Theory and Practice of Thermal Transient Testing
of Electronic Components

Márta Rencz • Gábor Farkas • András Poppe
Editors

Theory and Practice
of Thermal Transient Testing
# of Electronic Components

 Springer

Editors
Márta Rencz
Siemens Digital Industry Software STS
Budapest, Hungary

Budapest University of Technology
and Economics
Budapest, Hungary

András Poppe
Siemens Digital Industry Software STS
Budapest, Hungary

Budapest University of Technology
and Economics
Budapest, Hungary

Gábor Farkas
Siemens Digital Industry Software STS
Budapest, Hungary

ISBN 978-3-030-86176-6   ISBN 978-3-030-86174-2   (eBook)
https://doi.org/10.1007/978-3-030-86174-2

This Springer imprint is published by the registered company Springer Nature Switzerland AG
The registered company address is: Gewerbestrasse 11, 6330 Cham, Switzerland

This book is dedicated to the memory of Vladimír Székely, who was the founder of the network identification by deconvolution method, and who was the teacher of all the authors.
The authors wish to acknowledge the support of the Budapest University of Technology and Economics and Siemens Digital Industries, which were invaluable in the preparation of this book.

Gábor Farkas
András Poppe
Márta Rencz
Zoltán Sárkány
András Vass-Várnai

"Heat, like gravity, penetrates every substance of the universe, its rays occupy all parts of space."
Motivation for the research of heat equations,
by Joseph Fourier,
in The Analytical Theory of Heat, *1822 [1]*

"The force of the current in a galvanic circuit is proportional directly to the sum of all the tensions, and inversely to the entire reduced length of the circuit."
First formulation of an equation describing a discretized electric system,
by Georg Simon Ohm,
in The galvanic circuit investigated mathematically, *1827 [2]*

Contents

Chapter 1
Why Was Written and How to Read This Book

Márta Rencz and Gábor Farkas

Thermal transient measurements have become the most important characterization method of the thermal behavior of electronic systems in the last decades. This development is mainly due to the emergence of a new methodology, the *structure function method*, which is based on the network identification by deconvolution, introduced by V. Székely. This methodology in its mature form offers a "look inside the structure" of an electronic component with a single electrical measurement and the subsequent automated evaluation in software. It helps reveal data about the partial thermal resistances and capacitances inside the structure at all levels of an assembly, starting at a chip in a device package or module, through thermal interface and other material layers and various cooling mounts. The method may even provide temperature data on internal surfaces in the heat-conducting path which are otherwise not accessible for temperature measurements.

The users of the method soon understood that it is not pure magic, and to be able to fully exploit the capabilities of the methodology, a large amount of advanced knowledge is needed about the operation and the structure of the devices that are tested. In this book the authors, who are electrical engineers and university professors, tried to collect all information that is needed to fully understand the capabilities and the specialties of the thermal transient measurement technique.

The book is very timely now. The primary challenge in present engineering tasks is coping with the growing *power level* in electronic systems. Power controller units of electric cars and locomotives switch hundreds and thousands of amperes and forward many kilowatts toward the engine in order to bring tons of weight into

M. Rencz (✉)
Siemens Digital Industry Software STS, Budapest, Hungary

Budapest University of Technology and Economics, Budapest, Hungary
e-mail: rencz.marta@vik.bme.hu

G. Farkas
Siemens Digital Industry Software STS, Budapest, Hungary

motion. Solid-state lighting luminaires now operate in dozens and may dissipate hundreds of watts. Wind turbines and their power conversion units operate in the kilowatt to megawatt range, some high-voltage direct current electricity grid links are already in operation worldwide, and many new ones in Europe are under construction with a planned capacity of 1400 MW. The *power density* further increases in most of the systems in electronics. Processors run now at aggressive clock frequencies and dissipate hundreds of watts in a small box, video projectors which were formerly of suitcase size now resemble a pocketbook, and mobile phones produce although a few watts only but in a densely packed very thin case with no ventilation at all. These high-power levels represent an increased danger of overheating and damaging the devices.

Many of the power electronic systems work in extremely *harsh environments*. Automotive electronics, for example, must operate in the -30 to $+80$ °C external temperature range; this is similar for wind turbines, automotive lighting solutions, or street lighting luminaires.

Traditionally, the temperature of the internal semiconductor devices has been the principal factor which limited the system operability and influenced the system's reliability and lifetime. Due to the moderate cost and mature manufacturing technologies semiconductor power devices today are still mostly produced from silicon. Under the above-outlined environmental conditions, they sometimes reach their operation limits around 150 °C or 175 °C. With the advent of revolutionary wide band gap semiconductor materials, this is expected to change soon, and the structural materials of the device package will represent the new bottleneck in the system construction.

With increasingly sophisticated engineering and with the help of new thermal design methods based on *measurements* and *simulation*, the overheating of critical components can be prevented. Failure analysis shows that nowadays systems are correctly designed in this respect; the typical component breakdown is caused by the repeated *thermal transients*. Heating and cooling induce shear stress at the material interfaces in the structure, mostly at the die attach, or the solder joints, resulting in delamination, tear-off, etc. The poorer heat removal through a diminished surface can cause then thermal runaway.

The theory of heat propagation in materials was elaborated as early as the first decades of the nineteenth century. Since then, we know that the heat flows from the heat source toward the ambient, and the actual temperature of any point in between depends on the geometrical structure and the material properties of the parts where the heat flows through. Knowing the structure and the material parameters, the temperature distribution and the heat flow paths can be determined, if the heating sources are known. This calculation is done by *thermal simulation*. Thermal simulation can also reveal the time dependence of the temperature at any point in the structure.

In the case of thermal transient testing, the opposite is done. The time dependence of the temperature change is measured at a well-selected point in the system, resulting from a sudden change in the amount of the generated heat, and if the structure and the material composition of the system are known, the resulting and captured temperature transient enables determining the value of important structural parameters. This is of course an ill-defined problem with an infinite number of solutions, but if we have preliminary knowledge about the structure and we can control the direction of the heat flow, we can significantly limit the number of potential solutions. In the case of standard components in electronics, this is the case. In the methodology, we start a "heat signal" in an inner point of the system, measure the resulting temperature change in the same point of the system, and with mathematical calculations based on the theory of linear systems calculate the partial thermal resistances and capacitances of the heat flow path. The obtained system description that is called the *structure function* is a unique signature of the system. Any change in the structure results in a different "signature," rendering the method perfectly applicable for testing purposes. A special advantage is that the methodology is nondestructive, in contrast with the usual structure testing methods.

The technique can be used in all stages of design and manufacturing. The spectacular technological development described above has been enabled and achieved by *thermally aware system design*. In the last decades, thermal management has become an integral part of the design procedures, resulting in changing roles of different engineering disciplines in the overall design process.

Thermal managements usually start with simulation in the design phase and are followed by measurement on actual manufactured samples. Thermal transient measurements help in identifying the internal constituents of the system. Some parts in an assembly of an electronic system show high stability (die, ceramics, heat sink, etc.), while others, like die attach or thermal interface material (TIM) layers, may vary among samples, or during the lifetime of the assembly. With detecting the changing values in the structure during production testing, or from time to time in ageing monitoring, transient thermal testing also provides feedback on manufacturing stability.

During quality assurance in production, first, measurements are carried out, typically more of them at different boundary conditions, and then simulation can help in finding the root causes of eventual faults.

Two pillars of thermally aware design have been introduced above, namely, *measurement* and *simulation*. However, both implicitly rely on a third one that is *modeling* the thermal behavior of the system. When the measurement results are expressed in the form of thermal resistances and thermal capacitances or the peak temperature, a quite simple model of the analyzed system is inherently set up. More complex models can provide a better device or system description, can prevent unforeseen device failure, and help in avoiding overengineering. In up-to-date practice, *compact thermal models* and *reduced order thermal models* seem to be the promising direction for electronic datasheets and vendor-independent thermal models for simulation tools.

The precision of thermal testing depends on the *resolution, accuracy, repeatability*, and *reproducibility* of the measurement. It must be noted that the implicitly introduced models frequently use simplifications, such as "chip temperature" and "package case temperature." These system components are not point-like bodies but always have a certain finite size. This leads to an inhomogeneous temperature distribution on any surface unless the temperature is stabilized with and external force. These factors manifest as inherent *uncertainty* of the measurement that is usually much higher than the values one can enjoy in the case of, e.g., current or voltage measurements.

With our book we planned to fulfill various interests. It can be especially useful for engineering students who wish to understand the deeper physical and mathematical background of this broadly used new measurement technique. They should read the entire book from the beginning. Numerous examples throughout the book help in understanding the practical applicability of the theories.

> For improving the readability of the book, the examples are typographically separated from the main text body by gray shades, as here.

Chapter 2 outlines the underlying physical and mathematical notions, and their relationships needed to understand the thermal transient analysis technique. It contains the theory of thermal transient measurement evaluations. This chapter presents also the thermal metrics, used for the characterization of the thermal behavior of systems, and certain fundamentals about the standards that need to be considered when doing thermal measurements on electronics packages.

Chapter 3 presents the standardized and other thermal metrics used for the simplified characterization of packages and evaluates the advantages and disadvantages of the different methods used for determining them.

Chapter 4 summarizes the temperature-dependent electrical characteristics of the major semiconductor devices that may be used as heaters and sensors in the thermal transient measurements. After some introductory fundamentals of solid-state physics that is needed to understand the operation of these devices, the chapter presents the m features of the most frequently used semiconductor devices and their governing equations. From these equations the reader can see how the material properties and the structure of the devices that we wish to characterize or what we use as means to characterize the system influence the thermal transient measurements.

This book can be particularly useful for engineers who use the methodology in their everyday engineering practice and sometimes have difficulties in understanding or explaining the measured results. They can skip the details of the first mostly theoretical chapters, and after refreshing their knowledge on thermal metrics and semiconductor devices, they may jump soon to Chap. 5.

Chapter 5 presents the general scheme of thermal transient measurements and the usual additional tools that can extend the usability of the methodology. This chapter focuses on the current engineering practice of thermal transient characterization from the aspect of the measurement, evaluating also the different measurement methods.

For the practicing engineers, separate sections in Chap. 6 offer detailed information on the specialties of thermal transient testing of different electronic devices, with insight into the various sources of inaccuracies or potential mistakes in the measurements. This chapter is the most important part of the book for the novices of thermal transient measurement, as it presents all the details needed to accomplish the thermal characterization of the different devices.

Chapter 7 discusses the versatile applicability of the method, by highlighting the details of the most important use cases of the methodology from thermal qualification through the calibration of thermal simulation models, and reliability testing. These parts may bring innovative ideas also for the thermal engineers who currently use thermal transient testing for one purpose only.

Chapter 8 deals with the accuracy of the measurements, discussing all the factors that contribute to the uncertainty of the thermal transient measurements.

The chapters are written in such a way that enables reading them without reading the entire book. For this reason, some chapters start with some repetitions that enable the reader to understand the chapter without all the prior details.

At the end of the book, a well-structured Reference section provides links to information needed for a deeper understanding of the topics, as well as to details that expand the reader's knowledge in related curricula. References [1–17] refer to books, book chapters, or data compendiums which cover a wide range of physics background and thermal testing principles. References [18–53] specify standards and guidelines related to thermal or combined thermal and optical measurements. References [54–57] list software handbooks and test equipment manuals for thermal simulation and measurement control. In the last reference section, from [58] on journal and conference papers are listed which highlight related knowledge complementing the content of the book.

The authors hope that with the help of this book, the readers will gain a deeper understanding of the thermal transient measurements, and will find answers to all their questions related to their thermal transient measurement problems.

March 2022, Budapest

Chapter 2
Theoretical Background of Thermal Transient Measurements

Gábor Farkas, András Poppe, and Márta Rencz

In this chapter we collected the most important background knowledge that is needed to understand thermal transient measurements.

Speaking about measurements, we need to remember that a measurement is always accompanied by an inherent modeling step. Measuring the size of an object and claiming its length, width, and height is equivalent to replacing it with a model, which is a single (rectangular) block, and describing this model by these three quantitative parameters.

In thermal analysis the modeling of the system is much more challenging. All the physical quantities that play a role in thermal measurements must be precisely defined to avoid ambiguity.

2.1 Temperature and Heat Transfer

Temperature is the manifestation of the thermal energy of a finite size object. Thermal energy is the internal energy associated with the stochastic movement of particles in the object. These particles can be molecules in a fluid or gas, crystal lattice atoms in solids, or electrons in an electrically conductive material.

Heat is the internal energy, which is transferred between two or more finite size objects by various mechanisms at the level of particles (atoms and molecules). Heat transfer can be accompanied but does not need to involve transfer of matter.

G. Farkas
Siemens Digital Industry Software STS, Budapest, Hungary

A. Poppe (✉) · M. Rencz
Siemens Digital Industry Software STS, Budapest, Hungary

Budapest University of Technology and Economics, Budapest, Hungary
e-mail: Poppe.Andras@vik.bme.hu

The quantity of energy transferred as heat can be measured by its effect on the states of interacting bodies. Such effects can include the amount of matter participating in a phase change (e.g., amount of ice melted) or the change in temperature of a body dedicated to measuring the amount of transferred energy (temperature sensor, thermometer).

Power is the rate per unit time at which energy is applied externally on the system, or transferred between portions of the system.

Heat flux is the intensity of the transfer of energy per unit area per unit of time, that is, the power applied or forced through a unit of area.

The conventional symbol used to represent the temperature is T; the amount of heat transferred is denoted by Q. The SI unit of temperature is kelvin (K) or centigrade (°C); the unit of heat (also known as *thermal energy*) is joule (J, Ws).

In this work power is denoted by P and heat flux by φ. The SI unit of power is watt (W); the unit of heat flux is watt per square meter (W/m^2).

The primary mechanism of thermal energy transfer in electronic systems, as they are mostly solids, is *conduction*, at direct contact of objects, within molecular dimensions. The transfer occurs by the stochastic motion of particles, which can be "electrons" and "phonons" where the latter is the quantized lattice vibration.

Convection is a heat transfer mechanism in which one body heats another over macroscopic distances, through an intermediate circulating medium that carries energy from a boundary of one to a boundary of the other. The heat transfer on the surface of respective solid bodies towards the medium occurs by conduction.

In the related discipline of physics, in *fluid mechanics*, all media such as fluids or gases, aerosols, etc. are denoted as a generalized "fluid." Convection always involves the motion of matter. The internal energy of the medium is influenced not only by the stochastic motion of particles; it can be changed directly by *thermodynamic work*, by mechanisms that act macroscopically on the system, for example, by the motion of a piston.

Radiation is a heat transfer mechanism that occurs between separated or even remote bodies by means of electromagnetic waves. Accordingly, it requires no medium; it transfers heat over transparent matter or vacuum. All solid bodies emit radiation because of the stochastic motion of charged particles; this radiation grows in a "temperature to the fourth" manner.

The direction of heat transfer is always from the hotter to the cooler matter portions, as long as the temperature difference exists between them.

The heat transfer in solids is governed by local thermal properties: thermal conductivity and specific heat. These thermal parameters are temperature dependent in the semiconductor and package materials, which are most frequently used in electronics. However, the change of these parameters in the temperature range of the typical use is rather flat. This means that in many practical cases, the material parameters can be considered temperature independent, which simplifies the calculations, allowing in many cases the use of linear relationships.

Several thermal interface materials, such as thermal pastes and sheets, are anisotropic; they perform differently in different directions. Still, their orientation does not change during their operation.

In some thermal interface materials also phase change can occur at higher temperatures. Phase change accumulates or releases a large amount of heat, for example, phase change mechanisms enable intensive heat transfer in heat pipes.

2.2 Thermal Equilibrium, Steady State, and Thermal Transients

According to definition, two physical systems are in *thermal equilibrium* if there is no net flow of thermal energy between them when they are connected by a path permeable to heat. Extending this definition to *different portions of a system*, we consider a system being in thermal equilibrium when all parts of the system are at the same temperature. When one of the "systems" is the outer world, then the equilibrium is reached at the temperature of the ambient.

Steady state means that the temperature in different portions of the system does not change with time. Nonequilibrium states can be steady states if there is a source of energy to maintain the nonequilibrium condition. Without the source of energy, the system would settle into an equilibrium state after a certain time.

Thermal transient is a process through which the system or its portions transit from one temperature to another temperature.

In a heating process, the system moves from a lower temperature state to a higher one, and in a cooling process, the system starts from a higher and arrives at a lower temperature state.

Heating transients are always the result of adding energy to the system. This energy surplus is often applied on thermal systems as a time-dependent $P(t)$ power profile at one or more entry points, "driving points" over a time interval.

In electronic systems the energy that heats the system is in most cases the introduced electrical energy.

In cooling processes the energy leaves the system in the form of dissipation, that is, in the form of heat. Cooling can be a relaxation after revoking all power from the entry points for a prolonged time, or returning to a lower energy state at diminished power level.

The word "dissipation"[1] is used in a loose interpretation in the technical literature.

If the energy entry occurs at an area which is small compared to the size of the system, then that location is frequently called *junction*. The term means in thermal engineering a spot, which is considered to be isothermal and emitting homogeneous heat flux.

This name is inherited from power electronics where the heat source is in many cases a thin "dissipating," more precisely *heat-generating* layer near the upper

---

[1] In its original grammatical sense, "to dissipate" means *to scatter, to throw away, to dispose of* something.

surface of a semiconductor device. In many active power devices, this layer is in fact a *pn* junction, an area where semiconductors of different types join each other.

Proper distinction between different interpretations of this term will be made case by case in the subsequent chapters.

The development of the temperature change in time is highly influenced by the internal geometry and material properties of the system. Consequently, a systematic analysis of the transient process can yield relevant information on the structural composition of the system.

In the characterization of systems, those special transients play an eminent role where the system transits from one steady state to another steady state. Such "finished" transients yield the most complete information on the internal structure of the system. The structural details are determined with the best resolution in the vicinity of the power entry. From shorter transients where the steady state is not yet reached in all parts of the system, only limited information can be gained on the entire system structure.

2.3 Thermal Processes and Their Modeling in Electronic Systems

In electronic systems the most relevant heat sources are semiconductor devices; therefore, their junction temperature is a critical parameter influencing the reliability and the lifetime of the system.

The power generated in these devices can be calculated from the voltage and current values at their pins. These values change in time in a complex way and are determined by the electric characteristics of the devices, which are highly nonlinear, and *temperature dependent*. Typical device characteristics and their temperature dependence are discussed in detail in Chap. 6.

In many cases an electronic equipment operates in a relatively narrow temperature range, such as 0 °C–120 °C in laboratories, which is 273 K–393 K on the absolute temperature scale. In this range the temperature dependence of the thermal conductivity and specific heat of the used materials is usually negligible. In a system composed of materials of temperature-independent thermal properties, the flow of thermal energy is proportional to the temperature *differences*; the actual absolute temperature level has only minor influence.

For this reason, as long as the heat propagation takes the form of heat conduction and convection, the thermal behavior of electronic systems composed of semiconductor chips, their packages, and cooling mounts can be well described with the mathematical apparatus of the *theory of linear systems*. The linear approach can be used only with severe limitations in the case of systems with phase change materials, or systems, which operate at elevated temperatures where radiation from the outer surfaces becomes significant in the heat removal process.

T [°C]

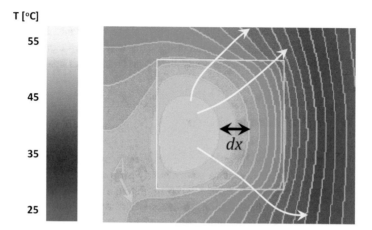

Fig. 2.1 Cross-sectional view of three-dimensional conductive heat transfer in solid material. Isotherms and heat-flow trajectories calculated in a thermal simulator are shown

The investigation of transient processes needs appropriate models of the thermal systems. These models can be of *continuous* or *distributed* nature, based on the differential equations governing the heat conduction in solids and the convection in gases and fluids. Such *detailed*, *continuous* models are usually analyzed numerically by finite element or finite difference software tools.

For analyzing all heat transfer mechanisms defined above, *computational fluid dynamics* (CFD) solvers, such as [56], offer a tool to simulate the conjugate heat and mass transfer.

The software tools display the simulation results in various forms. *Trajectories* demonstrate how the heat spreads in conductive regions and how the "fluid" flows in convective ones; *isothermal surfaces* denote where the temperature is equal at a certain time moment. It has to be noted that the geometric boundaries of physical objects and interface layers, which are often planar and rectangular in a real system, rarely coincide with isothermal surfaces, and the shape of the latter dynamically changes during a transient process.

Figure 2.1 visualizes the conductive heat transfer in a solid body. The different colors correspond to the temperature in the material, isothermal surfaces follow each other with an equidistant temperature difference, and the trajectories of the heat spreading are represented by curved arrows that are perpendicular to the isothermal surfaces.

The reference above to "finite element" or "finite difference" tools indicates that even the analytic equations that describe continuous (distributed) systems must be converted to their numerical counterparts, that is, a continuous system description must be *discretized*. This means that the continuously distributed material is lumped into small pieces for being analyzed by numerical methods in a computer. A net of adjoining lumps generated in the discretizing process of the model is often called a *mesh*. The lumps convey energy into each other in case of conduction and convey energy and matter into each other in case of convection. The abstract representation of the mesh that constitutes the numerical model of the distributed system is a set of *nodes* and *links* between the nodes. The graph of these nodes and links is called a *network* in the world of electrical engineering. This way, many techniques used in *network theory* are borrowed for the analysis of heat conduction problems in solids.

2.3.1 Equivalent Linear Models

The supposed linearity in the thermal domain implies that when increasing the power levels in the system, the temperature grows nearly proportionally. This assumption does not apply to cases with considerable nonlinear effects, e.g., systems with turbulent flow or significant radiation, or cases where the thermal conductivity of semiconductors exhibits strong temperature dependence, but these phenomena are usually negligible in the temperature range specified above.

This book investigates *time-invariant* thermal systems, in which the geometrical structure and material properties do not vary in time. Certain thermal interface types such as pastes tend to change their thickness during use, especially when pressure and power are applied on them the first time; successive transients yield slightly different results. Similarly, adhesives change their composition at initial curing or at their first use. Such effects are discussed in [63].

The effects of *wear* and *degradation* may cause significant alteration in the shape and thermal properties of some system components. These effects are treated in depth in the literature; some aspects that are closely related to the concept of thermal transient testing are discussed in Sect. 7.4. Still, the initial changes in system composition or later degradation can be observed rarely in the time span of a single transient; it is appropriate to use the time-invariant approach.

Linear time-invariant (LTI) systems can be analyzed in both continuous and discretized approaches. The discretization of a continuous body through a rectangular mesh is illustrated in Fig. 2.2.

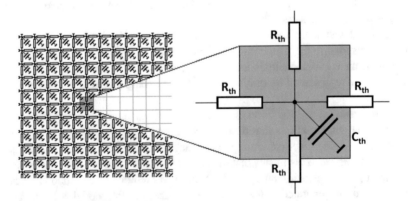

Fig. 2.2 Resistor-capacitor (RC) network model of a real physical structure

Taking the primary effect that is heat conduction, we experience a blend of two energy exchange mechanisms, heat propagation through an elementary portion of material and internal energy growth due to heat flow into that portion. The two mechanisms occur simultaneously in all system portions located anywhere in *space* and at any moment of *time*.

For simplicity, let us analyze first the two mechanisms separately. The energy storage and the temperature change can be formulated in an integral form on larger system regions and in a differential form for elementary portions.

In a *larger region* which contains no heat sources, the Q thermal energy grows *in time* as a result of φ heat flux flowing through all segments of its A surface. The heat flux integrated over the surface yields the P power applied on the region.

A changing $P(t)$ power causes a $Q(t)$ growth of energy in the V volume and m mass of the region. This energy growth manifests as $T(t)$ temperature elevation. The proportionality between thermal energy change and temperature change is expressed as C_{th} thermal capacitance.

Figure 2.1 demonstrates a case when an internal domain contains a heat source. Again, the P power can be interpreted as the sum of all heat trajectories, which intersect the A surface of a region enclosed by an isothermal boundary. The power is of the same P value on each A surface of isothermal shells containing the same heat source.

A *temperature difference between* two surface segments of a larger region induces a heat flow from the hotter towards the cooler one. Keeping this difference for a prolonged time steady state is reached; the heat flow stabilizes. The sum of all heat flux trajectories in a cross section of the region is now a steady P_{const} power, and each slice along the heat flow injects this power into the next region. The proportionality between the power and the temperature difference is expressed as the R_{th} thermal resistance.

The detailed temperature distribution and its change in time can be explored considering elementary portions. Suppose the elementary portion is cut out of the material between isothermal surfaces; it is of a small surface along which the thickness is of constant dx infinitesimal value.

Some parts in assemblies of power electronics correspond to a sandwich-like structure, in which all heat flows in a dedicated x direction. If uniform power is applied on a layer of such a laminate, a homogeneous heat flux will flow through all different material layers. In the following, the surface over which a homogeneous flux flows will be uniformly denoted by A, whether an elementary section or a whole laminate layer is considered, because the related equations are of the same format.

Figure 2.3 illustrates how the difference of the heat flux entering and leaving an elementary portion increases the thermal energy in it. The flux difference causes dQ energy growth and dT temperature growth in a dt time instant:

$$\Delta P = (\varphi_{in} - \varphi_{out}) \cdot A = dQ/dt = - c_p \cdot m \cdot dT/dt \qquad (2.1)$$

In (2.1) c_p denotes the specific heat of the material, and m is the mass of the section. The φ_{out} flux is forwarded into the next portion of matter, which behaves again in the same way.

Fig. 2.3 Heat flow into a material slice

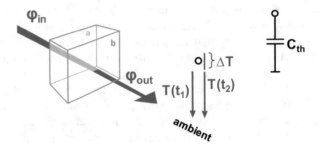

Fig. 2.4 Heat flow through a material slice

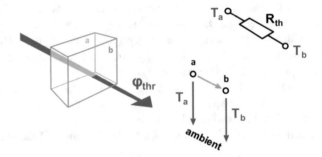

It is easy to measure the geometric size of an actual structural element, in this case the A surface and dx length. On the other hand, determining the mass of a segment of the structure is rarely feasible. For this reason it is more practical to introduce another material property, the c_V volumetric specific heat, also called volumetric thermal capacitance into (2.1):

$$\Delta P = dQ/dt = - c_V \cdot V \cdot dT/dt \tag{2.2}$$

$V = dx \cdot A$ is the volume of the slice; $c_V = c_p \cdot \rho$, and ρ is the material density.

Investigating the heat flow *through* the portion (Fig. 2.4), we find that in the continuous approach, it obeys the differential form of Fourier's law:

$$\varphi_{thr} = - \lambda \cdot \nabla T \tag{2.3}$$

where φ_{thr} is the heat flux across the section boundary, λ is the thermal conductivity in that section of material and ∇T is the gradient (spatial derivative) of the temperature.[2]

Equation (2.3) expresses that φ is a vector and the heat flows towards the cooler portions of a body.

---

[2]In the literature the thermal conductivity is denoted by k or λ. In this book we use λ; we reserve k for denoting the Boltzmann constant. The heat flux is often denoted as q, but in this book q is used for the elementary charge.

In a dx infinitesimal slice of A surface, (2.3) takes the form:

$$P = \varphi_{thr} \cdot A = -\lambda \cdot A \cdot dT/dx. \tag{2.4}$$

It can be observed in the figures and in the related equations that the dT quantity denotes in (2.4) a spatial temperature difference and in (2.1) a temporal difference. In (2.1) ΔP is a difference in space, the power difference between the two surfaces where the heat flow enters and leaves.

In electronics the heat dissipation is localized to the small volumes described as "junction" above. In other regions, the energy growth in a slice is the result of the trapped heat flux. Hence, the previous considerations can be combined into the classical heat equation, which is expressed in one dimension as

$$\frac{dT}{dt} = \left(\frac{\lambda}{c_V}\right) \cdot \frac{d^2T}{dx^2} = \alpha \cdot \frac{d^2T}{dx^2}. \tag{2.5}$$

The thermal diffusivity α, defined as $\alpha = \lambda/c_V$, is the measure of thermal inertia. In a material of high thermal diffusivity, heat moves rapidly; the substance conducts heat quickly relative to its volumetric heat capacity.

Discretizing the structure into small lumps of homogeneous material condenses (2.4) into thermal resistances. A lump has now dx length along the heat flow and A surface perpendicular to it; and a temperature drop of $T_a - T_b$ occurs between its a and b isothermal faces (Fig. 2.4). Now (2.4) takes the form:

$$T_a - T_b = P \cdot R_{th} = P\left(\frac{1}{\lambda}\frac{dx}{A}\right), \quad R_{th} = \left(\frac{1}{\lambda}\frac{dx}{A}\right) \tag{2.6}$$

In the discretized representation, the two faces of the section are connected by an R_{th} thermal resistance.

Similarly, in a material lump exposed to continuous φ heat flux over A area resulting in ΔP power difference between the entry and the exit, the energy change in a short $dt = t_2 - t_1$ time interval is from (2.1)

$$dQ = \Delta P \cdot dt = C_{th} \cdot (T_2 - T_1), \tag{2.7}$$

where $T_1 = T(t_1)$ is the temperature of the material at t_1 time and $T_2 = T(t_2)$ is the temperature of the material at t_2 time.

C_{th} is the thermal capacitance of the slice:

$$C_{th} = c \cdot m = c \cdot \rho \cdot dx \cdot A, \quad C_{th} = c_V \cdot V = c_V \cdot dx \cdot A. \tag{2.8}$$

In the discretized model of a complete system as represented in Fig. 2.2, also the topology in which the R_{th} and C_{th} constituents are connected is to be defined.

Temperature is a quantity, which is measured related to a reference value. When it is expressed, for example, in centigrade, then the temperature is related to the reference level of the internal energy of melting ice. In (2.6) the T_a and T_b temperatures are measured with respect to the reference level, and so are T_1 and T_2 in (2.7). Still, because of the linear approach, regardless of which reference level was chosen, it disappears from the equations when the temperatures are subtracted.

In thermal transient measurements, it is generally assumed that the near environment of the measurement arrangement such as the air temperature in the laboratory or the temperature of the circulated coolant in a cold plate device does not change during the measurement time. Similarly, CFD simulation always aims at analyzing a limited part of the universe only; the external world is often represented as a constant temperature on the system boundary. This constant temperature attributed to the near environment is called *ambient temperature* or *ambient* in short, usually denoted by T_A (or T_{amb}) in the literature. The ambient is the thermal counterpart of the electrical ground (zero reference potential, *datum reference* in other engineering disciplines).

The electrical networks are extremely abstract; a lump with its "volume" and "faces" appears as a node in the graph of a circuit scheme. The reference level in electronics is named "ground" and is represented by a $\perp$ sign.

Thus, as already shown in Fig. 2.2, the C_{th} thermal capacitance appears between the node representing the material portion and the ambient. The R_{th} thermal resistance connects two such nodes; the reference towards the ground disappears.

Equations (2.3)–(2.8) are of the same format as the descriptive differential equations of electronics, replacing the P power by the I electrical current, the T temperature by the U electrical potential, and the ΔT temperature difference by the V electrical voltage.[3] The thermal resistances and capacitances correspond to their electric counterparts of similar name.

In an electrical network, the current flows through the net of resistances and capacitances and causes a voltage drop (potential difference) between the nodes of the circuit. The current is measured in amperes and the voltage in volts; resistances and capacitances are measured in ohms (V/A) and farads (As/V), respectively.

In the thermal network, heat, that is power, flows through the net of thermal resistances and capacitances and causes a temperature drop between the nodes of the network. The power is measured in watts and the temperature in kelvins (K) or centigrade (°C); thermal resistances and capacitances are measured in K/W and Ws/K (J/K), respectively.

In network theory those networks in which excitation (powering) occurs at a single specific point can be characterized by their *thermal impedance*. This concept denoted as Z_{th} combines the effects of the C_{th} thermal capacitance and R_{th} thermal resistance into a single metric. Briefly, the thermal impedance represents both the temporal and spatial changes of temperature in a heat conduction path, that is, the

---

[3]The work of Joseph Fourier, *The Analytical Theory of Heat*, published in 1822 inspired Georg Simon Ohm to formulate his law in 1827. For a long time, the heat spreading and the change of temperature, both known for many thousand years, helped to understand the newly introduced concepts of voltage and current through their analogy.

Fig. 2.5 Foster (**a**) and Cauer (**b**) type representation of discretized structures

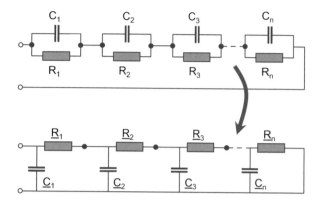

thermal impedance is the ratio of the dynamically changing temperature and the dynamically changing power. There are multiple, equivalent representations of the thermal impedance.[4] These representations will be introduced step by step throughout the different subsections of this chapter. We start the discussion with the *network model representations* of the thermal impedance.

The theory makes a distinction between self-impedances and transfer impedances. Self-impedances express the response of the linear system at the location where the excitation occurs, simply saying at the *driving point*. Transfer impedances describe the system response at a different location often referred to as *monitoring* or *(thermal) test point*.

The network theory states that arbitrary complex *RC* network in which the excitation occurs between the single driving point and the ground behaves in the same way as a reduced set of thermal resistances and capacitances arranged in one of the configurations shown in Fig. 2.5. This corresponds to a thermal network model of a single powered junction and an isothermal ambient.

It is important to note that the internal nodes in these models generally cannot be associated with the monitoring points of a complex *RC* network.

The *chain of RC stages* style (top of Fig. 2.5) is called Foster representation, and the *ladder of RC stages* style (bottom of the figure) is called Cauer representation, named after the inventors who introduced these canonic circuit topologies into linear filter synthesis. The conversion between the two models is a standard procedure in network theory (summarized in Annex C of the JEDEC JESD 51-14 standard [40]). An example of the conversion is presented in GNU Octave (MATLAB) code below.

---

[4]In the theory of linear systems, impedances are interpreted in a stricter sense in the frequency (f, $\omega = 2\pi f$) domain which is connected to the time domain through the Fourier transform or in the complex frequency ($s = j\omega$) domain which is connected to the time domain through the Laplace transform.

Example 2.1: The Foster-Cauer Transformation

A simple MATLAB code which carries out the Foster-Cauer conversion is as follows:

```
% Foster to Cauer transformation
function [Rc Cc] = foster2cauer(Rf, Cf)

%calculate Z= p/q polynomial

n = length(Rf);
p = Rf(1);
q =  [Cf(1)*Rf(1) 1];

for k= 2:1: n
     pn= [0 p] + Rf(k)*q + [Cf(k)*Rf(k).*p 0];
     qn= [0 q] + [Cf(k)*Rf(k).*q 0];
     p = pn;
     q = qn;
end
% calculate Cauer form

for k= 1:1:n
     Cc(k) = q(1)/p(1);
     q= (q - Cc(k)*[p 0]);
     q(1)=[];
     Rc(k)= p(1)/q(1);
     p = (p - Rc(k)*q);
     p(1)=[];
end
end
```

Running the MATLAB code for the values in Fig. 2.21, one gets on the output:

```
[Rc Cc] = foster2cauer([1 1 4], [100e-6 10e-3 100e-3])
Rc = 1.0220 1.1870 3.7909;
Cc = 0.000098912 0.009159927 0.095994437
```

   and

```
[Rc Cc] = foster2cauer([1 1 8], [100e-6 10e-3 100e-3])
Rc = 1.0220 1.1886 7.7894;
Cc = 0.000098912 0.009159713 0.093316320
```

2.3.2   *Energy Balance and Stability*

The simplest thermal model of a system consists of a single thermal resistance and a single thermal capacitance. This simplest model, shown in Fig. 2.6, consists of just two *thermal nodes*, the junction where power is applied and the ambient. The whole heat removal apparatus of the modeled system can be cumulated into a single and constant R_{thJA} junction to ambient thermal resistance. The energy storage is represented by a single C_{th} thermal capacitance.

The driving force of the heating and cooling processes is the thermal imbalance. Suppose there is a steady P_{gen} power generated in the system and fed into the "junction" node, which is at $T_J(t)$ junction temperature at t time. The temperature is always measured from a reference point; attributing any constant value to it in this simple model would not change the validity of the descriptive equations. In this section and in some further ones, we shall attribute $T_{amb} = 0$ temperature to the ambient in many cases, in order not to drag a constant value through all equations. This, of course, does not put any limitation to the validity of the equations.

We can state that before applying the P_{gen} power, the system is in a "low-power" thermal equilibrium, the energy stored in the system is zero, and so is the temperature of the T_J point. Applying P_{gen} power in the first instant elevates the internal thermal energy, $P_{gen} = P_{store}$ initially.

The continuous flow of P_{store} into the thermal capacitance increases the Q_{stored} thermal energy; thus, the T_J temperature elevates on C_{th}. We can recognize that P_{store} is the difference of P_{gen} and P_{diss}. Still, P_{gen} is constant, and the heat loss towards the ambient can be calculated as $P_{diss} = T_J/R_{thJA}$ at any time.

Solving the appropriate differential equations, we find that T_J grows exponentially:

$$T_J(t) = P_{gen} \cdot R_{thJA} \cdot \left(1 - e^{-\frac{t}{\tau}}\right), \tag{2.9}$$

where $\tau = R_{thJA} \cdot C_{th}$ is the characteristic *time constant* of the model.

Thus, during the transient process, the proportion of the heat loss through the junction to ambient thermal resistance grows, and the share towards the thermal capacitance diminishes.

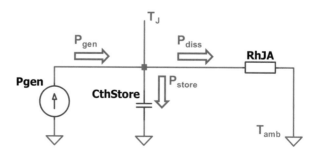

Fig. 2.6 Simple thermal model of a single RC stage driven by an external source. Generated P_{gen} and dissipated P_{diss} power indicated, their difference P_{store} elevates the Q_{stored} thermal energy

Integrating the heat flow (which is the power) over time the energy stored in and dissipated from the components can be also calculated. The integral of exponential functions until t time is straightforward, and the actual values can be directly derived from the electrical analogs.

The stored internal energy is $Q_{\text{stored}} = \frac{1}{2} C_{\text{th}} \cdot T_J^2$, the applied energy is $Q_{\text{in}} = P_{\text{gen}} \cdot t$, and the dissipated energy is $Q_{\text{in}} - Q_{\text{stored}}$, applying the principle of conservation of energy. The principle corresponds to the Kirchhoff current law at the junction node in the interpretation of network theory.

Example 2.2: A Simple Thermal Model of a Device in Its Environment

When a thermal system is built for transient testing, it can be divided into a "device under test" (DUT) part and a thermal environment, a test bench. In this example the model of the DUT is simplified to two thermal resistances and two thermal capacitances, and the thermal environment is represented by a single thermal R_{th} and a C_{th} (Fig. 2.7).

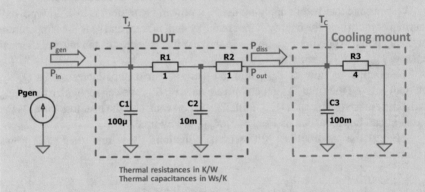

Thermal resistances in K/W
Thermal capacitances in Ws/K

Fig. 2.7 Simple thermal model: a power device represented by two thermal RC stages, and its thermal environment, simplified to a single RC stage. Component values of network elements are assigned for further calculations

The temperature of the entry point (driving point) is denoted as T_J; the temperature of the separation point is denoted as T_c.

It should be emphasized that this highly simplified model resembles only superficially a realistic device with an exposed cooling surface, which is typically denoted as "case." The calculated temperature of such a separation point is related only loosely to an actual complex temperature distribution on the case of a physical device.

As the use of "dissipation" is often ambiguous, let us denote below the P_{gen} generated power by P_{in} and the P_{diss} dissipated power, the heat loss by P_{out}.

(continued)

Example 2.2 (continued)

In steady state the thermal capacitances do not influence the heat flow. The system can be characterized by a total R_{thJA} junction to ambient thermal resistance, $R_{thJA} = R1 + R2 + R3$.

Steady state of the device is reached when $P_{in} = P_{out}$, and so the T_{Jss} steady state junction temperature is

$$T_{Jss} = P_{in} \cdot R_{thJA}. \tag{2.10}$$

In the case when T_J was below T_{Jss} when P_{in} was applied, $P_{in} > P_{out}$ and heating occurs until steady state is reached. In the opposite case, $T_J > T_{Jss}$, then $P_{in} < P_{out}$ and a cooling process governs the system towards steady state.

In some cases, the heating process may end in a steady state where the T_J device temperature is above the absolute maximum ratings of the semiconductor. This thermal condition is called *thermal overload*.

Thermal overload results either in longtime degradation or in a sudden breakdown of the device; consequently it has to be avoided in normal device operation. This overload, however, can be intentional for reliability/lifetime testing purposes. By systematic reliability tests and analysis of the degradation mechanisms, the thermally influenced safe operating area (SOA) of devices can be established. A more detailed treatment of SOA definition is given in Chap. 6.

So far only such cases have been considered in which the input power is stable during the transient, and the heat removal can be characterized by a constant R_{thJA} junction to ambient thermal resistance in steady state. In Chap. 6 we examine the transients of devices which are normally operated at constant current during the test, and their electrical characteristics have positive thermal coefficient, PTC. This latter term means that their voltage grows with elevating temperature at constant current bias, resulting in an increase of their own, internal power generation.

In addition, also the temperature coefficient of the thermal conductivity can be positive in crucial portions of the thermal environment, in the heat-conducting path. This can result in growing junction to ambient thermal resistance at higher temperatures.

Either one of these effects or the coincidence of them can cause thermal instability; due to a positive feedback loop, the temperature keeps growing in the powered state, which also elevates the power. When this situation finishes only at extreme temperature causing fatal damage of the device, then we speak about *thermal runaway*.

The effects of thermal instability and thermal runaway are treated in Sect. 6.1 for diodes and in Sect. 6.2 for MOS devices.

2.3.3 Heating and Cooling Curves

With the help of the theory of linear systems, there is no need to restrict the waveform of an actual power change. For any of them, the corresponding temperature change and the relevant system descriptors can be derived easily. Further on in this book, however, we put special emphasis on those specific power profiles, which have an eminent role in thermal testing, namely, heating and cooling at constant applied power for prolonged time, or applied in a periodic manner.

The time domain response of a linear system to an arbitrary excitation can be calculated if some specific descriptive functions of the system are known. These are the $g(t)$ Green function, the response to a Dirac-δ excitation (approximated by a very short pulse of known energy), and the $a(t)$ response function to unit step excitation.

In case of an actual thermal transient measurement, in response to a $P(t)$ input power excitation of some waveform, the system will react with a $T(t)$ temperature response. For example, in the typical case when a constant power is applied to a previously unpowered system (step function like excitation), a monotonous temperature elevation can be observed until a new, "hot" steady state is reached.

Figure 2.8 shows the temperature elevation in a distributed thermal system, composed of a MOSFET device in a common package, mounted on a cold plate. This example will be used in further sections in this chapter to demonstrate how structural information of the assembly can be extracted from a simple thermal transient and how the information can be best visualized.

To a single short pulse of Δt length and P_{p} "height" in power, the system will react with a

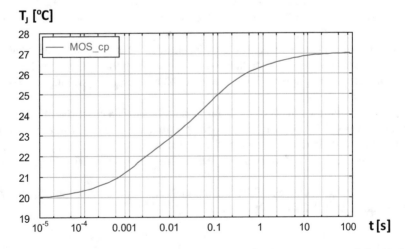

Fig. 2.8 Temperature change of a heated power device (MOSFET) on a water-cooled cold plate

$$T_p(t) = \Delta t \cdot P_p \cdot g(t) \qquad (2.11)$$

response. Any continuous $P(t)$ power profile can be sliced into a sequence of short $\Delta t \cdot P_p$ pulses. Linearity allows superposing the responses of the system, shifted in time as the pulses follow each other. Let us denote this series of pulses as $\Delta t \cdot P(t)$ where for the different t_i time instances $P_{p_i} = P(t_i)$. With $\tau = (\Delta t \rightarrow 0)$, the series of the discrete P_{p_i} pulses will become a continuous $P(t)$ function. With this, according to the superposition principle, the $T(t)$ temperature change can be written as

$$T(t) = \int_{-\infty}^{\infty} g(t) \cdot P(t - \tau)d\tau \qquad (2.12)$$

that is referred to as a *convolution integral*. Introducing the shorthand notation $\otimes$ for convolution, the above relation is written shortly as $T(t) = P(t) \otimes g(t)$. The formula in (2.12) expresses that at time t, the short pulse left behind by time τ is only present with weight $g(t)$ in the current temperature value.

There are strict conditions formulated for g and P in the theory of linear systems. These can be also expressed in ordinary terms. The $P(t)$ excitation has started at a certain time; the system was unpowered before that time. The retracting $P(t)$ function in (2.12) was zero before the start time; this can be reflected by changing the lower limit of integration in the formula from $-\infty$ to 0.

The convolution operation, $T(t) = P(t) \otimes g(t)$, also defines the components of the instrumentation needed for a transient measurement. A powering unit has to produce a $P(t)$ power profile during the measurement time. It is applied on the device under test, which is characterized by its $g(t)$ descriptive function. A data acquisition unit has to record the $T(t)$ temperature response. (As $g(t)$ embodies the relationship between the dynamic change of the temperature and the power, it can be considered as one possible theoretical representation of the thermal impedance.)

An obvious consequence of the above is that transient testing provides a direct means for system identification; the $P(t)$ excitation and the $T(t)$ response are known, in such a way the $g(t)$ system descriptor can be calculated:

$$g(t) = T(t) \otimes^{-1} P(t), \qquad (2.13)$$

where $\otimes^{-1}$ denotes the deconvolution operation.

As in most cases no analytic solution exists for convolution and deconvolution, the results of these can be calculated by numerical methods only. Generic implementations of these operators are available in software libraries, MATLAB codes, etc. The numerical algorithms for deconvolution require much sophistication

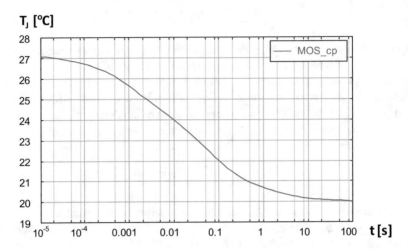

Fig. 2.9 Cooling of a power device (MOSFET) on a water-cooled cold plate, after heating is switched off in an abrupt, stepwise manner

and need careful considerations as advised in [58]. These considerations are typically overlooked in generic tools.

The linear approach implies that symmetric power profiles result in symmetric temperature changes. As a special case, heating and cooling processes are symmetric, if the previously mentioned nonlinearities do not apply.

The transient temperature change as illustrated in Fig. 2.8 or in Fig. 2.9 has a huge information content related to the structural details of the system, which is tested; but this information is not expressed in an obvious and apparent manner. The interpretation is much more evident in other equivalent representations highlighting one or other aspect of use. We can call these representations the "views" of a thermal transient.

In order to demonstrate the creation and use of the views, we shall present real transient measurements on real devices in actual thermal environments in the subsequent chapters. However, in some cases we found it more efficient to show the results in *simulation experiments* because it is easier to recognize the essence of a methodology in simplified structures. Moreover, these experiments can be easily repeated by the reader of this book.

Some simulated experiments will use the popular LTSpice analog simulator program [57] for analyzing both thermal and electronic properties of experimental arrangements. In these simulations the thermal equivalent of the electrical current is the heat flux ("flow quantity"), while the voltage carries the temperature values ("across quantity").

Example 2.3: A Simple Circuit Model of a Thermal System with Three Time Constants

A simple behavioral model of a thermal system is shown in Fig. 2.10, corresponding to the equivalent Foster network. In an electrical equivalent, the same current would flow through all stages in series; the voltage at the T_J driving point can be constructed as the sum of the voltage drops on the individual stages.

The thermal equivalent circuit in Fig. 2.10 has three time constants.

Running an LTSpice simulation with the circuit model, the "PULSE" directive forces $P_{in} = 10$ W input power at $t = 0$ time on the junction point. The simulation yields the T_J junction temperature transient response shown in Fig. 2.11.

We can observe three characteristic bumps in the curve; the reason of this will be discussed with more details in the next section.

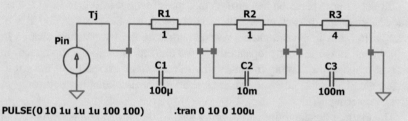

Fig. 2.10 Foster-type behavioral model of a simulated thermal system with three characteristic time constants ($\tau_1 = 100$ μs, $\tau_2 = 10$ ms, $\tau_3 = 400$ ms)

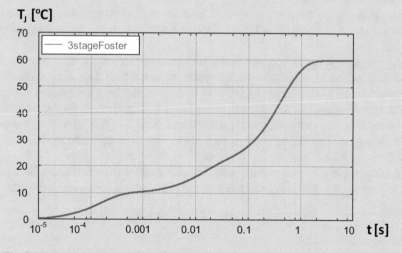

Fig. 2.11 Transient result from the LTSpice simulation defined in Fig. 2.10

Real thermal systems are sometimes simplified to similar circuits of three stages as a basic model. The first bump is interpreted as the heating of the chip with the die attach layer as obstacle in the heat spreading, the second bump corresponds to the heating of a package base with a thermal interface layer as bottleneck in the spreading, and the third one represents the characteristic heating of the cooling mount or the test environment. In the subsequent subsections, we shall present a more mature descriptive apparatus.

2.3.4 Z_{th} Curves

Examining Figs. 2.8 and 2.9 that present temperature transients measured on a real packaged MOSFET device, we can observe their "bumpy" nature.

These bumps are even more expressed in the simulated transient curve shown in Fig. 2.11 where the "system" has just three discrete time constants.

Engineering experience has proved that the structural information of the device under test and its thermal environment is encoded into the position and size of the bumps. In realistic measurements one can attribute the temperature change in the millisecond range to heat propagation in the die and through the die attach, in the second range to the cooling mount, in the minute range to heating of the circulated water, etc. This plot, however, characterizes the heat-conducting path only at the given powering.

The thermal conductivity and specific heat of the device components and of the measurement environment show only minor change in the typical temperature range of use. This implies that shifting the base plate temperature, we obtain similar recorded curves, and altering the applied power, we obtain again similar, proportionally magnified records.

Normalizing the temperature change with the applied power, we obtain the Z_{th} curves (Fig. 2.12). Sometimes the Z_{th} curve is referred to as the *thermal impedance curve*.[5]

At this point it has to be noted that it is common in the engineering practice that quantities which change over many orders of magnitude like time or thermal capacitance are plotted along logarithmic axes. Still the axes are labeled with the values of the quantity in the original, linear scale. In accordance with the customary representation, we refer to these quantities as $T(t)$, $C_{th}(R_{th})$, although the plot corresponds to $T(\log t)$, $\log C_{th}(R_{th})$, and so on. This also applies to the Z_{th} chart in Fig. 2.12 and all subsequent similar ones.

Z_{th} curves are always *monotonically increasing* due to their definition, because a heating curve (Fig. 2.8) is normalized with a positive and a cooling curve (Fig. 2.9) is

---

[5] In electronics the impedance is primarily defined in the frequency domain, not in the time domain as a step response function.

Z_th [K/W]

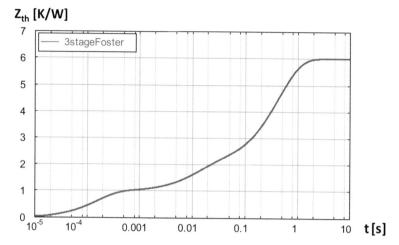

Fig. 2.12 Temperature change normalized by the applied power: the Z_{th} curve of the thermal system introduced in Fig. 2.10

normalized with its negative power step. The result is nearly the same Z_{th} curve in the two cases.

As a fairly accurate temperature transient for any power step can be produced if we multiply the Z_{th} curves by the actual power applied, this curve is used frequently for the characterization of the thermal behavior.

This concept of proportionality to power (i.e., linearity) is not fully accurate when measuring realistic systems. The actual shape of a cooling or heating curve depends on the temperature dependence of the material parameters as well. A more dominant factor is that at increased power level and at higher temperature elevation, the cooling mechanisms (turbulent convection, radiation) become more intensive; consequently the real temperature change is lower than the one extrapolated from the multiplied Z_{th} curve. As such, using Z_{th} for temperature estimation, let us remain on the safe side.

A deeper analysis of nonlinear effects is given in [64].

2.4 System Properties Calculated from the Thermal Transient

2.4.1 Time Constant Spectra

In a both theoretically and practically important case, a constant P_{on} power is turned on and kept on the tested system for a prolonged time.

Fig. 2.13 Time response of a single thermal RC stage to a step function excitation with its magnitude and time constant shown

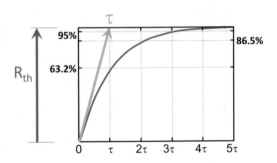

For unit step powering which is zero before switching on and unit value afterwards, based on the theory of linear systems, (2.12) will be of the form of

$$a(t) = \int_0^t g(x) \cdot 1 \, dx. \tag{2.14}$$

This means that the $a(t)$ unit step response is the integral of the $g(t)$ Green function. Producing a power step in reality is much easier than generating extremely short high energy pulses. Instead of directly measuring it, the $g(t)$ function can be derived from the measured step response. Eq. (2.14) implies that

$$g(t) = \frac{d\, a(t)}{dt}. \tag{2.15}$$

The unit step response function, for which the traditional notation in linear system theory is $a(t)$, is exactly the function that is called the $Z_{th}(t)$ normalized temperature transient curve in thermal engineering practice, presented earlier. In this book we use both notations.

When a constant P_{on} power is switched on at zero time, a single RC stage in Fig. 2.10 produces an exponential growth after switching on the power, adding a $T(t) = P_{on} \cdot R_{th} \cdot (1 - e^{-t/\tau})$ temperature term to the response of the entire series of the RC stages (Fig. 2.13). In the analogous electronic circuit, a steplike current is switched on, and exponential voltage growth is observed at the node driven by the current source.

The R_{th} *magnitude* denotes the thermal resistance of an elementary stage; $\tau = R_{th} \cdot C_{th}$ is the *time constant* where C_{th} is its thermal capacitance. Adding up the temperature drops of each subsequent stage in the series model, at the input (in this case at the junction), we get a *sum* of exponential functions:

$$T(t) = P_{on} \cdot \sum_{i=1}^{n} R_{th_i} \cdot (1 - e^{-t/\tau_i}). \tag{2.16}$$

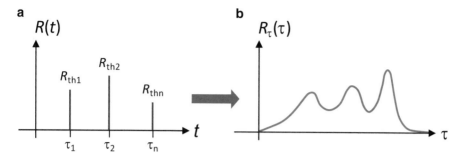

Fig. 2.14 Time constants in a lumped element system (**a**) and in a distributed parameter system (**b**)

Normalizing the $T(t)$ temperature response with the P_{on} power, we obtain the $Z_{th}(t)$ *thermal impedance*:

$$Z_{th}(t) = a(t) = \frac{T(t)}{P_{on}} = \sum_{i=1}^{n} R_{th_i} \cdot \left(1 - e^{-t/\tau_i}\right) \qquad (2.17)$$

that is nothing else than the $a(t)$ unit step response function, introduced earlier.

Different material slices have different characteristic thermal parameters; the resulting different R_{th_i} magnitudes and τ_i time constants make the $T(t)$ curve "bumpy," as seen in Figs. 2.8 and 2.11. The bumps are originated from the individual time constants (Fig. 2.14a) of the system.

The system can be fully characterized by a proper number of τ_i and R_{th_i} pairs; the sum in (2.16) restores the exact waveform of the temperature at the driving point. As the equation indicates, the dimension of magnitudes is K/W (kelvin/watt), and the dimension of the time constants is second.

If the thermal system is subdivided into a large number of thin slices, we reach a *continuous* model of elementary sections, forwarding energy into each other as indicated in (2.2) and (2.3). The sum formulae in (2.16) and (2.17) take the form of an integral:

$$T(t) = P_{on} \cdot \int_0^{\infty} R_\tau(\tau) \cdot \left(1 - e^{-\frac{t}{\tau}}\right) d\tau. \qquad (2.18)$$

$$Z_{th}(t) = a(t) = \frac{T(t)}{P_{on}} = \int_0^{\infty} R_\tau(\tau) \cdot \left(1 - e^{-\frac{t}{\tau}}\right) d\tau. \qquad (2.19)$$

We can easily realize that these relationships resemble the form of the convolution integral shown by Eq. (2.12).

Equation (2.18) corresponds to a measurement scheme again, in a similar but not equivalent way as it was found in (2.12) before. After the test equipment applies a P_0 constant power on the system under test at zero time, a measured $T(t)$ temperature response is recorded. The $R_\tau(\tau)$ time constant spectrum is now the characteristic system descriptor. As stated above for the discrete τ_i and R_{th_i} pairs, also the integral in (2.18) can be used for restoring the temperature waveform.

Although the Eqs. (2.18) and (2.19) seem to operate on continuous functions, the numerical procedures to obtain them produce discretized results. The primary parameter for the procedure is the intended $\Delta\tau$ time constant step.

Selecting small $\Delta\tau$ results in a high number of RC pairs in the discretized $R_\tau(\tau)$ time constant spectrum. The corresponding Foster-type network models properly the dynamic thermal behavior of the measured heat-flow path, seen from the T_J driving point (Fig. 2.15). The R_{th_i} magnitude element in the chain is composed as the product of the $R_\tau(\tau_i)$ spectrum value and the width of the τ_i time constant slice.

The C_{th_i} thermal capacitances can be calculated from the $\tau_i = R_{th_i} \cdot C_{th_i}$ relationship.

Real physical objects, which are to be tested as thermal systems, obey mechanical constraints. The heat propagates from a very thin active layer through tiny semiconductor chips, which are mounted into larger packages and modules. These devices are mounted on larger heat sinks, located in enclosures. Accordingly, thermal time constants of an actual electronic system may range from microseconds (thermal transient within the chip) to hours (temperature elevation of a cooling mount). For this reason are transient curves plotted in this chapter in logarithmic time scale; otherwise, the tiny details of early times could have been lost.

Similarly, thermal transient testers record the transients in a logarithmic time scale in order to store the transient results in a manageable amount of data. It is reasonable

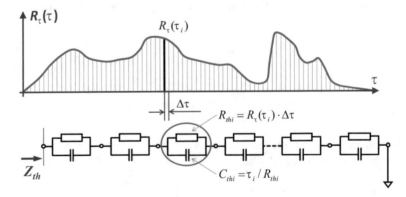

Fig. 2.15 Discretized time constant spectrum with equidistant $\Delta\tau$ time constant steps and the equivalent Foster chain

to process these data sets in *logarithmically equidistant* time increments, which correspond to a constant ratio of increments in linear time.

Introducing the $z = \ln(t)$ transformation for the time and the $\zeta = \ln(\tau)$ transformation for the time constants has a further advantage; it offers an easy transition between the convolution-style system descriptor in (2.12) and (2.13) and the "sum of exponentials" style system descriptor in (2.19).

The R_ζ time constant spectrum[6] defined on a logarithmic time scale can be obtained, indeed, in a convolution-type relationship with the first derivative of the unit step response. That is, the

$$\frac{d}{dz} Z_{\text{th}}(z = \ln(t)) = \frac{d}{dz} a(z) = R_\zeta(z) \otimes w_z(z) \qquad (2.20)$$

relationship holds where w_z is a fixed function: $w_z(z) = \exp[z - \exp(z)]$. (See further details in [58, 59].)

Equation (2.20) tells that from the unit step response of a (thermal) system, its time constant spectrum can be extracted by deconvolution as follows:

$$R_\zeta(z) = \left[\frac{d}{dz} a(z)\right] \otimes^{-1} w_z(z). \qquad (2.21)$$

At this point we have to emphasize that formulating (2.19) and transforming it in some steps to (2.21) is not aimless equation crunching; it is the essence of testing and system identification. Eq. (2.19) corresponds to the scheme of thermal testing; the thermal tester applies P_{on} on the tested system and records the $T(t)$ temperature response. Eq. (2.21) depicts a systematic process, which yields an R_ζ system descriptor, fully characterizing the thermal behavior of the system.

R_ζ is calculated from the $a(z = \ln(t))$ system response and a *fixed* auxiliary w_z function.

The deep mathematical background of the calculus and the fundamental concepts related to the time constant spectra of distributed RC systems are provided in [58–60].

The systematic use of the deconvolution which starts from the measured transient and results in the time constant spectrum calculated by (2.21) is the broadly used *network identification by deconvolution*, shortly the *NID method*.

The disadvantage of the Foster type of network model is that, although it is a *mathematically correct* model of the transient behavior, it cannot be used for building an equivalent of the real, *physical* thermal structure, because it contains *node-to-node* capacitances.

In the fundamental heat transfer equation (Eq. (2.7)), the T_1 and T_2 temperatures are measured from the ambient; a C_{th} thermal capacitance exists *between* a point representing the material portion and the *ambient* as underscored by (2.1).

---

[6]In contrast to the previous R_τ notation, with the index in the R_ζ function, we indicate that this is the version of the time constant spectrum that represents the time constant distribution on the logarithmic time scale.

Accordingly, the real thermal capacitances are always connected to the ground, since the stored thermal energy, which they represent, is proportional to the temperature elevation of *one node* with respect to the reference, and not to the temperature difference of two nodes as would be suggested by the Foster model. For this reason, the calculated Foster model has to be transformed into a Cauer model with a standard mathematical transformation, as outlined in [58, 61], and Annex C in the standard [40].

In order to facilitate understanding, let us examine the above-discussed functions of an artificial system.

Example 2.4: The Characteristic Functions of a Known System

A simple thermal system and its step response were presented in Fig. 2.10 and in Fig. 2.12. As the system is an artificial one with *exactly* three time constants, we can also produce the signal constituents with an appropriate simulation. The circuit and its separated subcircuits are shown in Fig. 2.16.

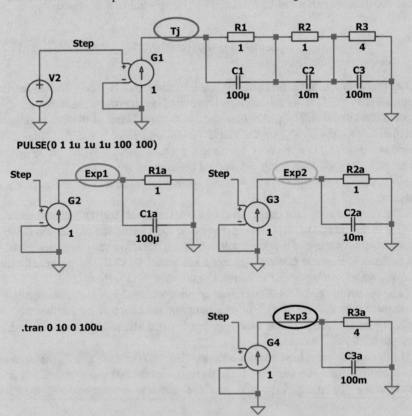

Fig. 2.16 Equivalent circuitry in LTSpice for producing the T_J temperature of the simulated system and for directly providing the three constituents of the resulting curve

(continued)

Example 2.4 (continued)

In Fig. 2.17 we present the simulated Z_{th} curve of Fig. 2.12 again, with the three bumps corresponding to the time constants and magnitudes, and also the constituting three exponential curves as *Exp1*, (R1 = 1 K/W, $\tau1$ = 100 μs); *Exp2*, (R2 = 1 K/W, $\tau2$ = 10 ms); and *Exp3*: (R3 = 4 K/W $\tau3$ = 400 ms).

It can be observed that the Foster representation gives some information on the nature of the Z_{th} curve, the time constants correspond to some extent to the position of the bump, while the magnitude refers to the curvature at that location.

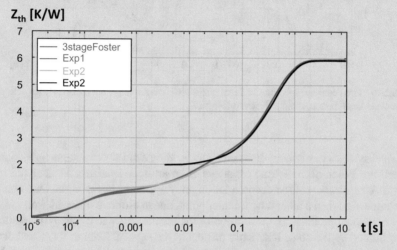

Fig. 2.17 Z_{th} curve with the three exponential components shown

If we calculate back the time constants from the simulated thermal transient curve with the NID method; we obtain the time constant spectrum shown in Fig. 2.18.

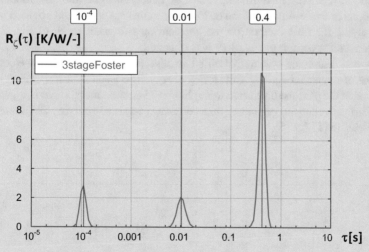

Fig. 2.18 Time constant spectra acquired by the NID method in the three-element lumped element system

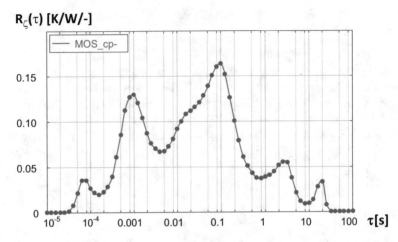

Fig. 2.19 Time constant spectrum of a real distributed parameter system (MOSFET on cold plate) calculated from the measured transient of Fig. 2.8

Real systems have many time constants forming a quasi-continuous time constant spectrum. When discrete time constants of an artificial system are calculated in an *iterative realization* of the deconvolution, they appear as peaks in the time constant spectrum, smashed a bit around a center point, due to different inherent limitations of any numerical process. The resolution of the results obtained by the NID method is also limited by the noise inherently present in the signals. Even in simulated curves (that can be extremely accurate), there is always a quantization noise present. Note however that despite limitations of the resolution, the locations of the maxima of the time constant spectrum such as the ones shown in Fig. 2.18 are exactly at the same time constant values as calculated directly from the element values of the Foster-type network model presented in Fig. 2.16. A deeper insight into the fundamental limitations in the resolution of the restored time constant spectra is provided in [58].

Applying the NID method on the transient measurement result of Fig. 2.8, the time constant spectrum of Fig. 2.19 was obtained.

The real distributed system (MOSFET on cold plate) consists of many elementary portions of matter forwarding and storing energy. The result of discretization can be observed in the plot; the time scale was split into 75 equidistant $d\zeta$ slices in logarithmic time. Further reading on generating time constant spectra and verifying their correctness is available in [174, 175].

2.4.2   The Structure Functions

Discrete time constants or a time constant spectrum can be produced from a measured $T(t)$ curve with the process presented so far. Still, the equivalent Foster circuit is a behavioral model[7] only.

In reality however, a slight change at a material layer in an actual physical object distorts many time constants in the chain. A physically sound approach can be based on the equivalent Cauer model of the system where we have a chain of RC stages with nodal capacitances, as illustrated in Fig. 2.20. In many cases elements of such a

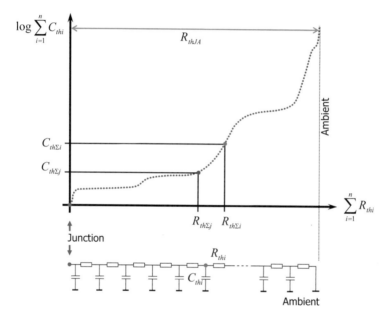

Fig. 2.20 Structure function: the graphic representation of the thermal RC equivalent of the system

---

[7] The behavioral nature of the Foster model can be endorsed by several arguments. As defined in Sect. 3.1, a thermal capacitance component in the network model of a thermal system represents the heat storage capability of a material region, lumped into a node of the network model. It describes the temporal changes of the temperature of that node with respect to the ambient; hence, it is referred as *nodal capacitance*. Accordingly, in an RC network model of a thermal system, it is represented by a thermal capacitance between a node of the model and the thermal reference point (ambient). Since the Foster-type models consist of *parallel* RC stages between adjacent nodes, neither the thermal resistances nor the thermal capacitances in it can be associated directly to the physical structure of the heat-flow path.

Equation (2.16) represents the Foster model in a mathematical formula. One can sum up the constituents in any order; the way in which sequence the individual stages are placed into the chain does not affect the overall temperature response at the junction. Based on the simple principle of mechanics that the junction is the tiniest "part" in the heat-conducting path and then smaller constituents are always mounted on larger ones, during the Foster-Cauer conversion, the stages are processed in growing order of $\tau_i = R_{th_i} \cdot C_{th_i}$ thermal time constants.

model can be attributed the different portions of the heat-flow path structure; changing the order of two RC stages in a Cauer model results in a different $T_J(t)$ junction temperature response. That is, a Cauer-type network model is not only another behavioral model of a thermal system, but it is also characteristic to the thermal system like a *signature*. Based on the so-called frequency domain calculus of the linear RC networks, there are standard procedures for the Foster $\leftrightarrow$ Cauer model transformations (see, e.g., Annex C of [40]).

Thus, a direct synthesis method for transforming the measured junction temperature transients into a *compact model* of the physical structure of a complex 3D thermal system is provided by accomplishing the deconvolution process, then discretizing the obtained thermal time constant spectrum and converting it into a Foster-type network model, and completing the Foster $\rightarrow$ Cauer transformation in a sequence. An early formulation of the concept was given in [137] for analogous mechanical problems, a modified method is presented in [136].

The obtained compact thermal model is a direct, physical representation of heat-flow path sections in which the heat spreading occurs in a true *one-dimensional* (1D) manner. Moreover, in cases where the spreading pattern can be expressed as a simple function of a single space coordinate, introduced as *essentially 1D* spreading in Sect. 2.5, the physical structure can be identified in a similar way. In all cases, regardless of the 1D, essentially 1D or complex 3D nature of the actual heat-spreading pattern, the Cauer-type models and their further representations are the *unique thermal signatures* of the physical structure of a semiconductor device package.

The first equivalent representation of a Cauer-type RC ladder model describing the heat-flow path is a graph, called *structure function* (shown in Fig. 2.20).

The quantities shown on the axes in the figure are the *cumulative thermal resistance*, defined as

$$R_{th\Sigma} = \sum_i R_{thi} \qquad (2.22)$$

and the *cumulative thermal capacitance*

$$C_{th\Sigma} = \sum_i C_{thi} \qquad (2.23)$$

In other words, starting from the driving point (the junction), we cumulate (sum) the partial thermal resistance and thermal capacitance values for of all subsequent heat-flow path sections. If we interpret the cumulative thermal capacitance as function of the cumulative thermal resistance, we obtain the *structure function* sometimes called *cumulative structure function*, often abbreviated as *SF*:

$$SF = C_{th\Sigma}(R_{th\Sigma}) \qquad (2.24)$$

The structure function is an excellent graphical tool to visualize the heat-conducting path. In accordance with the ladder of the figure, this chart sums up the thermal resistances, starting from the heat source (junction) along the x-axis and the thermal capacitances along the y-axis.

In *low gradient sections* of the structure function, a small volume, representing small thermal capacitance, causes large change in the thermal resistance. These regions have either *low thermal conductivity* or *small cross-sectional area*, or both. *Steep sections* in the function correspond to material regions of either *high thermal conductivity* or *large cross-sectional area*, or both. Sudden breaks of the slope belong to material or geometry changes.

Thus, thermal resistance and capacitance values, geometrical dimensions, heat transfer coefficients, and material parameters can be directly read on structure functions.

It is sometimes easier to identify the interface between the sections using the derivative of the cumulative curve: the *differential structure function*. Here peaks correspond to regions of high thermal conductivity like the chip or a heat sink, and valleys show regions of low thermal conductivity like die attach or air. Interface surfaces are represented as inflexion points between peaks and valleys.

From (2.6) and (2.8), we can say:

$$DSF = \frac{dC_{th\Sigma}}{dR_{th\Sigma}} = c_V \cdot dx \cdot A \cdot \left(\frac{1}{\lambda}\frac{dx}{A}\right)^{-1} = c_V \cdot \lambda \cdot A^2 \qquad (2.25)$$

The *differential structure function* (frequently abbreviated as *DSF*) yields information on the cross-sectional area along the heat conduction path. Further reading on producing structure functions and verifying their correctness is available in [176, 177].

In order to show how structural changes are represented in a structure function, we analyze below a simple artificial thermal model used in Example 2.3.

Example 2.5: The Structure Function of a Model Network

In order to demonstrate the easy usability of the structure functions, let us consider the following, still artificial example. In Fig. 2.21 we present two Cauer-type RC networks. The upper one is the converted Cauer-style version of Fig. 2.10. It can be observed that the thermal resistance and capacitance values slightly differ from the ones in the original Foster network, but based on the linear network theory, they produce equivalent thermal response. In other words, they have identical driving point impedance.

For making the example more plausible, we divided the three stages into two portions again: we assigned two stages to the device model; the third represents the test bench.

(continued)

Example 2.5 (continued)

Identification of structural elements in a system can be best facilitated with intentional material or geometry changes at relevant structural interfaces.

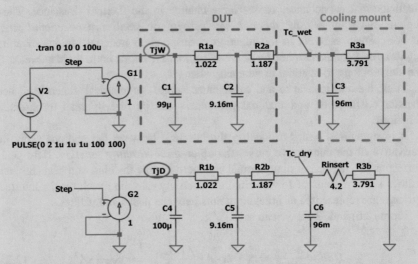

Fig. 2.21 The circuit scheme of Fig. 2.10 converted to Cauer equivalent. Two stages represent the device model; the third is the model of the test bench. Different thermal interface material quality is considered in the difference of the thermal resistance towards the ambient

To demonstrate this, the circuit scheme was altered to express the effect of different thermal interface materials between the device under test (DUT) and the test bench or cooling mount part. The different material quality was modeled by changing the 3.791 K/W thermal resistance towards the ambient to approximately 8 K/W in the second circuit.

The model of Example 2.5 resembles the concept of the broadly used JEDEC JESD 51-14 thermal measurement standard [40], also called the transient dual interface method (TDIM), developed for the testing of devices with an exposed cooling surface, such as packages with a cooling tab or modules with a baseplate. This separates the device under test from the cooling environment based on two thermal transient measurements, once on a cold plate without applied thermal paste ("dry" boundary) and then with appropriate wetting with a high-quality thermal paste material ("wet" boundary). This latter scheme corresponds to the actual intended mounting of power devices.

The inserted 4.2 K/W thermal resistance for the "dry" case can be interpreted as the thin air gap between the device and the cooling mount on a dry surface.

In the simulation example, 2 W power step was applied at $t = 0$ to the **TjW** and **TjD** input points. The simulated thermal transient curves are shown in Fig. 2.22.

T_J [°C]

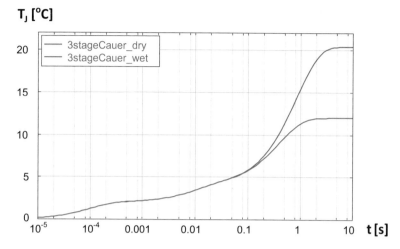

Fig. 2.22 Simulated temperature response of the two system variants at 2 W applied power

Z_th [K/W]

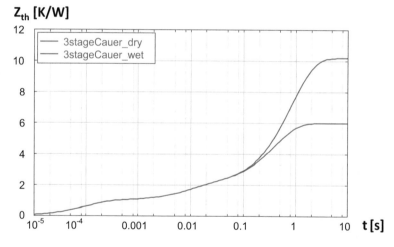

Fig. 2.23 Z_{th} curves of the two system variants

We can see in the figure that the two curves overlap up to about 0.2 s, but after that the junction temperature in the "dry thermal interface" case starts to increase much faster than in the "wet" case.

The same phenomenon can be observed of course also on the Z_{th} curves.

The time constant spectra of Fig. 2.24 are generated from the Z_{th} curves of Fig. 2.23 using the T3Ster-Master standard transient evaluation tool of a thermal tester equipment [54].

The simulated network represents a discretized system with three discrete time constants, while real systems are always continuous ones. The thermal transient evaluation software carries out mainly the steps defined in the previous sections.

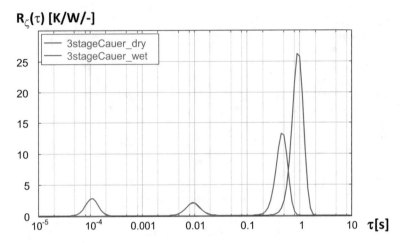

Fig. 2.24 Time constant spectra of the two system variants

Accordingly, it produces a continuous spectrum as defined in (2.21). A discretized time constant set can be produced from it by identifying the position of the three peaks and interpreting the area under the peaks as its magnitude.

There are various ways to accomplish the deconvolution process. A frequently used methodology is based on an iterative algorithm [58]. The actual deconvolution process resulted in the curve of Fig. 2.24 in 1000 iterative steps. Higher iteration number would result in sharper peaks around the time constants but with the same magnitude.

In Fig. 2.25 the structure functions of the thermal network of Example 2.5 are shown. The steep elevations correspond to the resistive elements and the flat plateaus to the capacitances in the Cauer-type representation. The readout of the values was performed with manual cursor positioning in the evaluation software [54] on the elevations and plateaus; an insignificant difference of these measured values from the original values can be observed.

It can be observed that the curves belonging to the "wet" and the "dry" boundary condition mostly coincide until the heat propagation in the "DUT" part in our example of Fig. 2.21 is represented, showing the fact that the two cases differ only in the last part of the heat-flow path.

The partial thermal resistances around 1 K/W each and the total R_{thJA} junction to ambient thermal resistance, 6 K/W for the "wet" and 10 K/W for the "dry" boundary, can be easily identified, and so are the appropriate thermal capacitances.

In real structures the steps in the structure functions are obviously less expressed, as demonstrated later in Example 2.7.

In Fig. 2.26 the differential structure functions of the thermal network and the manually measured peak positions, corresponding to the highest steepness in the cumulative structure function, are shown.

The next examples help in understanding the use of the structure functions.

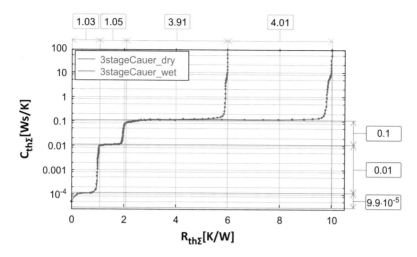

Fig. 2.25 (Cumulative) structure functions of the thermal network in Fig. 2.21. The steep elevations correspond to the resistive elements and the flat plateaus to the capacitances in the Cauer-type representation

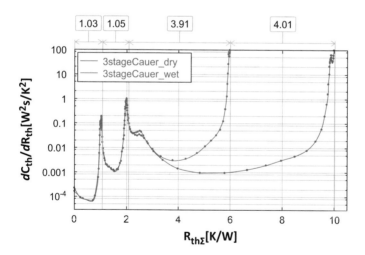

Fig. 2.26 Differential structure function of the thermal network in Fig. 2.21. The peak positions correspond to the resistive elements in the Cauer-type representation

---

Example 2.6: Analysis of the Heat Transfer in a Homogeneous Rod

A homogeneous rod with thermal boundary conditions is shown in Fig. 2.27. This rod can be considered as a series of infinitesimally small material sections as discussed above. Consequently, its discretized network model would also

(continued)

Example 2.6 (continued)

be a series connection of the single RC stages as shown in the figure. Thus, with this slicing along the heat conduction path, we create a ladder of lateral thermal resistances between two thermal nodes, and thermal capacitances between a node and the ambient.

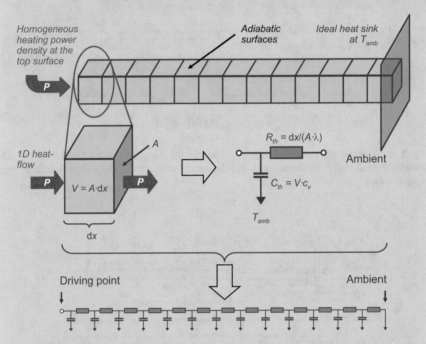

Fig. 2.27 The RC model of a narrow slice of the heat conduction path with perfect one-dimensional heat flow and the Cauer-type network model of the thermal impedance of the entire heat-flow path

Since homogeneity is assumed, the ratio of the elementary thermal capacitances and thermal resistances in the network model shown in Fig. 2.27 is constant. This means that the structure function of the rod is a straight line – its slope is determined by the C_{th}/R_{th} ratio of the network model and its differential structure function would be a constant – equal to the C_{th}/R_{th} ratio of the element values, as shown in Fig. 2.28.

This rod example demonstrates that the features of the structure functions are in a one-to-one correspondence with the properties of the heat conduction path.

Let us assume that in a given section in the middle of the rod, the C_{th}/R_{th} ratio is doubled. This results in a steeper middle section in the cumulative structure function (with the slope doubled) and in a peak in the differential structure function (which is twice as high as the constant value of the other sections). This is illustrated in Fig. 2.29.

(continued)

Example 2.6 (continued)

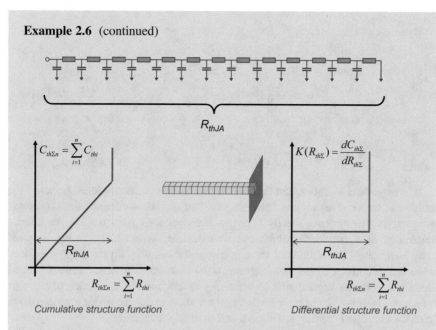

Fig. 2.28 The cumulative and differential structure functions of a homogeneous rod

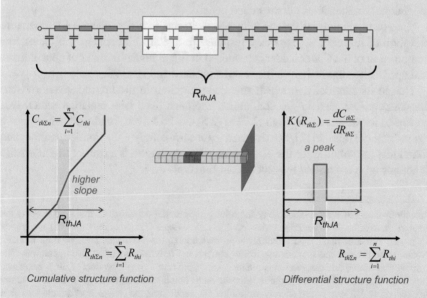

Fig. 2.29 The structure functions indicate the changes in the C_{th}/R_{th} ratio along the heat conduction path

(continued)

Example 2.6 (continued)

Obviously, this "doubling" of the C_{th}/R_{th} ratio can be the result of reaching a material section with different λ or c_V material parameters, or of larger cross-sectional area.

This discretized model offers a way to determine the thermal behavior of the rod. At any driving point excitation, an analog simulator tool, e.g., [57] solving the response of the RC ladder yields the temperature of any point within the rod, as it varies in time.

It is important to notice that there is no one-to-one correspondence between the number of material layers in a laminate structure and the number of time constants assigned to the system. Strictly taken, a homogeneous rod also has an infinite number of time constants, which can be calculated with (2.21). In the frequent case, however, when a material stack is composed of bulky layers of high conductance and thin layers of low conductance, a few characteristic time constants can be identified from the capacitance of the bulky layers and the resistance of the thin layers. A Cauer $\rightarrow$ Foster backwards transformation yields the major time constants from the identified thermal resistances and thermal capacitances.

Still, when the smallest thermal time constant of a system is to be ascertained, a generally good estimation can be the thermal capacitance of the chip multiplied by the thermal resistance of the die attach layer.

The RC chain normally starts with a capacitance, in order to avoid a temperature elevation of infinite steepness as a response to a sharp power step. An alternative composition of the Cauer ladder is proposed in [9]; at higher number of constituents, that approach is equivalent to the scheme outlined in Fig. 2.29.

Due to its simplicity, the heat spreading problem in the homogeneous rod has known solutions also in the continuous approach. The heat equation which was presented in various forms from (2.3) to (2.5) can be solved for any x position along the rod at any t time. In [1], Fourier used – not surprisingly – the Fourier method, and found that the solution for the $T(x, t)$ at any $P(t)$ excitation is a sum of trigonometric functions on x multiplied by exponential functions on t.[8]

---

[8] An alternative powerful methodology for solving differential equations of this kind is using the Laplace transform. This technique transforms the time-dependent equations into the s complex frequency domain, finds their solution there, and transforms the result back to time domain. Notably, the Fourier and Laplace transforms convert not only functions but also operations. The convolution of two functions in time domain is converted into a simple product of the transformed functions. Calculation of a temperature response on a complicated $P(t)$ function in the convolution integral of (2.12) can be effectively simplified by converting the $g(t)$ and $P(t)$ functions into corresponding $g'(s)$ and $P'(s)$ functions in the complex frequency domain, and then converting their $g'(s) \cdot P'(s)$ product back to time domain. A useful table of Laplace transformed forms of a bunch of functions can be found in [4].

The procedure of the Foster to Cauer-style conversion of RC circuits is also formulated in the s complex frequency domain in Example 2.1 and in Annex C of the measurement standard [40].

In the actual example, Figs. 2.28 and 2.29 represent a case with steady heat flux at the driving point side of the rod (Neumann boundary condition of heat transfer) and steady temperature on the ambient side (Dirichlet boundary condition of heat transfer). Solving the equation for the driving point, we get the already known sum of exponentials as outlined in (2.16) and (2.17).

A more compact analytical solution can be gained with further simplification of the problem. A result of practical importance was introduced in [18] followed by [62] and is used in [40].

Let us apply a P_0 power step on a rod of infinite length, that is, long enough in the sense that its far end remains at ambient temperature for the intended time of investigation. Solving the heat equation in the complex frequency domain at these boundary conditions [62], obtains for the time dependence of the T_J temperature of the driving point:

$$\Delta T_J(t) = \frac{P_0}{A} \cdot k_{\text{therm}} \cdot \sqrt{t} \qquad (2.26)$$

where the k_{therm} is

$$k_{\text{therm}} = \frac{2\sqrt{\alpha}}{\lambda\sqrt{\pi}} = \frac{2}{\sqrt{\pi \cdot c_V \cdot \lambda}} \qquad (2.27)$$

The thermal diffusivity α used in (2.27) again was defined as $\alpha = \lambda/c_V$ formerly at Eq. (2.5); it is the measure of thermal inertia. In a material of high thermal diffusivity, heat moves rapidly, and the substance conducts heat quickly relative to its volumetric heat capacity.

In practical constructions where the heat generates in a thin "junction" layer on the top of a block of homogeneous material, and starts spreading in that block, the initial section of the temperature transient can be precisely approximated by a square-root-time function.

When the heat spreading reaches the other end of the homogeneous block, then the temperature change takes another shape.

If the homogeneous block is of d thickness conditions, [62] claims that the square-root rule remains valid for short early times of t_v duration:

$$t_v < \frac{d^2}{2\alpha} \qquad (2.28)$$

Several thermal properties of typical materials used in the construction of power devices are listed in Table 2.1 (silicon at 25 °C and 125 °C, copper, solder die attach material).

It has to be noted that very thin layers of high thermal conductivity add a very small portion to the temperature elevation of a laminate composed of layers of several constructional materials. In these cases, the "square root of elapsed propagation time" style elevation of the temperature belonging to the next layer can be observed in measured thermal transients.

Table 2.2 lists the valid duration for the square-root-time approach for different materials and die (block) thickness. In the table typical semiconductor and die attach

Table 2.1 Thermal properties of typical materials in power devices

	λ [W/mK]	c_V [J/m³K]	α [mm²/s]	k_{therm} [m²K/W√s]	k_{therm} [mm²K/W√s]
Si@25 °C	125	$1.60 \cdot 10^6$	78.3	$7.99 \cdot 10^{-5}$	79.9
Si@125 °C	100	$1.60 \cdot 10^6$	62.7	$8.93 \cdot 10^{-5}$	89.3
Die attach	70	$1.66 \cdot 10^6$	42.1	$1.05 \cdot 10^{-4}$	104.6
Cu	390	$3.40 \cdot 10^6$	114.9	$3.10 \cdot 10^{-5}$	31.0

Table 2.2 Valid duration t_v values for the square-root-time approach

		Si@25 °C	Si@125 °C	Die attach	Cu
	d [mm]	t_v [ms]			
	0.1	0.063	0.056	0.048	0.161
	0.3	0.56	0.50	0.43	1.45
	0.5	1.56	1.40	1.20	4.03
	1				16.1
	2				64
	4				258

layer thickness values are shown. Copper is used in very thin layers on printed boards and direct bonded copper (DBC) constructions, and also as bulk material in cold plates. The table helps assigning the subsequent homogeneous spreading regions which can be observed in measured transients to material layers, based on the time range where the square-root-type temperature change occurs.

Example 2.7: Structure Functions of a Real Device

In Fig. 2.30 structure functions of a MOSFET device on a cold plate are shown. This assembly has been used in the former sections as an example for a distributed thermal system.

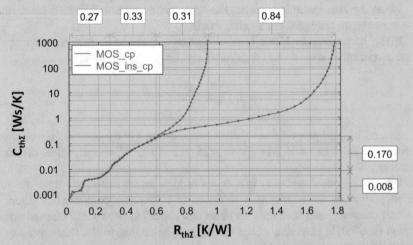

Fig. 2.30 Structure functions of a real distributed parameter system (MOSFET on cold plate, different TIM qualities) with characteristic R_{th} and C_{th} values

(continued)

Example 2.7 (continued)

Curve MOS_cp was derived from a thermal transient test when the MOSFET device was mounted on a water-cooled cold plate wetted by a high-quality thermal paste. From the cooling curve (Fig. 2.9), the NID methodology produced 160 RC stages in 1000 iteration steps; these are represented in the time constant spectrum of Fig. 2.19 in a quasi-continuous manner, without displaying each τ value separately. The Foster $\rightarrow$ Cauer transformation converted these into other 160 RC stages, for which the first 100 are shown as blue dots in Fig. 2.30. The remaining 60 stages are in the 1000 J/K to 10^{38} J/K thermal capacitance range and are not displayed because they are not relevant for the actual study.

In order to distinguish between the device and the test environment, the transient measurement was repeated inserting a ceramics sheet of 2.5 mm thickness between the device package and the cold plate. The transient measurement and the subsequent structure function calculation resulted in curve MOS_ins_cp.

This comparison of two structure functions of a device measured at different boundary conditions can be used for deriving standard thermal metrics, as expounded in Sect. 3.1.2 and standardized in [40]. A deeper analysis of structural details which can be recognized in the structure functions will be given in Sect. 7.1 in Example 7.1.

It can be observed that Z_{th} curves in previous figures do not disclose too much details of the structural composition; practically only the junction to ambient thermal resistance value or with multiple boundary conditions an approximate partial resistance until a divergence point, also called bifurcation point, can be read in them. The reason is that the equivalent thermal RC network of the system behaves as a low-pass filter; the sharp power step at its input is converted into the smoothed bumps of the thermal impedance function. On the other hand, the deconvolution algorithms, which produce the structure functions, are closely related to the image enhancement procedures which recreate lost fine details in a blurred picture.

In structure functions many details can be distinguished along with their partial thermal resistance and capacitance value. Still, it has to be mentioned that the structure function analysis is *not* a fully automated ("black box") technique.

There are three ways to assign actual assembly components to sections in the structure function. These are:

- The manufacturer of the device may know all internal geometries and material parameters. In such a way, a "synthetic" structure function can be built up, for example, superposing slices of material with given thermal resistance and capacitance in a spreadsheet tool, and comparing the measured structure functions to it.
- An approximate model can be built up in a finite element or a finite difference simulation tool, such as [56]. Thermal transients can be simulated in the tool and

structure functions can be composed of those. Geometry and material parameters can be fine-tuned until the simulated and measured structure functions match.
- Measured structure functions can be compared to an already identified "golden device." This technique is advantageous in production control.

In the case of Example 2.7, it was easy to measure the external dimensions of the standard TO-220 package which hosts the semiconductor chip. The size of the chip was determined by sectioning the package after the transient test. The assignment of the parts of the structure function to internal details of the same package is presented in Example 7.1 of Chap. 7, Sect. 7.1.

Without this thorough analysis, some characteristic portions of the assembly can be identified in the plot. The structure functions perfectly coincide until 0.6 K/W thermal resistance and 0.178 J/K thermal capacitance, hinting that until this point the heat propagates within the packaged device and the different TIM quality still did not affect the spreading.

The deeper investigation given in the example proves that the first section of the structure functions until 0.27 K/W and 8 mJ/K can be identified as a small silicon chip in the package and the die attach. The next section with 0.33 K/W partial thermal resistance and 0.17 J/K thermal capacitance can be attributed to the heat spreading in the copper tab of the package.

Beyond the identification of the structural elements within the package and the junction to case thermal resistance, also the thermal conductivity of the ceramics can be calculated from the chart. The inserted sheet with its 2.5 mm thickness added 0.84 K/W to the total junction to ambient thermal resistance. The effective cross-sectional area of the heat spreading was limited to the copper surface of the tab, which was 13 mm × 9 mm. According to (2.6), it follows from these geometrical data that the thermal conductivity of the ceramics is $\lambda = 25$ W/mK, a plausible value for sintered alumina material.

The structure function types introduced so far correspond to a one-dimensional mapping of the change of thermal resistance and thermal capacitance along the heat-conducting path. They depict how these local thermal quantities attributed to a section in an assembly change while advancing in the structure from the junction towards the ambient.

The (cumulative) structure function $C_{th\Sigma}(R_{th\Sigma})$ demonstrates the growth of the total cumulated thermal capacitance as a function of the total thermal resistance along the heat-flow path. In an alternative view, the differential structure function $dC_{th\Sigma}/dR_{th\Sigma}(R_{th\Sigma})$ was introduced, representing the change of the ratio of the thermal capacitance and thermal resistance, versus the total thermal resistance.

In both representations the thermal properties of structural elements can be identified; these carry the same information; still certain features are more perceptible in one or other form. For example, material interfaces induce a change of steepness in a cumulative structure function, but in its typical logarithmic portrayal, this change may become less apparent. The differential structure function enlarges these differences as obvious local maxima and minima, forming peaks and valleys. Volumes can be best measured in the cumulative version; material interface locations are often attributed to inflections in the differential one. In other words, the cumulative structure function gives answers on questions of "how much is what we look for?" the differential function rather identifies "where is what we look for?".

2.4.3 The Local Thermal Resistance Function

Some structural elements in an electronics assembly are of well-defined geometry and highly repeatable material properties. Such components are the semiconductor chips and the metal or ceramics parts of the package; their structure is stable in the manufacturing process and later during their lifetime.

As opposed to the above well-defined structural elements, the layers that connect them, the thermal interface materials (TIMs), may show high scatter. Solders and adhesives can be of different thickness and sometimes also of different structure after manufacturing, because of variations in processing steps and heat treatments. Thermal pastes change also later, depending on applied pressure and temperature fluctuations in the normal use of an assembly.

One can observe that the stable constituents are of higher thermal conductivity, resulting in lower thermal resistance. The TIM layers are often very thin, and they add accordingly a smaller portion to the thermal capacitance of a stack; still they contribute a large thermal resistance.

An important aim of structure testing is to find the location of the critical parts of high variation in the structure and to follow their change. A way to magnify the differences in the R_{th} values along the heat-conducting path can be drawing *local R_{th}* values as function of the cumulated thermal capacitances.

This $R_{th}(C_{th\Sigma})$ *local thermal resistance function* is also a graphical representation of the Cauer ladder: on the horizontal axis, the sum of the C_{th} elements and on the vertical axis the next R_{th} element of the chain are shown.

The $C_{th\Sigma}$ values grow monotonously from the origin of the heat towards the ambient. Due to the steadiness of the stable components and the low share in the total thermal capacitance of the TIM layers, the horizontal axis can be considered to correspond to the geometrical location.

Figure 2.31 presents the local thermal resistance functions of four packaged LED samples, soldered to aluminum starboard and mounted on a temperature-controlled cold plate. Sample 1C_W has a serious delamination problem, which can be identified as a high thermal resistance peak in the 0.2 mJ/K–30 mJ/K range of the local thermal resistance function.

A further example on the use of the local thermal resistance function is presented in Sect. 7.1, Example 7.3. The die attach delamination problem in Fig. 2.31 is treated in depth in Example 7.4.

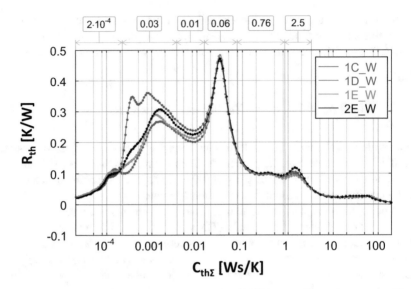

Fig. 2.31 Local thermal resistance functions of packaged LED samples on a temperature-controlled cold plate. Sample 1C_W has a serious delamination problem, as it can be observed in the $C_{th\Sigma} = 0.2$ mJ/K–30 mJ/K range

2.5 Heat-Spreading Patterns in Regular Geometries and Their Appearance in the Structure Functions

In this section we examine some practical cases in which different heat-spreading types can be recognized in the structure functions.

Heat Spreading in a Generalized Tube, Essentially 1D Heat Spreading

Numerical methods can solve the equations of heat spreading for arbitrary shapes and material composition. In the first 150 years of the 200-year history of studying the laws of heat spreading, only analytical methods were available. They can still be used very effectively even today because they yield universal solutions; inserting a few parameters describing the geometry and the material properties, the results are instantly available.

The thermal resistance is always to be measured between two isothermal surfaces in a solid body. In many practical cases, the body can be constructed as a "tube" or "beam," with a surface of varying cross-sectional area shifted along a line or curve (Fig. 2.32).

If the area of the surface is $A(x)$ at position x on the curve, then the thermal resistance of a short dx section is $dR_{th} = (1/\lambda) \cdot (1/A(x)) \cdot dx$, and the added thermal capacitance of the section is $dC_{th} = c_V \cdot A(x) \cdot dx$. Both dR_{th} and dC_{th} are to be added to the cumulated quantities summed up until the point where the dx section started.

Between points x_0 and x_1, the increase of the thermal resistance and thermal capacitance can be calculated as

Fig. 2.32 Two isothermal surfaces connected by a heat flux tube

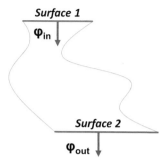

$$R_{th} = \int dR = \int_{x_0}^{x_1} \frac{1}{\lambda A(x)} dx \quad \text{and} \quad C_{th} = \int dC = \int_{x_0}^{x_1} c_V A(x) dx \qquad (2.29)$$

As suggested by the formulae in (2.29), for the heat transfer in a structure shown in Fig. 2.32, the spatial distribution of both the thermal resistance and the thermal capacitance can be represented as a continuous function of the same independent variable. If this is the case, one can say that the heat spreading in the investigated region is *one dimensional* (see the example of a homogenous rod shown in Figs. 2.27 and 2.28) or *essentially one dimensional*, as discussed below. If the actual dependence of the thermal resistance and thermal capacitance on that common independent variable is known, then the $C_{th}(R_{th})$ relationship, i.e., the structure function, is also known, in certain cases given also by analytic formulae.

The Classic 1D Solution: 1D Longitudinal Spreading

As already analyzed in previous sections, the solution of (2.29) is obvious when the area of the isothermal surfaces is constant between the starting and final positions. The object can be considered to be a block or a cylinder, not necessarily a right circular one.

Such a cylinder can be a stand-alone object with a heater on one of its base surfaces, but can also be interpreted as a cylindrical protrusion on a larger object, which emits heat flux into a base surface (Fig. 2.33).

The thermal resistance and capacitance along an L length between x_0 and x_1 positions can be calculated inserting the constant A area as

$$R_{th}(L) = \frac{1}{\lambda A} \cdot L, \quad C_{th}(L) = c_V A \cdot L \qquad (2.30)$$

Expressing the L length from the first formula and inserting it into the second, we get the analytical expression for the structure function of a cylindrical body:

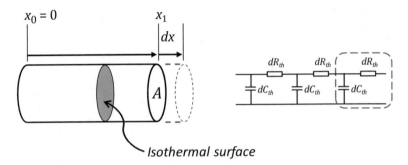

$x_0 = 0$ x_1

Isothermal surface

Fig. 2.33 Longitudinal heat spreading in a cylindrical object with constant A isothermal surface area

$$L = \frac{1}{\lambda \cdot A} \cdot R_{th}, \quad C_{th}(R_{th}) = \lambda \cdot c_V \cdot A^2 \cdot R_{th} \tag{2.31}$$

The differential structure function is $dC_{th}/dR_{th} = \lambda \cdot c_V \cdot A^2$ again, but this formula is always valid as previously shown in (2.25).

The Classic 2D Solution: Essentially 1D Radial Spreading

When a smaller heat source is mounted on a heat spreader plate of w thickness, then the heat propagates radially, with concentric isothermal surfaces around the heater position, at least after a certain distance from the position of a heater.

Such a situation can also be represented as a sort of one-dimensional heat spreading, requiring only a transformation of the spatial distribution of the thermal resistance and thermal capacitance to a radial coordinate system. Those cases of heat spreading when a similar transformation from a single space coordinate to another unique space coordinate is possible are considered *essentially one dimensional*.

In a hollow cylinder defined by annular base and w thickness, the dR_{th} and dC_{th} increases can be formulated easily again.

As Fig. 2.34 hints, at radius x the perimeter of a ring is $2\pi x$; thus, in a layer of thickness w, the isothermal lateral surface area around the ring is $A(x) = 2\pi x w$. Adding a thin annular shell of dx infinitesimal thickness to the propagation profile, the lateral thermal resistance growth will be $dR_{th} = 1/[\lambda \cdot A(x)] \cdot dx$, and the growth in the thermal capacitance when adding an annulus to the existing profile will be $dC_{th} = c_V \cdot A(x) \cdot dx$.

Integrating between x_0 and x_1 positions, one gets for R_{th}

$$R_{th} = \int_{x_0}^{x_1} \frac{1}{\lambda \cdot w \cdot 2\pi} \cdot \frac{1}{x} dx = \frac{1}{\lambda \cdot w \cdot 2\pi} \cdot \ln\left(\frac{x_1}{x_0}\right) \tag{2.32}$$

and for C_{th}

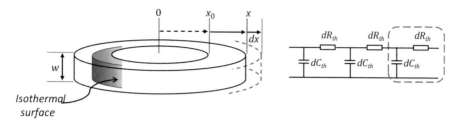

Fig. 2.34 Radial heat spreading in a plate of w thickness

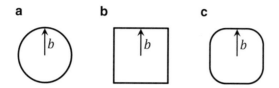

Fig. 2.35 The perimeter of various 2D profiles can be calculated as a characteristic "feature size" b multiplied by an s_p ratio. For the circle (**a**), the ratio is $s_p = 2\pi$, for the square (**b**) $s_p = 4$, while for an arbitrary rounded rectangle (**c**) such a ratio cannot be given easily

$$C_{th} = \int_{x_0}^{x_1} c_V \cdot w \cdot 2\pi \cdot x \cdot dx = c_V \cdot w \cdot \pi \cdot (x_1^2 - x_0^2) \qquad (2.33)$$

In (2.33) one can recognize the area of the annular base as $\pi \cdot (x_1^2 - x_0^2)$. The descriptive formulae for the change of thermal resistance and thermal capacitance are analogous to (2.32) and (2.33) also in other similar cases, when the spreading occurs in a material sheet of w thickness on profiles of *geometrical similarity* along a radial coordinate x in a polar coordinate system. The 2π ratio between the perimeter and the radius of the circle will be replaced by a different s_p factor for other profiles characterized by a b "feature size" (Fig. 2.35).

An example for spreading in circular sectors of growing radius is shown in Chap. 3, Sect. 3.1.2 (Example 3.1 and Figs. 3.3, 3.4, 3.5, 3.6, 3.7, and 3.8).

Expressing x_1 from (2.32) and inserting it into (2.33) in two steps, one gets:

$$\frac{x_1}{x_0} = e^{\lambda \cdot w \cdot 2\pi R_{th}} \qquad (2.34)$$

$$C_{th} = c_V \cdot w \cdot \pi \cdot x_0^2 \left(\frac{x_1^2}{x_0^2} - 1 \right) = c_V \cdot w \cdot \pi \cdot x_0^2 \cdot \left(e^{4\pi\lambda \cdot w \cdot R_{th}} - 1 \right) \qquad (2.35)$$

The (cumulative) structure function is typically plotted in a lin-log coordinate system as exposed in Sect. 2.4.2.

At larger radii of the heat spreading where $x_1 \gg x_0$, the $\ln(C_{th}(R_{th}))$ function will become a straight line, as (2.35) indicates.

Equation (2.35) is of $D \cdot \exp.(4 \pi \lambda w R_{th})$ form; thus

$$\ln(C_{th}) = \ln(D) + 4\pi\lambda \cdot w \cdot R_{th} \tag{2.36}$$

Taking two points of the structure function plot on a straight section of the $\ln(C_{th}(R_{th}))$ function:

$$\ln(C_{th2}) - \ln(C_{th1}) = \ln(C_{th2}/C_{th1}) = 4\pi\lambda \cdot w \cdot (R_{th2} - R_{th1}) \tag{2.37}$$

and λ can be determined as

$$\lambda \cdot w = \frac{1}{4\pi} \cdot \frac{\ln(C_{th2}/C_{th1})}{(R_{th2} - R_{th1})} \tag{2.38}$$

An example of determining the thermal conductivity of a substrate based on (2.38) is presented in [65].

The Classic 3D Solution: Essentially 1D Conical Spreading

A coordinate transformation can also map a true three-dimensional heat spreading into a corresponding one-dimensional one, if the isothermal surfaces of the spreading conform the principle of *similarity*. In these cases the $A(x)$ surfaces along an x space coordinate grow with a scale factor, or ratio of similarity. For simplicity let us denote the constant scale factor as K.

For determining the scale factor of surfaces in space, first consider a cross section of two typical spreading patterns as shown in Fig. 2.36.

Figure 2.36a corresponds to the spreading in a truncated cone, or truncated pyramid, a standard concept of heat spreading, which is broadly treated in textbooks and also in the literature [60, 65]. In the figure the length of the y vertical sections is always proportional to their distance from the origin, $y(x) = m \cdot x$; the parameter m is the slope of the $y(x)$ function. The area of the flat surface will be $A(x) = y^2\pi = m^2x^2\pi$ if the isothermal surface is circular and $A(x) = 4y^2 = 4m^2x^2$ for a square surface. In

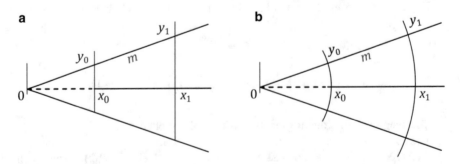

Fig. 2.36 Similar triangles and circular sectors characterized by a 1D growth parameter **m**

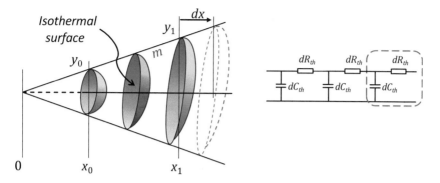

Fig. 2.37 Heat spreading in a "generalized cone," through isothermal surfaces characterized by similarity

both cases the surface grows with an $A(x) = K \cdot x^2$ formula. K can be also interpreted for other shapes, too, like the ones in Fig. 2.35c.

The formulae remain of the same style for spherical spreading, depicted in Fig. 2.36b. In this case a similar m scale factor determines the ratio of the y arc length and the x radius. In geometry m is called central angle, expressed in radians.

For a full sphere, $y = m \cdot x = 2\pi x$ and $A(x) = 4x^2\pi$. For an $m < 2\pi$ central angle value, the surface of the spreading is proportionally smaller:

$$A(x) = 4x^2\pi \cdot \left(\frac{m}{2\pi} \right)^2 = \frac{m^2}{\pi} x^2 \qquad (2.39)$$

which is a $K \cdot x^2$ formula again. Analogous formulae depict the growth of the isothermal surfaces of arbitrary shape as illustrated in Fig. 2.37.

The integral between x_0 and x_1 positions yields R_{th} as

$$R_{th} = \int_{x_0}^{x_1} \frac{1}{\lambda \cdot K} \cdot \frac{1}{x^2} dx = \frac{1}{\lambda \cdot K} \cdot \left(\frac{1}{x_0} - \frac{1}{x_1} \right) \qquad (2.40)$$

and C_{th} as

$$C_{th} = \int_{x_0}^{x_1} c_V \cdot K \cdot x^2 \cdot dx = \frac{c_V \cdot K}{3} \cdot (x_1^3 - x_0^3) \qquad (2.41)$$

It can be observed that this generalized "conical" spreading scheme corresponds to several realistic heat propagation patterns, including cones and pyramids.

Selecting the appropriate K scale factor the heat spreading can be calculated in a dome, starting from a small spot in an infinite half space, which is a valid approximation for chips with a hot spot, or bulky heat sinks farther from the mounted device. Similarly, in a larger distance from the actual investigated thermal system, the spreading in the ambient can be considered spherical, obeying (2.40) and (2.41).

The "factor of 1/3" in (2.41) is justified by geometrical considerations, the volume of a cone or pyramid is a third of the enclosing cylinder or block, and the volume of the dome or sphere is two thirds of the circumscribed cylinder. These ratios can be well recognized in actual measured thermal capacitances in real structure function.

In a 3D spreading, the "actual position to ambient" thermal resistance is of finite value; (2.40) yields for "infinite" conical and spherical spreading, from position x_0 towards the ambient $R_{thJA} = 1/(\lambda K x_0)$. For truncated cones and pyramids, and for spherical shells, (2.40) provides the textbook formula of

$$R_{th}(L) = \frac{1}{\lambda \cdot K \cdot b_0 \cdot b_1} \cdot L, \qquad (2.42)$$

where b_0 and b_1 are the measured "feature size" at the beginning and at the end of the generalized "cone." As defined in Fig. 2.35, b is the radius of a circular shape and half of the edge for a rectangle; K is π for the circle and 4 for the rectangle. It can be seen that (2.42) describes the heat propagation over growing surfaces along an x coordinate as a "generalized rod" where the surface in (2.30) is replaced by the product of the linear b_0, b_1 dimensions at the beginning and at the end. Still, the junction to ambient thermal resistance remains finite, as the L/b_1 ratio converges to the constant m factor for large lengths.

Further research results on the spreading shapes are published in [66, 67, 139].

As illustrated in Fig. 2.38, in real package structures, one often finds actual heat-flow paths with different sections that can be characterized with 1D or essentially 1D heat-spreading patterns discussed above.

In summary, we can state that in many cases, there is a one-to-one correspondence between the physical sections of the junction-to-ambient heat-flow path and the structure functions. There are cases, however, where there is a real, complex heat-spreading pattern within a package where one cannot identify any dominant heat-flow path. In such a situation, the structure function (as one of the representations of a junction-to-ambient thermal impedance) is the *thermal signature* of the

The *junction* as heat source

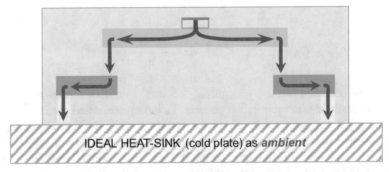

Fig. 2.38 Real package structures often can be represented as a series of different heat-flow path sections characterized by (essentially) 1D heat spreading

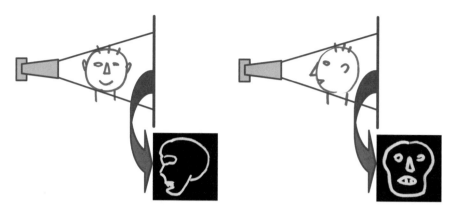

Fig. 2.39 Structure functions can be considered as "1D projections" of complex 3D heat-spreading patterns like X-ray images are 2D projections of three-dimensional bodies

system; it can be considered only as a one-dimensional "projection" of the complex 3D pattern. Note that this signature is still perfectly applicable to test the integrity of the structure.

Measuring the thermal impedance of such packages on cold plates helps un-blurring the image, though the heat is directed mainly into one major path, towards that surface of the package that is in direct contact with the cold plate during the test, while all other surfaces of the package are thermally isolated. Changing the test setup such that another surface of the package is in contact with the cold plate while the previous one is thermally isolated from the ambient directs most of the heat into another path.

The structure functions obtained this way are somewhat similar to one's taking X-ray images from different directions, as illustrated in Fig. 2.39. Thermal "CAT-scan images," though, cannot be constructed from measured junction temperature transients.

The DELPHI methodology [68] defines four different boundary conditions with various high and low heat transfer coefficients on different package surfaces for validating simulated package models by measurements. The actual boundary conditions are presented in Chap. 7, Sect. 7.7.3.

2.6 The Concept of the Heat Transfer Coefficient

In the previous section, the thermal properties of structural elements were formulated in Eqs. (2.30)–(2.42), and more generally in (2.29). The thermal resistances were calculated assuming that the geometry of the heat-conducting path is known in details.

Table 2.3 Typical values of convective heat transfer coefficient

Convection type	h [W/m^{2}K]
Air, free	2.5–25
Air, forced	10–500
Liquids, forced	100–15,000
Boiling water	2500–25,000
Condensing water vapor	5000–100,000
Surface of a cold plate	1000–5000

In many cases it is more practical to use the concept of the h heat transfer coefficient, the proportionality factor between the φ heat flux through and the ΔT temperature between two isothermal surfaces.

The h coefficient is in common use as a thermal figure of merit of commodities manufactured in specific thickness and composition. It is widely used for TIM sheets or coatings in electronics or similarly for characterizing thermal insulation or glazing in building industry.[9]

Another field of use is to characterize the termination of the heat-conducting path in an assembly. In case of convective cooling, many formulae are used for determining the cooling capability of a gas or fluid flow on a solid surface. The formulae are mostly empirical and take into consideration the nature of the fluid flow (natural, laminar, turbulent flow) and material-related coefficients (Prandtl, Rayleigh, Nusselt numbers, etc.). Table 2.3 lists approximate heat transfer coefficients for some convective cooling solutions used in power electronics. Boiling water and vapor condensation were included in the table because of the growing importance of heat pipes in electronics cooling.

As the heat transfer coefficient is defined as $h = \varphi/\Delta T$, the thermal resistance of an inserted material or a heat exchanger of A surface can be calculated as

$$R_{\text{th}} = \frac{1}{h \cdot A} \tag{2.43}$$

2.7 Driving Point and Transfer Impedances: Self-Heating and Transfer Heating

So far only such cases were investigated where a single point was heated, its own temperature was measured, and the whole system was reduced into a single thermal RC network model. There are many practical situations, however, where (due to different reasons not discussed here) one cannot measure the temperature response at the same location where the heating was applied. Typical examples are laterally arranged *multi-die packages* (such as illustrated in Fig. 2.40), stacked die packages,

---

[9]The heat transfer coefficient is denoted as U or K factor in the building industry.

Fig. 2.40 Axonometric view of the major parts of a package of a dual-chip power device with two heat sources (driving points) and three temperature monitoring points. T_1, T_2, and T_3 represent the temperatures of the two chips and the temperature of one of the pins of the device, respectively

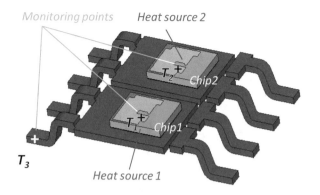

thermal test chips [32], LED packages with multiple LED chips, or PCB-assembled LED modules with a thermal test point on the board. In case of a multi-die package or an RGB LED module one is interested, how the heat dissipation of one semiconductor chip affects the temperature of the other dies in the system.

To allow a proper distinction and precise description of such cases, the concepts of *driving points* and *temperature monitoring points* were introduced in the technical literature [69].

The *driving points* are the *locations of the heat sources in the system* (i.e., the junctions of the chips).

The *temperature monitoring points* are the *locations where the temperature responses are measured*.

In most practical cases, as in the examples discussed previously, the *junctions* of the chips are both driving points and temperature monitoring points. The thermal impedance obtained from measurements when the power step is applied and the temperature response is captured at the same location is called *driving point thermal impedance* or *self-impedance* in short.

When the driving point and the temperature monitoring point are separated in space, the thermal impedance obtained is called *thermal transfer impedance*, in this book also referred to shortly as *transfer impedance*.

Example 2.8: Self-Heating and Transfer Heating in a DDR RAM Module

As an example for the driving point and transfer impedances, a DDR RAM module, mounted into the socket of a PC motherboard, was simulated. Applying 0.2 W power on the internal device `Chip1` in the leftmost RAM package `RAM_1`, the temperature distribution, which develops on the module surface in steady state, is shown in Fig. 2.41.

(continued)

Example 2.8 (continued)

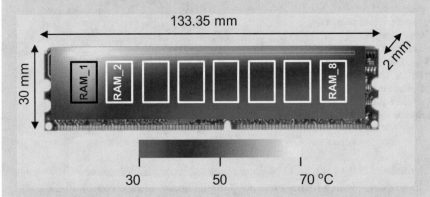

Fig. 2.41 Simulated temperature distribution in a DDR RAM module in PC socket, 0.2 W applied on the leftmost RAM package RAM_1

As a further example, Fig. 2.42 presents the simulated transients on the internal chips in Chip1, Chip2, and Chip8 encapsulated into packages RAM_1, RAM_2, and RAM_8. The self-heating curve of the driving point Chip1 starts growing at early time, a delay of 1.1 s can be observed in the transfer curve of Chip2, and it takes nearly 10s until the heat propagates towards monitoring point Chip8.

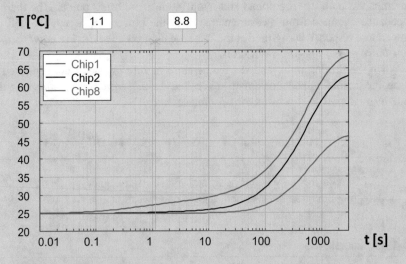

Fig. 2.42 Simulated temperature response of the DDR RAM module in PC socket, 0.2 W applied on the leftmost RAM package. Transients at the driving point Chip1 and monitoring points Chip2 and Chip8 are shown

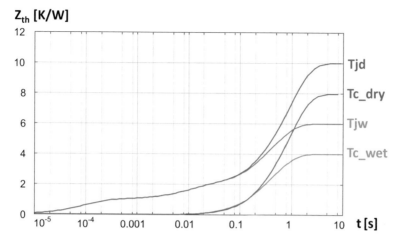

Fig. 2.43 Self-impedance curves Tjd and Tjw and transfer curves Tc_dry and Tc_wet of the two simple Cauer models represent the dry and wet boundary conditions in Fig. 2.21 of Example 2.5. The naming of the curves corresponds to the node names in the model

Physical considerations imply that the temperature response at a *single driving point* is always monotonous; the temperature constantly grows when power is applied and decreases when it is revoked. Similarly, also transfer curves remain monotonous if the structure is nearly one dimensional between the driving point and the monitoring point.

The simulation of the two simple Cauer-style models in Fig. 2.21 of Example 2.5 yields the Z_{th} curves of Fig. 2.43. The curves denoted with Tjd and Tc_dry correspond to the simulated self-impedance and transfer impedance of the model at dry boundary condition, represented by higher thermal resistance from the internal Tc_dry location towards the ambient. The curves Tjw and Tc_wet show the results for the wet boundary condition.

The *propagation delay* between the driving and the monitoring points is characteristic to the transfer impedances; it is approximately 10 ms in this assembly.

The propagation delay in the time domain curves of the transfer impedances results in some negative magnitude values in the time constant spectra.

The monotonous nature of the transfer curves cannot be assured in systems with multiple heat sources or multiple heat-conducting paths within the thermal system. The root cause of this non-monotonous behavior and its practical conscequences will be treated in detail in Sect. 3.5.

A thermal system with multiple driving points and temperature monitoring points (such as a multi-die system; see Fig. 2.44) can be fully represented if all its possible self- and transfer impedances are known, i.e., all of these thermal impedances are measured or simulated.

On the right side of Fig. 2.44, the Z_{ii}^* and Z_{jk}^* symbols representing all possible thermal impedances are arranged in a matrix. In [69] such matrices are called *thermal transfer impedance matrices*. This is the dynamic extension of the concept

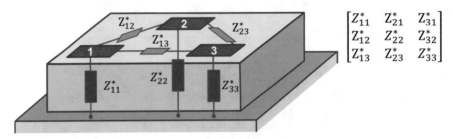

Fig. 2.44 A substrate with three heat sources (i.e., driving points) that are also used as temperature monitoring points and the illustration of the corresponding self- and transfer impedances

of the *thermal resistance matrix* introduced among others in [70]. Nowadays such a matrix is called *thermal characterization matrix* when steady-state values are included in the matrix (usually denoted by R_{th}^*) or *dynamic thermal characterization matrix* (usually denoted by Z_{th}^*) when thermal impedances are therein. The features and issues related to these thermal characterization matrices will be discussed later in Chap. 3 dealing with the so-called thermal metrics.

Note that the elements included in these thermal characterization matrices do not represent the element values of a so-called multi-port compact thermal model of a multi-die system. Steps of obtaining a steady-state thermal model from the R_{th}^* matrix are described, e.g., in [71]; an application for dynamic compact thermal modeling of digital IC chips based on the Z_{th}^* matrix is described in [72].

2.8   System Descriptors for Periodic Excitations

Power electronics applications are mostly exposed to periodic excitations. Some appliances, such as motors, generators, or the input side of power supplies, are directly connected to the power grid, which operates at sinusoid alternating current. Other applications, such as car and locomotive electronic traction control units (ECUs), PWM controls in LED lighting, and internal circuitries of switching power supplies, operate at pulsed direct current.

It was stated in the previous sections that the linear network theory enables the calculation of the system response on arbitrary excitations from the measured step response. The methodologies offered by the linear approach are particularly suitable for directly deriving system responses on periodic excitations.

Below two cases of practical importance are discussed. First, the concept of the *complex locus*, a system descriptor that enables the direct production of the response of a thermal system on arbitrary periodic power change, will be presented. The methodology expects the spectral decomposition, that is, the frequency, amplitude, and phase of the constituents of the power as input, and yields the temperature response in the frequency domain. The apparatus of the Fourier transform and inverse Fourier transform connects the excitation and the response to their view in time domain. The technique is equally suitable for calculating the response of driving and monitoring points.

The use of complex loci is indispensable in the investigation of the stability of systems with thermal feedback, for tuning the control loop of thermostats at a given thermal mass, etc.

The concept of *pulse thermal resistance diagrams* has a simpler and less universal use. The thermal system is represented as a parametrized set of curves from which the peak temperature in the stationary state can be read when an excitation of repeated pulses of known period time and duty cycle is applied.

2.8.1 Complex Loci

The frequency domain representation of the thermal impedance can be calculated from the time domain $Z_{th}(t)$ function. The Fourier transform yields the $Z_{th}(\omega)$ function as

$$Z_{th}(\omega) = \int_0^\infty Z_{th}(t) \cdot e^{-j\omega t} dt \qquad (2.44)$$

where ω is the angular frequency of the excitation. The resulting $Z_{th}(\omega)$ complex thermal impedance function can be visualized, e.g., by means of a *complex locus*, also known as Nyquist diagram.

In Example 2.5 an artificial Cauer model consisting of three RC ladder stages was presented; the time domain Z_{th} curves characterizing the self-impedance and transfer impedance were shown in Fig. 2.43.

Applying (2.44) on the self-impedance curves Tjd and Tjw corresponding to the "dry" and "wet" boundary condition, the complex loci of Fig. 2.45 can be gained.

The complex loci depict how a thermal system responds to a unit-size sinusoid power of the f frequency and the $2\pi f$ angular frequency.[10] In the chart, the Re real part of the thermal impedance corresponds to dissipation, and the Im imaginary part expresses heat storage. The amplitude of the temperature response is the length (absolute value) of the vector between the origin and the corresponding point in the plot; the phase shift between the power and the temperature is the angle between the x axis and the vector, representing the delay in the temperature response. As the temperature always lags behind the power, the angle is negative; consequently the thermal impedance curves belonging to a driving point are in the fourth quadrant of the complex plane.

At a single sinusoid excitation, the thermal system behaves as a thermal resistance of Re $Z_{th}(\omega)$ and a thermal capacitance calculated from the formula Im $Z_{th}(\omega) = 1/$

---

[10]The linear network theory also uses an alternative representation of the same information. The Bode diagrams display the absolute value and the phase of the system response in separated charts.

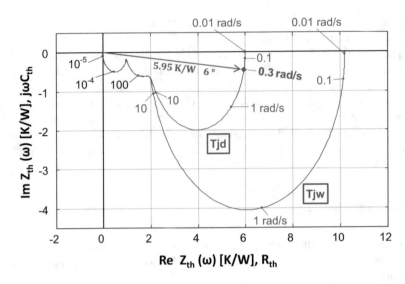

Fig. 2.45 Complex loci calculated from the time domain Z_{th} curves of Fig. 2.43, characterizing the self-impedance in two simple systems of different boundary conditions, as presented in Example 2.5

($j\omega C_{th}$). The physical meaning of the smaller temperature change on the same power amplitude is that the periodically changing heat can be locally stored and released by the structures near to the junction; it does not reach the ambient.

A single parallel RC circuit has the locus of a half circle, as its points can be calculated from the fixed R_{th} thermal resistance value and the shrinking $j\omega C_{th}$ thermal admittance. In Fig. 2.45 the portions of the three half circles can be clearly recognized showing the three discrete time constants of the system.

The zero-frequency value expresses the R_{thJA} junction to ambient thermal resistance at the two boundary conditions. When a periodic power signal is applied on the thermal system, which can be decomposed into several single frequency components of various amplitude and phase angles, the system responds in the frequency domain with a temperature response composed as the sum of the $Z_{th}(\omega)$ vectors.

Figure 2.46 shows the self-impedance and transfer impedance, calculated from the time domain Z_{th} curves Tjd and Tc_dry in Fig. 2.43, belonging to the "dry" boundary. In the enlarged excerpt of Fig. 2.47, it can be observed that the temperature change at the Tc_dry monitoring point is of a phase shift higher than 90 ° at frequencies above 40 rad/s which corresponds to 12 Hz. In general, complex loci corresponding to transfer impedances extend both to the third and fourth quarters.

This can be also interpreted in such a way that because of the propagation delay between the points Tjd and Tc_dry, the temperature change can be in an opposite phase as compared to the power excitation.

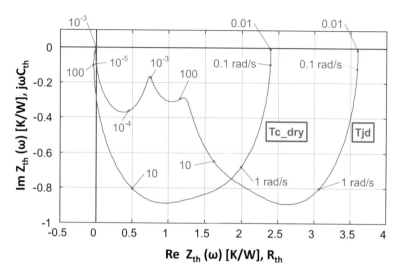

Fig. 2.46 Complex loci calculated from the time domain Z_{th} curves in Fig. 2.43 belonging to the "dry" boundary, characterizing self-heating (Tjd) and transfer heating (Tc_dry)

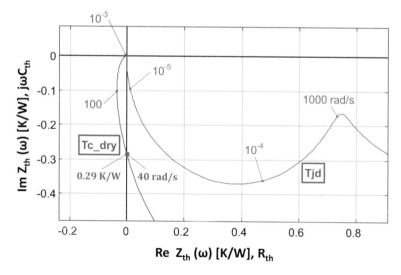

Fig. 2.47 Excerpt of Fig. 2.45. With positive feedback and high enough gain, thermal oscillation can be expected above ~40 rad/s corresponding to ~12 Hz

In this system if the coupling between the electric side of the powering and the Tc_dry point has a high enough gain, the positive feedback can induce a thermal oscillation above 12 Hz.

Fig. 2.48 Periodic power
pulse sequence

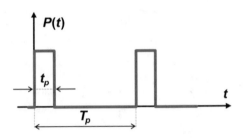

Complex loci are a very powerful representation of the component and its environment when analyzing periodic excitations. An application example is the single valued "AC thermal impedance" of LEDs [73], detailed in Sect. 6.10.

2.8.2 Pulse Thermal Resistance Diagrams

So far we have discussed the thermal characterization of a system based on its response to a single power step. In an important class of practical applications such as switching power supplies and motor drives, the power excitation can be described as a series of repeated power pulses. Similar pulse sequences play an important role, e.g., in reliability testing (Chap. 7, Sect. 7.4).

A periodic pulse sequence can be characterized by the T_p period time and the t_p length of the "on" state (Fig. 2.48), or by the T_p period time and the $D = t_p/T_p$ duty cycle.

With an excitation of repetitive pulses at a certain duty cycle, some heat will be stored in the internal thermal capacitances during the "on" state of the pulse and will be released during the "off" state. In a more detailed view, if the pulse is applied at a certain location of the system, the thermal energy is first stored in the material sections in close vicinity of the excitation, and farther sections are filled up with thermal energy at longer times as the heat propagates.

If T_p is longer than the shortest relevant time constant of the system, the thermal capacitances cannot be fully discharged in the "off" state; the average temperature continuously elevates in the system until stationary state is reached.

Example 2.9: Momentary Temperature Change and Peak Temperature at Pulse Excitation

Two versions of a simplified heat-conducting path were presented in Fig. 2.21, representing a DUT on a cooling mount. Different thermal interface qualities were denoted as "wet" and "dry" boundary condition.

In the following example, the "dry" scheme was driven in three subsequent simulations by different pulse sequences (Fig. 2.49).

(continued)

Example 2.9 (continued)

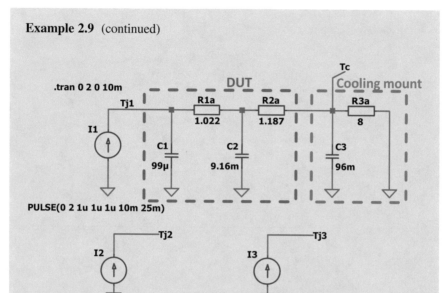

Fig. 2.49 Equivalent RC ladder scheme of a heat-conducting path, corresponding to the "dry" boundary condition in Fig. 2.21, pulsed power excitation at different period time and duty cycle

The transients were simulated until 2 s. In the first simulation, the circuit was driven by 2 W power, with 40% duty cycle, 25 ms period time, and 10 ms "on" time. The result of the simulation is shown as the red Tj1 line in Fig. 2.50. The second simulation was carried out with the same duty cycle but with 250 μs period time (Tj2, green line). A "long" step excitation was applied in the third simulation, at least longer than the total 2 s simulation time.

It can be observed in Fig. 2.50 that, as a consequence of the linear approach, the average temperature of the transients taken at 40% duty cycle is around 40% of the transient temperature in the step response. However, for thermal overload and lifetime prediction, the *peak temperature* reached at pulsed powering is of primary interest.

This peak temperature is highest in a pulsed heating process when stationary state is reached. This occurs when the elapsed heating time exceeds a few times the largest time constant of the system; in Fig. 2.50 this is reached at a few seconds.

(continued)

Example 2.9 (continued)

a

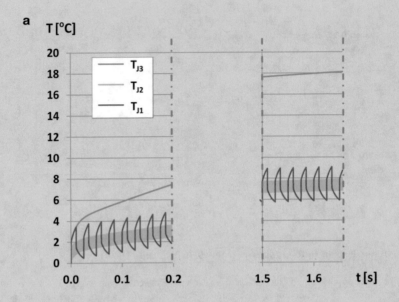

b

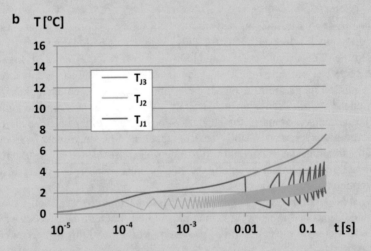

Fig. 2.50 Simulated results with pulsed power excitation at 40% duty cycle and different period time: Tj1, $T_p = 25$ ms; Tj2, $T_p = 250$ µs; Tj3, step excitation (**a**) linear time scale (**b**) logarithmic time scale

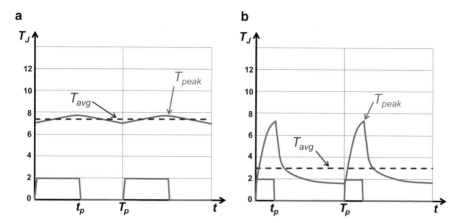

Fig. 2.51 Junction temperature waveform at different heating pulse series:
(a) $t_p = 0.1$ ms, $T_p = 0.2$ ms, $D = 0.5, f = 1/T_p = 5$ kHz, $dT_{peak} = 7.8$ K, $dT_{avg} = 7.3$ K. (**b**) $t_p = 1$ ms, $T_p = 5$ ms, $D = 0.2, f = 1/T_p = 200$ Hz, $dT_{peak} = 7.9$ K, $dT_{avg} = 2.9$ K

In stationary state the junction temperature follows a periodic function of stable waveform, similar to the ones shown in Fig. 2.51.

In an actual thermal testing of a real system, one way to establish the peak value of the periodic temperature response is applying different pulses of different length and amplitude and measuring the peak temperature directly. For practical reasons this can be done in the "off" state just after the falling edge of the power pulse as suggested in Sect. 5.4.1.

Carrying out a series of measurements with square wave excitation at many frequencies and duty cycles is rather tedious. Instead, a plot representing the peak temperature at several T period times and D duty cycles can be derived from a single measurement with a sole step function excitation, followed by some mathematical calculations. This plot, called *pulse thermal impedance* plot, can be calculated from the time constant spectrum.

The actual temperature waveform can be determined by measurement and simulation. First, one of the RC models of the device defined in previous sections is to be constructed from a thermal test with step excitation, and then a SPICE-like circuit simulator can produce the waveforms. This technique is often used for creating the pulse thermal resistance diagrams for data sheets; however, a direct methodology based on LTI theory can produce the plot in a single convolution step.

Equation (2.21) yields the $R_\zeta(z)$ time constant spectrum from a measured step response. In a linear approach, a single power pulse can be interpreted as a pair of consecutive step functions; a negative power step after t_p time extinguishes the first positive one. The corresponding thermal responses can also be superposed, deferred by t_p time and with opposite sign.

This concept can be extended to a series of pulses which follow each other by a T period time.

The direct calculation of the convolution integral of (2.12) in the time domain would need numerical approximate formulae. However, periodic functions like the pulse sequence discussed so far have a compact form in the (complex) frequency domain. The solution becomes of manageable complexity carrying out the following operations again:

- Transforming the periodic excitation and the (logarithmic) time constant spectrum by Fourier (or Laplace) transform into the (complex) frequency domain
- Multiplying the transformed functions
- Applying the inverse Fourier (or Laplace) transform to get the (logarithmic) time domain solution

In case of a periodic pulse excitation with t_p pulse width, T_p period time and $D = t_p/T_p$ duty factor (Fig. 2.48), the curves of the pulse thermal resistance diagram can be calculated from the time constant spectrum by the following convolution operation:

$$Z_{th}(z_p = \ln\ t_p, D) = R_\zeta(z) \otimes \frac{1 - \exp\ [-\exp\ (z)]}{1 - \exp\ [-\exp\ (z)/D]} \tag{2.45}$$

From the $Z_{th}(z_p, D)$ result, substituting z_p with $\ln\ t_p$ a $Z_{th}(t_p, D)$ function can be constructed again. Keeping the convention of engineering practice used so far, in charts $Z_{th}(z_p, D)$ will be plotted, but the horizontal axes will be labeled with t_p and the plot will be referred to as $Z_{th}(t_p, D)$.

The result of (2.45) is far from being mere theory; it is the mathematical expression of a very practical algorithm which can convert the result of a single transient measurement (test response) into the pulsed thermal impedance, without the need to test a structure by pulse patterns of various t_p, T and D parameters. An example of this calculation is presented in [60]. An actual software tool which realizes the calculation is part of the tester configuration of [54].

The calculated $Z_{th}(t_p, D)$ pulse thermal resistance plots for the scheme of Fig. 2.49 are shown in Fig. 2.52. Several duty cycles are plotted in the 5%–50% range. The curves were distilled from the response of the *whole RC ladder*, including both the DUT part and the cooling mount part. Accordingly, the chart can be called as junction to ambient pulse thermal resistance plot and can be denoted as $Z_{thJA}(t_p, D)$; it reflects the temporary energy storage on both the device and the cooling mount sections. The peak temperature of the stationary state can be calculated for power pulses of P height as $T_{peak} = P \cdot Z_{thJA}(t_p, D)$.

The $D = 0\%$ curve corresponds to the Z_{th} curve of the "dry" boundary scheme, drawn as 3stageCauer_dry plot in Fig. 2.23. At long period times, the temperature can reach its full $T_{peak} = P \cdot R_{thJA}$ value. At high repetition frequencies, that is at low t_p the peak junction temperature equals approximately the average temperature, $Z_{th} = D \cdot R_{thJA}$ and $T_{peak} = P \cdot D \cdot R_{thJA}$. In Fig. 2.52 the junction to ambient

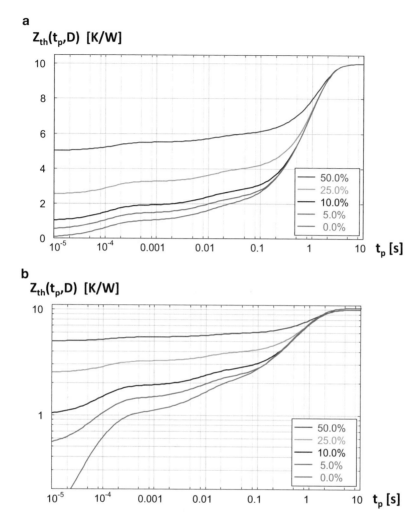

a

$Z_{th}(t_p,D)$ [K/W]

b

$Z_{th}(t_p,D)$ [K/W]

Fig. 2.52 $Z_{thJA}(t_p,D)$ junction to ambient pulse thermal resistance of the heat-conducting path scheme of Fig. 2.49; (**a**) linear, (**b**) logarithmic pulse thermal resistance scale

thermal resistance is $R_{thJA} = 10$ K/W; we can observe at short pulses the pulse thermal resistance value of 5 K/W, 2.5 K/W, and 1 K/W for duty cycles 50%, 25%, and 10%, respectively.

Data sheets typically present simulated charts in logarithmic pulse thermal resistance scale. These charts are mostly "reduced" to the "case" surface of a packaged device or module, in various, and mostly doubtful ways.

In a more sophisticated technique, a simulation is carried out on the detailed geometry of the device at constant uniform temperature on the baseplate, and the transient results are used to compose the pulse thermal resistance chart.

In a simpler approach, a Cauer RC network is composed from a measured thermal transient of the device in a conductive test environment (defined in Chap. 5, Sect. 5.1), practically on a cold plate. Based on some assumptions, a single internal node in the Cauer ladder is denoted as the "case" and that point is connected to the ambient. An analog circuit simulator software tool is used to compose the pulsed response of the shortened ladder.

In an even less justified approach, a Foster equivalent of a few stages is presented as junction to case thermal model. A seen before, contrary to the Cauer model, when a Foster model is terminated with different thermal impedances, the element values in the chain vary, as the Foster model is only valid for a single boundary condition. In other words, the Cauer model is a boundary condition-independent (BCI) model for the cases, when there is only one heat-flow path from the junction to the ambient, while the Foster model is always a boundary condition-dependent model.

A $Z_{thJC}(t_p,D)$ "junction to case" pulse thermal resistance chart is presented in Fig. 2.53. The plot was composed using the scheme of Fig. 2.49 again; the T_C node in the Cauer ladder was assumed to be the interface between the models of the device and the cooling mount. For mimicking the broadly used but theoretically wrong procedure, T_C was connected to the ambient (ground), and the pulse thermal resistance was constructed with (2.45). The chart suggests that the heat spreading in the modeled device reaches the assumed "case" surface at approximately 50 ms; the partial thermal resistance between the junction and the case node is 2.2 K/W, as also known from the circuit scheme.

This process for deriving the junction to case pulse thermal resistance is quite ill-defined; in reality no isothermal "case" surface exists, and the complicated trajectories of heat spreading cannot be reduced to a single one-dimensional RC network with a dedicated case node. These ambiguities will be treated more in detail in Chap. 3, where a standard R_{thJC} junction to case thermal resistance metrics for devices with a single major heat-conducting path is defined.

For all these reasons, the $Z_{thJC}(t_p,D)$ plots reduced to an assumed case surface have very limited practical meaning; they provide solely a rough estimation on the thermal behavior of a device in an actual assembly.

In most power electronics constructions, the portion of the heat-conducting path within the packaged device is of lower thermal resistance than that of the cooling mount. Taking the $Z_{thJC}(t_p,D)$ plots as system model for pulsed excitations would postulate that the package case or module baseplate is at a fixed and known temperature, which practically never occurs; it could be realized with infinite cooling capability on the case surface.

Assuming the other extreme, considering the cooling mount as a mere additional thermal resistance, the actual cooling performance of the assembly at pulsed excitation is severely underestimated. A bulky heat sink can absorb thermal pulses well in the minute range; its contribution to the transitory storage of the heat can be taken into consideration in several steps.

An estimation on the performance of a given heat sink at pulsed excitations can be based on the analytical Eq. (2.51), presented later in Sect. 2.11. As a further step, a

a

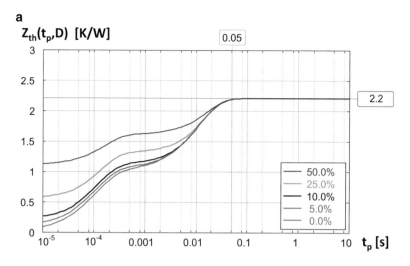

b

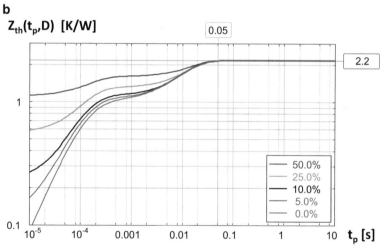

Fig. 2.53 $Z_{thJC}(z,D)$, junction to case pulse thermal resistance of the heat-conducting path scheme of Fig. 2.49, grounded at the Tc node; (**a**) linear, (**b**) logarithmic pulse thermal resistance scale

thermal transient simulation on a detailed model can confirm the suitability of the selected cooling solution.

When the full realized assembly composed of the device with thermal interface and cooling mount is available, it is essential to carry out *single power step measurements* at different powering. These can give an insight into the thermal performance of *all parts* in the assembly, in a wide power, frequency, and duty cycle range, relying on the concept of the pulse thermal resistance.

A more detailed study on pulse thermal resistances is given in [74].

2.9 Relationship Among the Different Representations of Thermal Impedances

To summarize the various representations of the thermal impedance, Fig. 2.54 provides an overview chart that shows also how they are related to each other.

The boxes with lighter blue background are generic to both driving point and transfer impedances.

The time constant spectrum is also generic to both types of the impedances. It has to be emphasized that in case of *transfer impedances*, real, physically meaningful time constant spectra may also include *negative magnitude (R) values*. Practical implementations of the numerical deconvolution algorithm defined by Eq. (2.21) may or may not yield these negative time constant values. This implementation dependence is indicated by the darker blue color of the corresponding box in Fig. 2.54. To be on the safe side, it is better to derive further representations of the self-impedances only from time constant spectra, if the unique properties of the used deconvolution algorithms are not known.

Dark red backgrounds indicate the representations that are defined or used only for driving point thermal impedances. Though the procedures used to obtain the Foster models or the pulsed thermal resistance diagrams do not pose mathematical problems even when the time constant spectra contain negative magnitude values, these representations are mostly used for driving point thermal impedances. Note

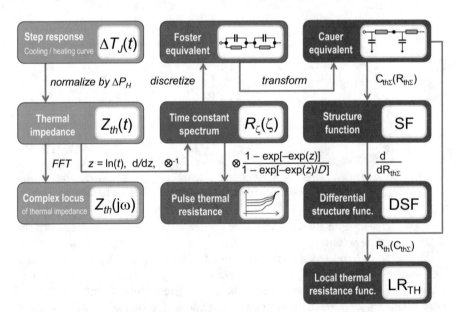

Fig. 2.54 Summary chart of the different representations of the driving point thermal impedances and the transformation paths among them. In light blue boxes, generic representations both for self-impedances (aka driving point impedances) and transfer impedances are shown. In dark red boxes, representations that are defined solely for driving point impedances are presented

that Foster-type network models with negative thermal resistance values can be well handled by circuit simulation algorithms. Note that if the numerical deconvolution algorithm used for the implementation of Eq. (2.21) provides time constant spectra with positive magnitude values only even in case of transfer impedances, the representations shown with any red background in Fig. 2.54 can be numerically calculated. Still, one has to be aware of the fact that parts of the structure functions obtained this way from thermal transfer impedances do not represent physical reality. As the calculation of the complex loci of thermal impedances bypasses the calculation of time constant spectra, a quick look at these loci allows to check if the thermal impedance is a pure driving point impedance or there is some "transfer effect" included therein. If a locus has a section in the quarter of the complex frequency plane with negative real part, that curve represents a transfer impedance, such as the red curve in Fig. 2.47, and the corresponding $Z_{th}(t)$ function cannot be represented by any version of the structure functions.

2.10   Distributed Heat Transfer on a Surface Towards a Convective Environment

Previously in Sect. 2.5 closed formulae have been constructed for obtaining the thermal resistance and capacitance of finite length beams or tubes. It was assumed in all cases that the heat transfer occurs exclusively at the two ends of the tube; the heat flux enters the tube at one end, at the "driving point" of the equivalent thermal network, and leaves on the other end, at an isothermal surface.

In a convective environment, the cooling of actual heat sinks and cold plates occurs on their whole surface, in a distributed way.

From the viewpoint of thermal transient testing, this distributed heat loss can be an intentional part of the test arrangement terminated by a cooling mount, or it can be an undesirable parasitic effect which distorts the measured thermal quantities.

Such parasitic effects can be parallel heat-conducting paths from the test setup through the surrounding air, or the distortion of the temperature field caused by sensor probes attached to a hot surface. An analytical treatment of the parallel heat-conducting paths and a methodology for a partial reconstruction of the primary path are given in [75–77].

Large chapters of mechanics deal with heat convection, and empirical formulae are listed in the literature which take into consideration the speed of the coolant, possible turbulent effects, and surface roughness and similar. The unintentional parallel cooling in thermal transient testing typically occurs towards the ambient air, and considering a constant h heat transfer coefficient for calculations is mostly satisfactory.

Simulation of detailed system models yields the distribution of temperatures and fluxes for arbitrary geometries and the heat loss at various surfaces towards a convective environment, but the result is valid only for a given geometry and powering.

The analytical formulae of distributed cooling are obtained in a semiempirical way, amalgamating the equations of heat spreading in Sect. 2.1 with empirical correction factors. The treatment of these effects is beyond the scope of this book. However, we illustrate in a short example that the results can be also obtained starting from the discretized RC approach of the previous sections.

Example 2.10: Heat Spreading in Wires and Long Fins in a Convective Environment

In an important practical case, when a temperature sensor is attached to a hot surface, the leads of the electrical connection cause an additional cooling at the measured spot, and in such a way, they distort the measured temperature value. The heat loss occurs on the whole wire surface; the cooling at the very end, at the instrument, can be neglected.

Similarly, plate or pin fins of a heat sink can be considered "long" when most of the heat flux leaves on the surface, before reaching the far edge or tip of the fin.

Assuming infinite length of these structures, simple analytical formulae of their thermal resistance can be obtained again.

Suppose in a Cauer-type ladder network (Fig. 2.55) the series Z_s elements represent the thermal resistance of a section in the fin or wire which impedes the heat propagation towards further similar sections. The parallel Z_p elements correspond to the heat transfer towards the air or other convective environment.

Adding a further Z_s-Z_p pair to an infinite ladder does not change its Z_{in} driving point impedance:

$$Z_{in} = Z_s + \frac{Z_p \cdot Z_{in}}{Z_p + Z_{in}}, \qquad (2.46)$$

rearranging and solving for Z_{in} the equation yields

$$Z_{in} = \frac{Z_s + \sqrt{Z_s^2 + 4 \cdot Z_s \cdot Z_p}}{2} \qquad (2.47)$$

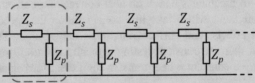

Fig. 2.55 Infinite Cauer ladder

(continued)

Example 2.10 (continued)

Along a dx length of the wire or a pin fin, the infinitesimal thermal resistances can be calculated as

$$Z_s = \frac{1}{\lambda \cdot A_s} dx = \frac{1}{\lambda \cdot r^2 \pi} dx \quad \text{and} \quad Z_p = \frac{1}{h \cdot A_p} = \frac{1}{h \cdot 2r\pi \cdot dx}, \qquad (2.48)$$

where r is the radius of the cylindrical element, A_s and A_p are the cross-sectional and lateral surfaces and h is the heat transfer coefficient towards the air.

The physical and geometrical parameters in (2.48) can be cumulated into respective M and N multiplying constants: $Z_s = M \cdot dx$ and $Z_p = N/dx$. For this infinitesimal geometry, $Z_s \cdot Z_p = N \cdot M$, and the result from (2.47) can be written as

$$Z_{in} = \frac{Mdx + \sqrt{(Mdx)^2 + 4NM}}{2}, \qquad (2.49)$$

The limit of the driving point impedance becomes $Z_p = \sqrt{N \times M}$ as dx approaches zero:

$$Z_{in} = \sqrt{\frac{1}{\lambda \cdot r^2 \pi} \cdot \frac{1}{h \cdot 2r\pi}} = \sqrt{\frac{1}{\lambda \cdot h \cdot 2r^3 \pi^2}} \qquad (2.50)$$

Equation (2.50) is known as "heat loss from infinite fin"; it is a useful formula for calculating the heat removal from a surface when measured by thermocouple or PT100 sensor. The λ thermal conductivity is known from the material composition of the thermocouple or of the connecting wire; similarly, the r radius (or "gauge") is provided by the manufacturers.

The $1/r^{3/2}$ dependence of the thermal resistance in the formula suggests that the thickness of the thermocouple is of eminent importance. Likewise, thermal insulation on wires can diminish the heat removal effect.

A correction algorithm to restore the structure functions of an equivalent heat-spreading scheme, subtracting the parallel heat loss towards the ambient, is presented in [75, 76], and is a realized feature in the software toolset of [54].

Figure 2.56 illustrates the simulated temperature distribution in a K-type (chromel-alumel) thermocouple. The diameter of the welded ball at the tip was supposed to be 0.6 mm; the diameter of both wires is 0.2 mm (gauge 32). The tip touches a cold plate of 30 °C temperature; the upper half of the ball is embedded in thermal grease of 40 W/mK thermal conductivity. The ambient temperature was set

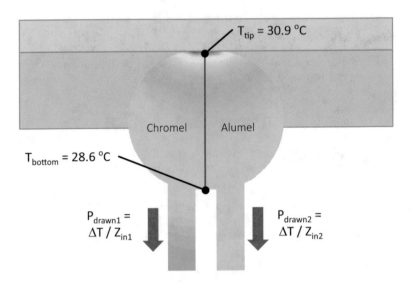

Fig. 2.56 Simulated temperature distribution in a K-type (chromel-alumel) thermocouple. The tip touches a cold plate of 30.9 °C temperature; the upper half of the welded ball is embedded in thermal grease. The temperature at the point where the wires adjoin becomes $T_{bottom} = 28.6$ °C at $T_A = 20$ °C ambient temperature and $h = 10$ W/m^{2}K heat transfer coefficient towards the ambient. The ΔT temperature difference in the figure is $T_{bottom} - T_A$

to $T_A = 20$ °C in the simulation, and the heat transfer coefficient towards air was taken as 10 W/m^{2}K.

The thermocouple "reports" the temperature of that point where the two wires adjoin, which is in this case $T_{bottom} = 28.6$ °C. A larger distance from the surface, typical PTFE insulating coating, and similar other factors may diminish further the accuracy of thermocouple measurements.

2.11 Temporary Heat Storage in the Cooling Mount

In an interesting way, a formula similar to Eq. (2.50) can be used for calculating the thermal impedance of a rod, blade fin, or pin fin at pulsed excitation. The pulsed waveform can be decomposed into an average steady power and a series of power waveforms of frequency f, angular frequency $\omega = 2\pi f$ alternating around the average value.

In this case the Z_s series elements of the Cauer ladder in Fig. 2.55 correspond to the R_{th} thermal resistance of an infinitesimal section of the fin; the parallel component corresponds to the thermal capacitance as $Z_p = 1/sC$ or $Z_p = 1/j\omega C$. Composing the $Z_s \cdot Z_p$ product, the equivalent thermal impedance of a fin at angular frequency ω will be

$$Z_{in} = \sqrt{\frac{1}{\lambda A} \cdot \frac{1}{j\omega c_V A}} = \frac{1}{A} \cdot \sqrt{\frac{1}{\lambda j\omega c_V}} \qquad (2.51)$$

During system design an estimation on the performance of a selected heat sink can be made based on (2.51).[11] All blade or pin fins contribute to the cooling, as a first assumption their cross-sectional area can be summed up in the formula. As a further step, a thermal transient simulation on a detailed model can confirm that the cooling solution fits the purpose.

A detailed treatment on the effects of external air or liquid cooling on heat sink fins or wires is presented in [79].

2.12 The Limits of the Linearity Assumption

In the previous sections of this chapter, so far the characteristics and behavior of thermal systems have been treated in a fully linear approach. This approach is justified when the investigated system operates in a temperature range where the nonlinearities can be neglected.

The root cause of nonlinearity in the thermal behavior is the temperature dependence of the thermal parameters, namely, of the thermal conductivity and the specific heat in the material layers which are the most exposed to temperature change. Typically, these structural parts include the semiconductor chip, the die attach, and the ceramics or metal base to which the chip is attached.

Farther elements in the heat-conducting path are mostly at lower temperature because of the applied external cooling. The typical cooling solution in electronic systems is convective heat transfer on dedicated cooling surfaces, assured by either air or liquid cooling. The convective cooling mechanisms have inherent nonlinearities, but the detailed discussion of these effects is beyond the scope of this book.[12]

Nevertheless, thermal testing standards dealing with the environmental conditions of the measurements are aware of these, especially in case of natural convection [31] and forced air cooling[13] [34]. For example, in case of a standard natural convection cooling environment, the air temperature has to be measured at the temperature monitoring point of the test chamber in order to assure that the test environment remained stable during a measurement.

---

[11] This formula is related to the "RMS heat storage for the thermal skin effect," useful for calculating the seasonal temperature change in different depths of the earth. The phase change effect with a periodic thermal signal is also used to measure thermal conductivity, because it is proportional to the thermal diffusivity (3ω method).

[12] These effects are related to the nature of the fluid flow of these media (such as laminar flow turning into turbulent, etc.); discussion of these is the subject of fluid mechanics.

[13] As the CFD-based thermal simulation tools emerged and their use became daily engineering practice in electronics cooling design, the relevance of physical testing of semiconductor device packages under forced air cooling conditions has significantly decreased.

Some special electronic appliances operate at high temperatures, such as vacuum tubes in broadcasting systems or silicon carbide rectifiers in locomotive applications. At these temperatures the investigation of radiation has to be involved, which follows the Stefan-Boltzmann law:

$$\varphi(t) = \varepsilon \cdot \sigma \cdot T(t)^4, \tag{2.52}$$

which tells that the emitted φ heat flux from a surface portion of a hot body is proportional to the fourth power of its T temperature. In (2.52) σ is a physical constant, the Stefan-Boltzmann constant. The material composition and the surface quality also influence the emitted power. This is represented by the ε *emissivity* of the surface; it is $\varepsilon = 1$ for an "absolute black body" and lower for real materials. Shiny metal surfaces have an emissivity below 0.1, while ε of the anodized aluminum is above 0.7. Paints are typically above 0.9, independently from their color, as they have typically the same "color" in the infrared spectral range where emissivity really counts.

Although the radiative heat transfer from the circuit boards or hotter package or heat sink surfaces improves slightly the cooling of regular electronics, these surfaces are typically below 100 °C, where radiation has a minor role. The spectacular blackening of heat sinks serves mainly marketing purposes.

Time-dependent variation of the material properties is also out of the scope of this current discussion, especially since these changes (e.g., thermal conductivity change due to the dry-out of TIM pastes) are slow. Slow in this context means that that the pace of such changes is slower by multiple orders of magnitudes than the lengths of the temperature transients that we aim to measure. This huge difference in the pace of changes allows one to use structure functions to monitor the degenerative (aging) processes in certain structural elements of semiconductor device packages. Such applications of thermal transient testing are discussed in Sect. 7.4.

Accounting for possible nonlinearities is the most relevant for the early parts of driving point thermal impedances since these originate mostly from the temperature dependence of the materials used inside a semiconductor device package [69]. Though these nonlinearities slightly effect the thermal transfer impedances as well (such as the *Chip1-Chip2* transfer impedance for the arrangement presented in Fig. 2.40), they are most affected by the properties of materials outside the package structure – as it will be shown later, e.g., in Chap. 3.

Thermal transient testing may aim at different targets. One purpose can be the determination of the temperature change in time, especially finding the maximum temperature as a crucial factor which influences lifetime. Another target can be checking internal structural details, partial resistances, and assembly integrity.

Therefore, during thermal transient testing, it is worthwhile to use different levels of heating power for different purposes:

1. If the applied temperature sensors and the measurement apparatus are sensitive enough to record tiny signals at proper resolution, lower levels of heating power can be applied that result in a low junction temperature rise. Remaining below

10 °C temperature change, the nonlinearities of the material properties are negligible; thus, using the linear system theory for the postprocessing of the measured thermal transients is well justified.

2. If the purpose of the thermal transient testing is to characterize a device under conditions close to the relevant field applications, the heating power levels should approximately match the levels of the foreseen levels in use. In many cases typical junction temperature elevations are still expected to stay below 100 °C, because of reliability considerations. A case study of exceeding this temperature change limit and stepping into the range where nonlinearities of the material parameters matter will be presented below in Example 2.11. Still, the outcome of the study is that the linear apparatus can manage thermal changes in the 150 °C range.

3. In reliability and accelerated lifetime tests, a common practice is to apply heating power levels beyond the ones usual in field applications. These result in high temperature elevations when nonlinear effects become significant. Such effects may not necessarily hamper the NID method-based postprocessing of the measured transients, but need to be known and properly accounted for, e.g., applying right correction formulae. Even in accelerated tests when the device is continuously stressed with high-power pulses, the variations in the structure functions still can properly reveal when and how fatal device degradations appear and develop. Such applications are treated in Sect. 7.4 of Chap. 7.

Further on in this section, we focus on the second case, on nonlinearities encountered during the measurements in typical operating conditions.

2.12.1   The Most Common Nonlinearities: Temperature-Dependent Material Parameters

The major source of the nonlinearity of the thermal systems is the temperature dependence of the λ thermal conductivity and of the c_v volumetric heat capacity of the materials used in the structure. Different materials show rather different temperature dependence, but the alteration of thermal parameters is low in the usual range of the operation of electronics devices (200–400 Kelvin). In [64] a thermal transient technique is presented which yields the temperature dependence of the thermal conductivity of materials used in electronics packaging.

In the usual temperature range of operation, these dependences can be of different nature, but for typical materials, they can be described with an exponential formula:

$$\lambda = \lambda_0 \cdot \exp\left[\alpha_\lambda \cdot (T - T_0)\right], \tag{2.53}$$

where T_o is the reference temperature, λ_o is the thermal conductivity of the material at the reference temperature and α_λ is the coefficient of temperature dependence (CTD) of λ. The α_λ value is nearly equal to the relative change of λ for 1 °C

Table 2.4 Values the λ_0, α_λ, c_{v0}, and α_c for some packaging materials (averages in the 300–400 Kelvin range)

Material	λ_0 W/mK	α_λ 1/K	c_{v0} Ws/m^{3}K	α_c. 1/K
Cu	401	−0.0001	3.44·10^6	0.0003
Ni	90.7	−0.00012	3.95·10^6	0.0008
Ag	429	−0.000094	2.47·10^6	0.00017
Inconel	11.7	0.0014	3.74·10^6	0.00075
Al$_2$O$_3$	36	−0.0031	3.04·10^6	0.002
Si	148	−0.004	1.66·10^6	0.001

temperature rise. For small temperature changes, (2.53) can be well approximated with the

$$\lambda = \lambda_0 \cdot [1 + \alpha_\lambda \cdot (T - T_0)] \tag{2.54}$$

linear relationship. The values for some common materials of packages are presented in Table 2.4, taken from [64].

In the dynamic behavior, the temperature dependence of the heat capacitances may also play a role. Fortunately, this effect is rather small and often negligible in the 0–150 °C range. For the description of the temperature dependence of the heat capacity, a function similar to (2.53) can be used as

$$c_v = c_{v0} \cdot \exp\left[\alpha_c (T - T_0)\right], \tag{2.55}$$

where c_{v0} is the volumetric heat capacity value at the reference temperature and α_c is the coefficient of the temperature dependence. Similarly to the thermal conductivity, the linear approximation holds also here.

Values of c_{v0} and α_c are also presented in Table 2.4; their temperature dependence is shown in Fig. 2.57. As the table shows, these parameters change only slightly in the usual temperature range of operation.

Various authors have examined the significance of nonlinearities in the thermal behavior of electronics packages [64, 69, 80]. They agree that for small temperature changes (<25 °C), the error of the linear approximation is negligible. For temperature changes within the 0 °C, 100 °C range the error is about 2–5%, depending of the materials used in the structure. In case of larger temperature changes, the error can be higher, depending of course again on the α parameters of the different materials in the structure.

As mentioned previously in this section, the temperature dependence of the structural materials of the heat conduction path from the semiconductor junction affects mostly the driving point thermal impedances at early times, as emphasized by D. Schweitzer et al. in [69].

In Eqs. (2.26) and (2.27) in Sect. 2.4.2, the time evolution of the junction temperature transient was expressed with a $\Delta T_J(t) = k_{therm} \times \sqrt{t}$ formula, where both the λ thermal conductivity and the c_v volumetric heat capacity appear in the definition of the k_{therm} coefficient.

Fig. 2.57 Temperature dependence of the volumetric heat capacity of a few materials frequently used in semiconductor device packaging [3]

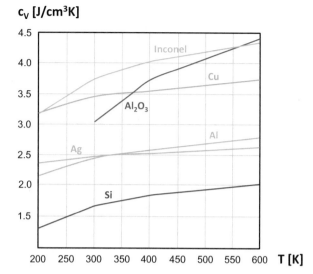

As shown in [69], if the $Z_{th}(t) = \Delta T_J(t)/P_H$ driving point thermal impedance is known for an initial T_1 temperature of the chip, then for an elevated T_2 chip temperature, the value of the thermal impedance can be rescaled as follows:

$$Z_{th}(T_2,t) = Z_{th}(T_1,t) \cdot \sqrt{\frac{\lambda_{sem}(T_1) \cdot c_{v-sem}(T_1)}{\lambda_{sem}(T_2) \cdot c_{v-sem}(T_2)}} \qquad (2.56)$$

where λ_{sem} and c_{v-sem} are the temperature-dependent thermal properties of the semiconductor chip.

The above equation and Eqs. (2.26) and (2.27) were derived with the assumption in [69] that the semiconductor chip is thick enough to consider as infinite in size for a prolonged time. Moreover, it was assumed that the heat leaves the junction through the chip, towards the bottom of the structure, and there is no heat flux towards the top. Extending this concept, A. Alexeev et al. recently derived in [77, 78] a new formula for $\Delta T_J(t)$ that is valid for the cases, when the heat generated at the junction flows in both directions, for example, in case of LEDs where some heat also leaves through the lens. Similarly to D. Schweitzer's correction for nonlinearities, also in this case, early transients follow the $\Delta T_J(t) \sim \sqrt{t}$ time dependence; the square-root approach for $Z_{th}(t)$ is maintained.

Based on (2.56), the correction procedure in [69] accounts for these temperature dependencies during thermal simulations. The models of the temperature dependence can be similar to (2.53) and (2.55) or to the simpler (2.54) for the thermal conductivity.

Figure 2.58 presents finite element-based simulation results for a device driven with a short heating pulse of 200 W [69]. The blue curve represents the "real" device behavior with considering the temperature dependence of the thermal conductivity

Fig. 2.58 Simulated response of a real multi-chip power device structure (similar to the one in Fig. 2.40) for a short power pulse of 200 W

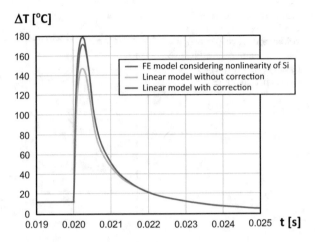

and volumetric heat capacity of the heated silicon chip in an exact way. The green curve was obtained by a purely linear model-based calculation (thus, with temperature-independent, constant material properties), while the red curve was obtained by calculations where a linear thermal model completed by the correction procedure based on Eq. (2.56) was used [69]. By comparing the green and the blue curves, one can clearly see that at junction temperature elevations above ~150 °C, neglecting nonlinearities results in large (>10%) error.

Note, however, that according to the study reported in [64], if the junction temperature elevations do not exceed ~100 °C, the errors due to neglecting the temperature dependence of the material properties remain below 5%. Thus, the postprocessing of the measured thermal transients using deconvolution or other apparatus based on the linear system theory is justified from an engineering perspective.

Temperature dependence of the effective thermal conductivity of thermal interface layers is another reason why one may observe temperature-related changes in structure functions.

Example 2.11: Nonlinearity of the Thermal Behavior of a Packaged SiC Power Device

A CFD simulation with the tool of [56] was carried out on the detailed model presented in Fig. 2.59. In the arrangement a power MOSFET device in a TO220 package was placed on a cold plate with an inserted alumina sheet of 1 mm thickness.

The chip was modeled as a SiC block of 1.3 mm × 1.3 mm × 0.3 mm size, with a 1 mm × 1 mm × 0.005 mm dissipating junction on its surface. The die attach layer was of 0.025 mm thickness. The simulation was carried out in the 10 °C–90 °C cold plate temperature range, in 20 °C steps.

(continued)

Example 2.11 (continued)

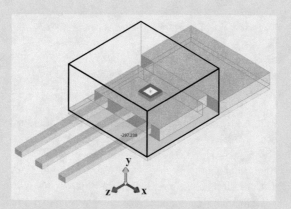

Fig. 2.59 Detailed model of a SiC power transistor for thermal simulation in the SIEMENS SIMCENTER Flotherm tool [56]

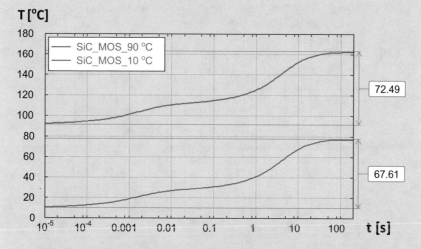

Fig. 2.60 Temperature change of the MOSFET device on cold plate, during heating at applied 10 W power, at 10 °C and 90 °C cold plate temperature

In the simulation partly the material parameters built into the tool, partly data from the literature were used.

The temperature range to be considered was quite broad; at 10 W applied power, the simulated device temperature varied between 10 °C and 160 °C, depending on the cold plate temperature; see Fig. 2.60.

In this range the simulation tool uses a piecewise linear approach for the thermal conductivity of the SiC chip material ($\lambda = 330$ W/mK until 125 °C and

(continued)

Example 2.11 (continued)

$\lambda = 214$ W/mK between 125 °C and 225 °C). We added further data to the material library of the simulation tool for the copper and ceramics layers in the DBC structure.

Copper was characterized with the thermal conductivity value $\lambda = 401$ W/mK along with a negative temperature coefficient of $\alpha = -0.00011$ /K. The die attach material was modeled with the thermal parameters of $\lambda = 40$ W/mK; $\alpha_\lambda = -0.007$ /K. The ceramics layers were represented by Al_2O_3 material, with $\lambda = 36$ W/mK; $\alpha_\lambda = -0.003$ /K values. The c_V volumetric specific heat (volumetric heat capacity) was specified as 3.45 J/cm^3K for copper, 1.65 J/cm^3K for the die attach, and 3.03 J/cm^3K for the ceramics.

The simulated transients were converted to structure functions, as shown in Fig. 2.61. Several regions could be identified in the structure functions based on their thermal capacitance calculated from the volume in the module and the assigned specific heat.

The sections below 1 mJ/K were identified as the SiC chip and until 8 mJ/K as the die attach, denoted as DA. Two arrowed lines correspond to the volume of a copper block based on the dimensions of the package base in the model (line Cu) and to the third of the volume (line Cu/3). The end of the steep section which can be attributed to the high conductivity of copper lies between the Cu and Cu/3 positions. This may refer to a heat spreading in a truncated pyramid between the small chip and the wide package base bottom touching the ceramics, as introduced in Sect. 2.5.

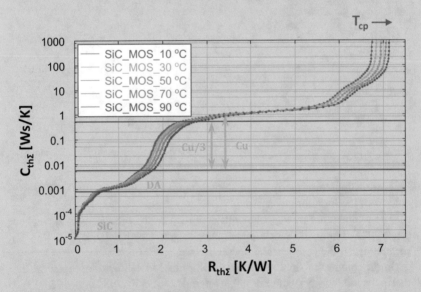

Fig. 2.61 Structure functions of the MOSFET device on cold plate, at applied 10 W power in the 10 °C–90 °C cold plate temperature range

(continued)

Example 2.11 (continued)

It has to be noted that the structure functions gained from real thermal transient measurements are often of lower steepness in the die attach region. This is typically due to die attach voids or delamination. A study on the effect of die attach voids of different coverage is presented in Example 7.5 in Sect. 7.3.

As all constructional materials were supposed to have a negative thermal coefficient, accordingly, growth of the thermal resistance at higher temperature can be observed in all portions in the heat-conducting path. Figure 2.61 proves that the temperature-related difference in the thermal resistance starts building up in the SiC and die attach regions. Fitting the structure functions at the ambient (Fig. 2.62), we can see that the copper and the subsequent layers have a minor share only in the growth of the thermal resistance.

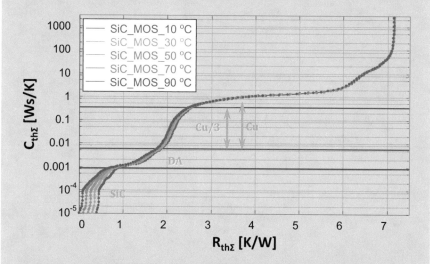

Fig. 2.62 Structure functions of Fig. 2.61 fitted at the thermal resistance of the ambient

Figures 2.63 and 2.64 present the calculated structure functions for the lowest and highest cold plate temperatures, when both heating and cooling transients were simulated. The orange and red curves denote the results of *heating*, for 10 °C and 90 °C cold plate temperatures, respectively. The results calculated from cooling transients are represented as blue and gray plots.

(continued)

Example 2.11 (continued)

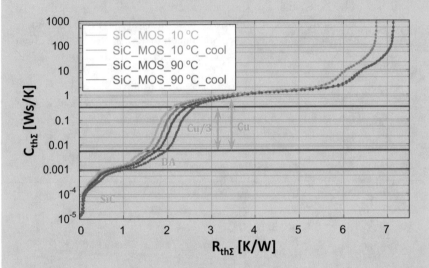

Fig. 2.63 Structure functions derived from simulated heating transients (red and brown curves) and cooling transients (blue and green curves), at cold plate temperature 10 °C and 90 °C

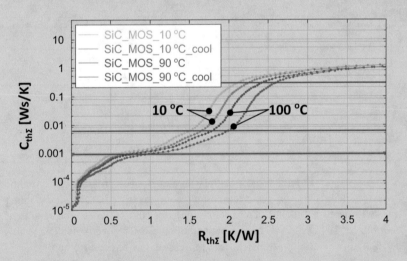

Fig. 2.64 Excerpt of Fig. 2.63. Change of the thermal conductivity in the SiC, die attach, and copper region can be observed

In the previous example, the structure functions calculated from heating and cooling measurements match well along the heat-conducting path. This suggests that for this device, cooling or heating transients yield nearly the same data for structural analysis, despite the assumed nonlinearities. Accordingly, using the linear system theory for data processing is well justified.

It can be observed in the plots that as all material layers were expected to have a thermal conductivity of negative thermal coefficient, all regions in all structure functions shift towards higher thermal resistance values at higher temperatures. This indicates a possible instability, which may result in thermal runaway. In a thermal assembly design, increased cooling capability of the cooling mounts at higher temperatures has to be ensured in order to guarantee thermal stability.

This highlights the fact that in case of identifying the so-called standard thermal metrics of packages, the test environment (applying a layer of thermal interface material in this example) and the test conditions[14] (the cold plate temperature here) have a significant effect on the overall junction to ambient thermal resistance we measure. This also suggests that one needs to apply well-defined and easily repeatable and reproducible procedures to separate the structure function portions corresponding to the package under test and corresponding to the test environment.

2.12.2   Measurement Artifacts Appearing as Nonlinearities

In our discussions so far, we assumed that ideal measurement apparatus was used, i.e., all the captured signals represent the true temperature transients. In practice, this is not always the case. It strongly depends on the way how the temperature is measured by the thermal transient test equipment. Details of usual realizations of such equipment are discussed in Chap. 5, followed by Chap. 6, providing descriptions of measurement basics of different types of electronic components. The discussion here is restricted solely to the possible artifacts caused by the actual temperature measurement method. These – if not known – are wrongly appearing as the nonlinear behavior of the thermal system realized by the device package.

In daily practice direct measurement of temperature is replaced by indirect methods, matched to the actual temperature range of interest. Early thermometers used to measure temperatures in everyday human environment were based on the physical effect of thermal expansion, converting it to a length scale. Modern, electrical thermometers convert the temperature to an electrical signal, e.g., to voltage. The accuracy of the practical temperature measurement depends on how accurately a known temperature change is calibrated against the change of the electrical signal of the electrical thermometer.

---

[14]The so-called standard thermal metrics and issues of the related thermal test conditions will be discussed in Chap. 3 in detail.

In thermal testing of electronics, the temperature change is converted to electric signal by active semiconductor devices used as temperature sensors using one of their temperature-sensitive parameters (TSP), or by other dedicated electrical temperature sensors attached to accessible surfaces of the measured system. A few important ones are:

- *Diode sensors*

  We usually say that the T_J junction temperature dependence of the V_F forward voltage of a diode driven by a constant forward current is linear, but this is also just an approximation, which is true only for small temperature excursions ($<50\,^\circ$C). In Chap. 4 the actual sensitivity is derived from semiconductor physics, and the resulting equation shows that, e.g., in a range of 200 °C, this dependence is far from linear. The nature of nonlinearity of the $V_F(T_J)$ relationship is more pronounced, e.g., for III–V compound semiconductors than for silicon.

- *Resistor sensors*

  In resistor-based temperature sensors, the temperature dependence of the electrical resistance is utilized. Such thermometers also frequently show nonlinearities. For example, the resistance of metal sheets is of exponential temperature dependence. For example, a PT100 platinum sensor has 100 Ω resistance at 0 °C, and it grows by 385 ppm/K with the temperature. In a small temperature range, this means that applying 10 mA electric current to sense the temperature, the obtained voltage will be 1 V, with the sensor exhibiting 3.85 mV/K sensitivity, but in a broader range, the sensitivity will change, corresponding to the sensor's exponential temperature dependence.

In both cases, in simple measurements, a single sensitivity parameter is used that is a good approximation of the sensors' real characteristics only for a relatively small temperature range (e.g., $\Delta T < 50\,^\circ$ C). If the temperature elevations are beyond the validity of linear approximation of the temperature sensors' characteristics, but still a single sensitivity parameter is used for the temperature-voltage conversion, the measured $\Delta T_J(t)$ transients and the corresponding $Z_{th}(t)$ thermal impedances will be distorted by the measurement error, and the measured thermal system would appear as if it was nonlinear. Such a nonlinearity of a thermal system is obviously a *measurement artifact*.

If during thermal characterization of a semiconductor device larger temperature ranges are involved (e.g., $\Delta T > 50\,^\circ$ C), a careful calibration of the sensors and the exact calculation with the temperature sensors' actual characteristics is a must to avoid the above measurement artifacts. For example, the software of [54] supports such careful sensor calibrations and performs polynomial or exponential fitting on the set of measured points; the resulting calibration files can be used by the data postprocessing software to yield $Z_{th}(t)$ thermal impedances free of the abovementioned artifact.

Example 2.12: Forward Voltage: Temperature Mapping of a Power LED Device

Figure 2.65 shows a calibration curve of the voltage-to-temperature mapping (calibration curve) of a Royal Blue Cree XP-E2 medium power LED device driven by 10 mA constant forward current. The temperature range covered by the presented set of calibration data is 160 °C. In the diagram we also present results of two different linear approximations for the low and high temperature ranges.

If the sensitivity parameter derived by linear regression for the low temperature range is used, e.g., for a junction temperature above 50 °C (resulting in $\Delta T_J > 60\ ^\circ C$) from the reference, the low temperature region), significant errors will develop. To avoid artifacts in the postprocessing of the measured $\Delta T_J(t)$ transients of the LEDs, a quadratic approximation of the $V_F(T_J)$ relationship, as shown in the figure, is already sufficient [81, 82].

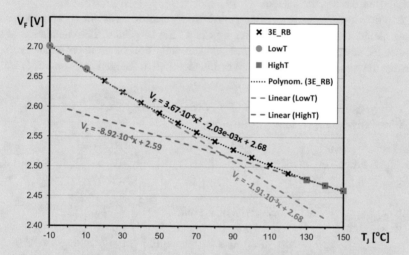

Fig. 2.65 Voltage-to-temperature calibration curve of a Royal Blue Cree XP-E2 medium power LED device at 10 mA sensor current. (Based on Refs. [81, 82])

2.12.3   Limits of the Validity of the Linear Approach in Actual Measurement Results

So far this chapter has elaborated the theoretical aspects of heating and cooling of a thermal system. In further chapters the implementation and evaluation of actual measurements will be discussed.

Nowadays all thermal measurements are based on recording temperature-related electric signals of the chips in an appliance or of dedicated temperature sensors. The recorded signal has to be examined in all time intervals for the following criteria:

- Is the electric transient in a time interval related to a thermal change of some parameters or is it caused by other electric effects?
- If it is of thermally induced nature, does the related thermal change occur within the investigated thermal subsystem such as a tested device or module, or in a broader environment out of the device?
- If it is within the investigated thermal subsystem, is its behavior linear in the thermal domain?

In subsequent chapters, especially in Chap. 5 it will be expounded how high power is generated in different device categories such as transistors, diodes, or integrated circuits and why is some power maintained on these devices also during cooling measurements. Without going into details, at this point it can be stated that a common way of powering is to force a higher electric current through the device, and a general way of measuring the temperature of an active device is to record some electric device parameter at a low current bias, called the measurement current. The applied power can be always determined from known currents and measured voltages in the electric system.

Multiplying a measured, often called "raw" electric transient by an appropriate scale factor, a sort of "quasi thermal transient" can be gained. The calculation of the power for several device categories and the definition of an appropriate "voltage change to temperature change" conversion factor are presented in Chaps. 5, 6, and 7.

Example 2.13: Measurement of a Power MOSFET Device at Different Powering and Boundary Conditions

Forced current of various levels such as 1 A, 1.5 A, and 2 A was applied on a packaged power MOSFET (IRF540) for 10 seconds. The resulting power on the device was 0.74 W, 1.14 W, and 1.6 W at the three current levels. After revoking the heating power, the change of the voltage on the device was recorded at different measurement currents (100 μA, 1 mA, 10 mA). The measurements were repeated with the device placed on dry and wet cold plate.

Figure 2.66 presents the "quasi temperature" curves recorded at a few selected combinations of heating current, measurement current, and boundary. Three curves in the figure, denoted with a key starting as wet_D_, were measured on the wet plate, and the other three on the dry one. The voltage-to-temperature conversion factor was determined for all measurement currents in a calibration process as defined above in Example 2.12 and later in Sect. 5.6.2.

As introduced before, the thermal impedance is defined as the temperature change in time divided by the applied power. The Z_{th} curves of a linear thermal system are identical at different power levels and starting temperature.

(continued)

Example 2.13 (continued)

quasi T

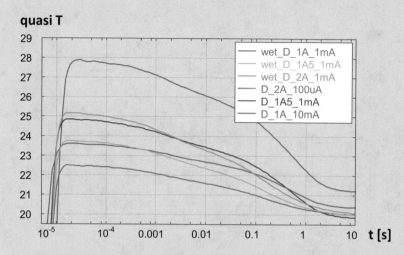

Fig. 2.66 "Quasi temperature" curves calculated from thermal transients of a power MOSFET at various powering and at two boundary conditions. At applied heating current of 1 A, 1.5 A, and 2 A, the power which developed on the device was 0.74 W, 1.14 W, and 1.6 W, respectively

A sort of "quasi thermal impedance" can be produced dividing the transient change of the "quasi temperature" by the applied power. In Fig. 2.67 the curves of this "quasi thermal impedance" are shown, gained from Fig. 2.66 with dividing by the corresponding power. It can be observed that between 30 μs and 100 ms seconds, all curves coincide, regardless of the boundary condition; then, until 10 seconds those curves coincide which were measured at the same "wet" or "dry" boundary. This indicates that the variation in the recorded "quasi thermal impedance" was of purely thermal root cause after 30 μs; it was proportional to the applied power only. The difference in the thermal quality of the "wet" or "dry" cold plate causes the divergence of the curves after 100 ms; this can be used for identifying the role of the device as opposed to the test environment in the thermally induced temperature change.

Figure 2.68 shows the early section of the "quasi Z_{th}" curves enlarged. All presented "wet" curves in the previous figure were taken at 1 mA measurement current, as it was impossible to find any difference between them at the present resolution of the charts; for better clarity only one of them is shown now.

(continued)

Example 2.13 (continued)

quasi Z$_{th}$

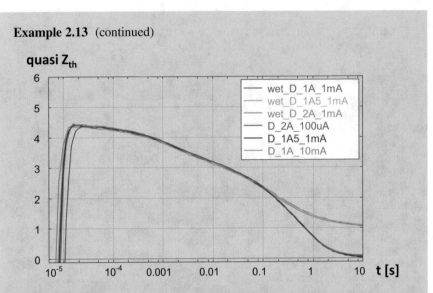

Fig. 2.67 "Quasi Z$_{th}$" curves calculated from thermal transients of a power MOSFET at various powering and at two boundary conditions. Pure thermal nature of the variation can be seen between 30 μs and 10 seconds where all curves measured at the same boundary condition coincide

quasi Z$_{th}$

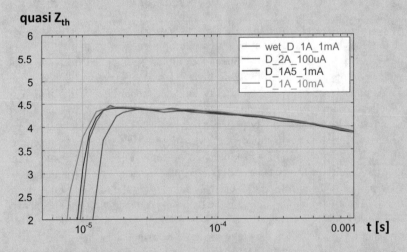

Fig. 2.68 Excerpt of Fig. 2.67. The voltage change on the device is of purely thermally induced nature when the curves belonging to different heating and measurement currents coincide

The range in which the transient change is of the thermally induced nature can be determined with the comparison of transient curves at different powering and measurement currents, as shown in Example 2.13. Omitting the part of nonthermal root cause, typically described as "electric transient"; the valid, true Z$_{th}$ thermal impedance curves can be gained from the "quasi" Z_{th} curves. Based on Eq. (2.26) even the restoration of the thermal change covered by the "electric transient" becomes possible, as presented in Chap. 6, Sect. 6.1.4.

The distinction between thermal and nonthermal changes in the "quasi" thermal impedance curves is especially complex in the testing of devices based on GaN material (Chap. 6, Sect. 6.9.2). Depending on the measurement technique, these devices can produce electric transients in the many millisecond range which can be mistakenly attributed to thermal effects. A study on the separation of transient changes of different nature and correction procedures is given in [83].

Effect of the Voltage Drop on Wiring

A sort of "procedural" nonlinearity can be observed in measured Z_{th} curves of large power modules at high heating current. As defined in several standards, linear thermal descriptors of a device such as thermal impedance curves and structure functions are composed from recorded transient curves normalized by the applied power. However, the power value is calculated from the forced heating current multiplied by the voltage, measured on the external pins of the module. The wiring within the module is farther from the active devices and is not heated in the same manner as those are, still, the power fraction due to the voltage drop on the wires is added to the power used for normalizing. This increase in the calculated power which actually does not appear in the heating of the active devices causes a characteristic shrinking of the Z_{th} curves and the structure function at larger heating currents. Although this effect seems to be an artifact at the first glance, it is rather a feature caused by the definition of the thermal measurement standards. The phenomenon is discussed in this context, because it affects the structure functions in a similar way as the nonlinear effects caused by the temperature-dependent material parameters.

Effects of Additional Energy Transport

A similar "procedural" nonlinearity can be attributed to devices with multiple energy transport, like LEDs in which the applied electric power is partly dissipated as heat, partly emitted as light. The emitted optical power does not contribute to the temperature elevation of the device. The LED efficiency that is the share of the emitted optical power compared to the electric power fed into the device strongly depends on the current and temperature. Accordingly, in case the thermal transient measurement aims at the structural analysis of the device, the optical fraction in the power has to be measured and subtracted. Similarly, when an external cooling mount is added to the device, only this "optically corrected," reduced heating power is to be taken into consideration.

This concept is illustrated in Fig. 2.69 showing the structure functions of a Cree MCE LED module on the temperature-regulated device holder plate of an optical integrating sphere. The voltage-to-temperature mapping of the LED was similar to the one shown in Fig. 2.65; it was an obviously nonlinear one.

Still, converting the temperature-induced forward voltage change to temperature by the exact mapping and subtracting the optical power, all structure functions taken at different cold plate temperature fit perfectly in region (2.1) of the figure, belonging to the LED and its metal core heat distributor board. The thermal system of the LED

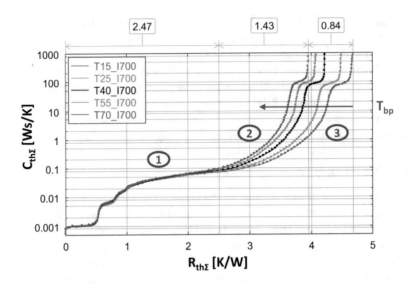

Fig. 2.69 Optically corrected structure functions of a 10 W LED module on the temperature-regulated device holder plate of an optical integrating sphere. Plate temperature T_J is elevated from 15 °C to 70 °C

module can be considered linear. The shrinking of the structure functions in regions (2.2) and (2.3) corresponds to the thinning of the thermal interface between the module and holder plate, and the higher heat sinking capability of the temperature-controlled holder at higher temperature.

Details of the optical correction procedure are presented in Chap. 6, Sect. 6.10.

Summary of Nonlinear Effects

A complete steady-state to steady-state $\Delta T_J(t)$ junction temperature transient comprises all information about the junction to ambient heat transfer, including the heat transfer processes in the test environment, also beyond the conduction path(s) of the package or module under test. Such mechanisms are, e.g., convective cooling from the thermal test board holding the package or, in extreme cases, radiative heat transfer from hot package surfaces. These mechanisms exhibit temperature dependence: the natural convection from hotter surfaces is more intensive; the radiative heat transfer exhibits very strong temperature dependence according to the Stefan-Boltzmann law. In both cases the nonlinearities appear at the largest time constants of the system that can be clearly identified as changes at the very ends of the structure functions and, thus, can be well separated from the parts that are characteristic of the package under test [156].

Treating the thermal effects which are characteristic to the test environment is beyond the scope of this book. It has to be mentioned that providing a correction formula to account for these effects in a linear model-based approach is not straightforward.

Chapter 3
Thermal Metrics

András Poppe and Gábor Farkas

As seen previously, measuring the Z_{th} thermal impedance of a packaged semiconductor chip provides all thermal information that is theoretically available; thus, it can provide input also for generating *standard thermal metrics*.

The thermal metrics of single die packages are single numbers that represent the quality of a certain package or package type, allowing easy comparison among them during everyday thermal engineering practice. These aggregate numbers do not resolve structural details of the packages, publishing them on product data sheets does not reveal proprietary information of semiconductor vendors or packaging houses, and, therefore, they are widely used in the electronics industry. Simplification, however, if applied without the understanding of the underlying physical and measurement concepts, may lead to improper use in daily practice. Therefore, in this chapter, we introduce the fundamental physical considerations that with good engineering approximations may lead to standardized thermal metrics. Primarily, we use the notations established earlier [18] and the terminology of the most widely used international industrial standards for thermal testing of packaged semiconductor devices, the JEDEC JESD51-x family of standards [29].

A. Poppe (✉)
Siemens Digital Industry Software STS, Budapest, Hungary

Budapest University of Technology and Economics, Budapest, Hungary
e-mail: Poppe.Andras@vik.bme.hu

G. Farkas
Siemens Digital Industry Software STS, Budapest, Hungary

© The Author(s), under exclusive license to Springer Nature Switzerland AG 2022
M. Rencz et al. (eds.), *Theory and Practice of Thermal Transient Testing of Electronic Components*, https://doi.org/10.1007/978-3-030-86174-2_3

3.1 Standard Thermal Metrics for Single Die Packages

First, we start with the basic concepts of the thermal metrics defined in these standards; then, we introduce both the classical way of measuring them with simple steady-state methods, and the description on how these metrics can be identified from thermal transient measurements in such a way that is compatible with the original concepts of JEDEC's JESD51-x family of standards.

3.1.1 A Few Thoughts on the Thermal Resistance in Relation to Deriving Standard Thermal Metrics

So far the concept of the lumped thermal resistance has been introduced in a very abstract way, representing the temperature drop between two *nodes* in a single heat-flow path as a response to a steady, constant heat flux φ_{thr} (see Fig. 2.4). As suggested by Eq. (2.4), integrating the heat flux over the surface it passes through, we obtain the P power entering one node and leaving through the other one. As shown in Eq. (2.6), the ratio of the temperature drop and this power results in a single quantity, called *thermal resistance*. In this theoretical introduction, the thermal resistance is defined between two abstract locations, represented by zero-dimensional nodes. It is not obvious how one can define the thermal resistance or thermal impedances between arbitrary locations of a real heat-flow path related to a real, 3D physical semiconductor package structure.

Device packages are comprised of different material regions with larger volumes, surrounded by *complex surfaces*. Despite well-defined standards, there has been lot of misunderstanding before the basic concepts got clarified in the technical literature on the definition of a single thermal resistance [84, 85].

In product data sheets, the thermal data presented are rather a sort of a *characterization parameter* representing the strength of the *thermal coupling* between two locations in a thermal system, than a real "thermal resistance"-like property.

In steady-state (static) conditions when thermal capacitances do not play any role, the following formal definition of a thermal resistance holds: the thermal resistance of a closed volume of material (see Fig. 3.1) is:

"*the temperature difference between two isothermal surfaces divided by the amount of heat that flows between them is the thermal resistanceThermal resistances of the materials enclosed between the two isothermal surfaces and the heat fluxHeat flux tube originating and ending on the boundaries of the two isothermal surfaces.*" [86]

The essential point to understand is that a thermal resistance can never be based on measuring or calculating from values of two *points*, *unless* the planes are isothermal. Additionally, the heat flux φ_{in} entering the closed volume at *Surface1* should be equal to the heat flux φ_{out} leaving it through *Surface2* as shown in Fig. 3.1.

In reality, the above conditions can never be exactly realized, but good approximations for the above conditions can be provided. This is true for standardized test

Fig. 3.1 Two isothermal
surfaces connected by a heat
flux tube

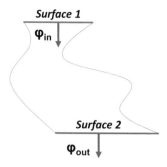

environments, and in many cases, it is also true for real-life operating conditions,
which are approximated by such test environments.

Many thermal measurement standards aim at realizing nearly isothermal physical
surfaces and deduce thermal metrics from a temperature measurement at two
dedicated points. The procedures defined in these standards are collectively referred
to as the *two-point methods*. In another approach the characteristics of the heat flux
are changed at the terminating surfaces; the related technique is called *transient dual
interface method* (TDIM).

3.1.2 The Concept of the Junction-to-Ambient and the Junction-to-Case *Thermal Resistances*

For semiconductor device packages, the R_{thJA} junction-to-ambient thermal resistance is a generally used thermal performance indicator, which is to be measured in a
natural convection test environment as shown in Fig. 3.2.

Though the temperature exhibits a given spatial distribution over the surface of
any active semiconductor die, measuring the junction temperature by the electrical
test method defined in [29] and [30] results in a well-repeatable average temperature
value which is called T_J junction temperature in the JESD51 series of standards; in
other words, the *junction* is considered as an isothermal source. It is an approximate
model that works fairly well in everyday practice. This way the R_{thJA} junction-to-
ambient thermal resistance is defined such that the junction is surrounded by the
ambient being forced to the constant T_A ambient temperature. Figure 3.3 demonstrates that in this case, we have two isothermal locations between which there is no
heat loss, so conditions of the formal definition of the thermal resistance are met. As
seen in Fig. 3.2, the still-air chamber is only one part of the measurement environment; the used standard test board is also part of the test environment of the package.

Focusing first on thermal characterization on package level, one can generally
state that packages of power semiconductor devices have been designed to be cooled

Fig. 3.2 An IC package on a certain type of standard thermal test board in a JEDEC standard natural convection test environment. In this photo, a practical realization of a so-called still-air chamber (built according to the recommendations of the JEDEC JESD51-2A standard [31]) can be seen. Details of standard thermal test environments will be discussed later in this chapter

Fig. 3.3 Illustration of the R_{thJA} junction-to-ambient thermal resistance according to the formal definition of the thermal resistance

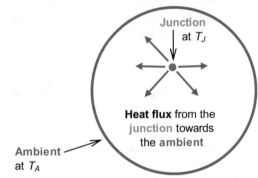

through a dedicated cooling surface, called *case* in several testing standards. In most package designs, a single heat-flow path exists between the chip's junction and the case. Therefore, for thermal characterization of power device packages, the R_{thJC} *junction-to-case thermal resistance* is the recommended thermal metric, and cold plate cooling is the recommended test environment. In this approximation it is assumed that most of the heat generated in the junction leaves the package through the mating surface of the case and the cold plate, which is considered isothermal as shown in Fig. 3.4.

Though in a junction-to-case thermal resistance test setup (see Figs. 3.4 and 3.5a) it is fair to assume the cold plate to be isothermal, under real application conditions

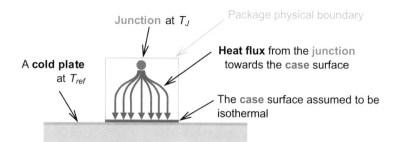

Fig. 3.4 Illustration of the R_{thJC} junction-to-case thermal resistance of power semiconductor device packages according to the formal definition of the thermal resistance

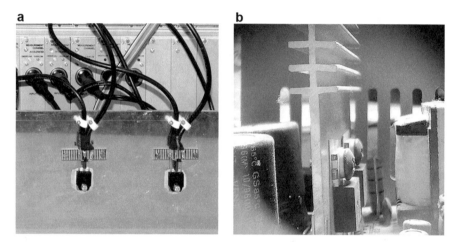

Fig. 3.5 (**a**) Actual test setup for the thermal measurement of transistors in power packages on a water-cooled cold plate. (**b**) The transistors in a real product environment: package case surfaces are attached to a heat sink

where a heat sink substitutes the cold plate (Fig. 3.5b), neither the case surface nor the mating surface of the heat sink is at uniform temperature.

For both the R_{thJA} and the R_{thJC} metrics, one can state that the actual test conditions have strong influence on their measured value. In case of the R_{thJA} junction-to-ambient thermal resistance, for example, the type of the test board has a great influence.

In case of the R_{thJC} junction-to-case thermal resistance, when measured with the junction as a temperature sensor and with a thermocouple attached to the case surface, the temperature profile at the case–cold plate or case–heat sink interfaces poses a problem. This nonuniform temperature profile is influenced by the quality of heat transfer from the actual package case surface towards the cold plate during test, or to the heat sink in real-life conditions, determined mostly by the applied thermal interface material, if any. See below in Example 3.1.

In the relatively new JEDEC standard for thermal characterization (JESD51-14 [40]), the use of thermal transient testing and structure functions helps mitigate the above practical problems in measuring R_{thJC} thermal resistances. A similar concept can be also applied to the R_{thJA} junction-to-ambient thermal resistances. Nevertheless, when reporting the measured thermal resistances values, the test report should indicate the actual test method and the details of the test conditions.

Example 3.1: R_{thJA} Measurement Results for Different Thermal Test Boards

For illustrating how the R_{thJA} thermal resistance measurement results are affected by the applied test conditions, we present two different results obtained for the same device type (Infineon's TLE6236G device in an SO-28 package), using a high thermal conductivity and a low thermal conductivity thermal test board as shown in Fig. 3.6. In these two setups, the heat transfer from the heated junction towards the package leads is identical, but from the leads on, towards the ambient, two different heat transfer patterns are realized by the boards. Section A shown in Fig. 3.6a represents enhanced conductive heat transfer provided by the high conductivity board towards the edge connector of the board; section B represents the heat transfer from the board towards the ambient, mostly by natural convection.

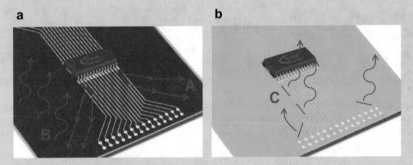

Fig. 3.6 IC package mounted on (**a**) high thermal conductivity, (**b**) low thermal conductivity test board for R_{thJA} test in still-air chamber. Photographs of such boards are shown in Fig. 5.3 in Sect. 5.1 of Chap. 5

In the case of the low conductivity board shown in Fig. 3.6b, the conductive part through the traces and the board itself is much lower, though some cooling takes place also by natural convection towards the walls of the still-air chamber. This results in almost twice as high thermal resistance value (81 K/W) as in case of the high conductivity board (50 K/W), as shown in Fig. 3.7. As one can see, the thermal impedance curves coincide up to ~10s, corresponding to about 33 K/W thermal resistance between the junction and the package leads.

(continued)

Example 3.1 (continued)

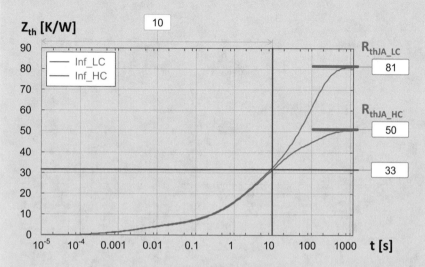

Fig. 3.7 Measured thermal impedance curves of the setups shown in Fig. 3.6, curve `Inf_HC` belongs to the high thermal conductivity, `Inf_LC` to the low thermal conductivity test board

Note in Fig. 3.7 that the steady-state values of the measured thermal impedances provide values of the R_{thJA_LC} and R_{thJA_HC}, the junction-to-ambient
thermal resistances, measured using the high and the low effective thermal
conductivity test boards, respectively. There is no sharp divergence between
the Z_{th} curves belonging to the two boundary conditions; the value of 33 K/W
indicated in the chart was taken from the structure function analysis presented
in Fig. 3.8.

With the structure functions shown in Fig. 3.8, one can better understand
the root cause of differences between the junction-to-ambient thermal resistances R_{thJA_LC} and R_{thJA_HC}; the thermal resistances for the high thermal
conductivity and low thermal conductivity test boards, respectively. The heat
transfer structure changes, when the heat leaves the package at the leads and
enters the test boards. This is clearly indicated by the location where the two
structure functions start diverging. Up to this divergence point, the cumulative
thermal resistance is 33 K/W; this is the R_{thJB} "junction-to-board thermal
resistance" of the package[1].

---

[1] This metrics is denoted as junction-to-lead or junction-to-pin thermal resistance in several
works. We do not investigate how close this "thermal resistance" satisfies the formal
definition of the thermal resistance given earlier; that is why we use quotes.

(continued)

Example 3.1 (continued)

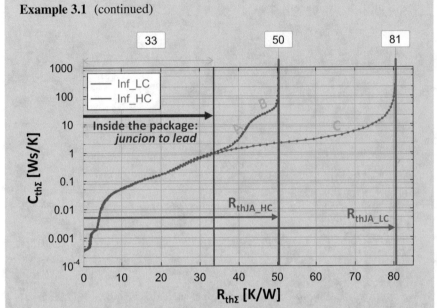

Fig. 3.8 Structure functions obtained from the thermal impedance curves of Fig. 3.7, curve `Inf_HC` belongs to the high thermal conductivity, `Inf_LC` to the low thermal conductivity test board

The perfectly matching initial parts of the structure functions represent the heat conduction paths inside the two identical packages, from the junction through the leadframe until the bottom of the leads.

In case of the high thermal conductivity board (`Inf_HC` curve), sections A and B are clearly distinguishable. Section A represents conductive heat spreading by the board itself. Its straight shape in the lin-log representation of the structure function suggests that the heat is spreading in concentric rings in the two solid internal copper layers of the high conductivity board. Similarly, the long straight section in the structure function before the divergence point indicates a ring-style spreading in the inner leadframe.

Section B represents the overall effect of convective heat transfer from the board surface towards the ambient. In case of the low effective thermal conductivity test board in section C, the effect of the conductive heat spreading by the board itself is negligible; we cannot separate it from the effect of the natural convection of the air inside the test chamber.

In the case of the high effective thermal conductivity board, we have an excess cumulative resistance of 17 K/W adding on top of the 33 K/W initial section, representing 51% of the total R_{thJA_HC} junction-to-ambient thermal resistance. In the case of the low effective thermal conductivity board, the excess thermal resistance originating from the test environment itself is 48 K/W which is about 145% of the own contribution of the package.

As a summary, it can be stated that the R_{thJA} thermal resistance, though relatively easy to measure, is more characteristic for the test environment itself than for the package under test. The thermal engineering community was only partially aware of this until structure functions and reliable thermal simulation tools became widely available.

Another conclusion is that the test environment with the shortest possible overall junction-to-ambient heat-conducting path is to be chosen for device characterization. The measured R_{thJA} value is more characteristic to the package tested than to the test environment keeping the contribution of the test environment to R_{thJA} the smallest. From this perspective, measurements performed on cold plates offer obvious advantages. This example also highlights why it is of paramount importance to report the details of the test conditions/test environment together with measured values.

3.1.3   The Concept of Thermal Characterization Parameters

Classical thermal testing standards, such as the JEDEC JESD51 overview document [29] and the JESD51-1 standard [30], define the generic concept of "junction-to-environment X" thermal resistance type steady-state thermal metrics as

$$R_{thJX} = \frac{T_J - T_x}{P_H} \qquad (3.1)$$

where T_x denotes the steady-state temperature in a given location of environment X (such as the "ambient" or the "case") and P_H denotes the steady-state heating power at the location of the J "junction." As long as the entire P_H heating power is completely dissipated to the environment through the temperature monitoring location X, Eq. (3.1) meets the strict definition of the thermal resistance provided in the previous subsection. In reality, however, the X surface is usually not a plane; a small variation in the location of the external sensor causes significant change in the measured R_{thJX} value. A usual, alternate notation of the junction to reference environment X thermal resistance is Θ_{JX}.

There are test scenarios, when the accessible location X does not reside at an isothermal surface which is perpendicularly crossed by the entire heat flux leaving the junction. In a widely used practical test setup, the temperature of the printed circuit board holding the powered device is measured, at a dedicated thermal test point near to the package pins. For example, in case of power LEDs mounted on MCPCB substrate, this is the only temperature change that one can monitor; see Fig. 3.9a.

In such cases when only part of the heat flux originating from the junction affects the temperature of the temperature monitoring point X, Eq. (3.1) takes the form:

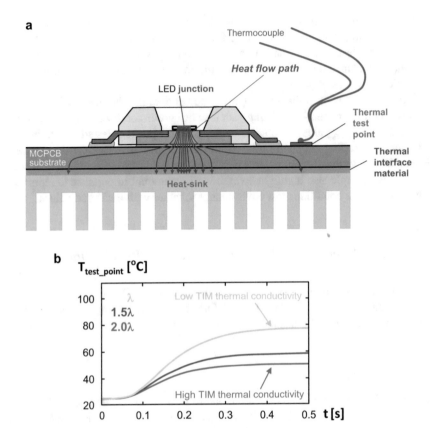

Fig. 3.9 (a) Power LED package on a metal core printed circuit board substrate attached to a heat sink; (b) simulated transient of the thermal test point temperature with varying λ thermal conductivity of the TIM between the MCPCB and the heat sink. (Based on [87])

$$\Psi_{\mathrm{JX}} = \frac{T_{\mathrm{J}} - T_{\mathrm{x}}}{P_{\mathrm{H}}} \qquad (3.2)$$

and the Ψ_{JX} quantity defined this way is called the junction-to-X *thermal characterization parameter* [39]. Though its dimension is also K/W like that of the thermal resistance, one has to be aware of the fundamental difference between the R_{th} thermal resistances and the Ψ thermal characterization parameters especially since no direct physical meaning can be attributed to the latter.

The test scenario illustrated in Fig. 3.9a is typical in field applications when the end user cannot have an access to the junction temperature. In such cases, only the temperature of the thermal test point can be monitored or recorded. The junction temperature can be estimated only from this temperature data, using the P_{H} heating power at the junction and the $\Psi_{\mathrm{J\text{-}to\text{-}testpoint}}$ thermal characterization parameter provided by the manufacturer.

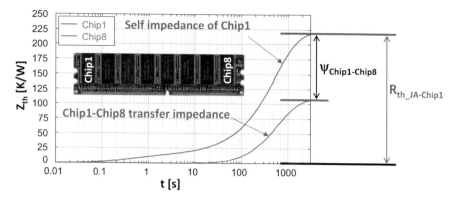

Fig. 3.10 The thermal transients of the RAM module of Example 2.8 shown in Fig. 2.42 turned into thermal impedances by dividing them with the power applied at Chip1

On package level, typical thermal characterization parameters provided on product data sheets are the Ψ_{JL} *junction-to-lead*, Ψ_{JB} *junction-to-board*, or Ψ_{JT} *junction-to-top* (of the package) thermal characterization parameters [39].

One has to note that the measured thermal characterization parameters are highly dependent on the actual thermal environment of the package or assembly, and this limits their use. As an example, a simulation study for a power LED assembly such as shown in Fig. 3.9a revealed that the temperature transient recorded at the thermal test point of the assembly was strongly influenced by the thermal conductivity of the thermal interface material (TIM) applied between the module substrate and the heat sink [87]; see Fig. 3.9b.

Note that if the temperature shown in Fig. 3.9b is normalized by the heating power, we obtain the junction-to-thermal test point *transfer impedance*[2].

Note also that applying Eq. (3.2) to the $T_J(t)$ and $T_X(t)$ temperature transients results in a time function that has no physical meaning, though the final steady-state value would be the Ψ_{JX} thermal characterization parameter.

This means that a thermal characterization parameter is always a difference of the steady-state values of a driving point impedance and a transfer impedance, as illustrated in Fig. 3.10.

This explains why *no real physical meaning can be attributed to the Ψ values* (not to mention that the way these values are calculated does not meet the strict definition of the thermal resistance given is Sect. 3.1.1).

Last, but not least, it has to be noted that due to the propagation delays of the transfer impedances, the $\Delta T_X(t)$ transient would reach steady-state later than $\Delta T_J(t)$.

---

[2]The concept of the thermal transfer impedance was introduced in Sect. 2.7; see, e.g., Fig. 2.41. there.

Therefore, when measuring R_{thJX} thermal resistances by switching on the heating power P_H, in Eq. (3.2), the final, steady-state value of the environment temperature T_x must be used. This also applies to the measurements of multi-die packages discussed in Sect. 3.2.

3.1.4 The Standard Thermal Metric of Single Die Packages with Exposed Cooling Surface: $\mathbf{R}_{thJC}$

In many cases, the X surface is an exposed cooling surface of a power device or module, the "case." In the simplest approach, the junction-to-case thermal resistance, R_{thJC}, can be derived from the definition given by Eq. (3.1) as follows:

$$R_{thJC} = \frac{T_J - T_C}{P_H} \tag{3.3}$$

where T_J is the junction temperature of the device under test and T_C denotes the "case" temperature measured in the junction-to-case thermal resistance measurement environment.

According to the original concept of the classical thermal testing standards, the measurement based on (3.3) is a two-point measurement, involving two thermometers separated spatially. The first thermometer measures the junction temperature, T_J. In most cases this thermometer is the pn junction driven by a constant forward current, using its forward voltage, V_F as a temperature-sensitive parameter (TSP).

The second thermometer, attached to the "case" surface of the package under test as an external device, is typically a thermocouple. As mentioned earlier in Sect. 2. 12.2 in Chap. 2, precise calibration of these temperature sensors is needed in order to avoid false nonlinearities of the thermal measurement itself.

Although the test setup illustrated in Fig. 3.4 provides some approximation of the "isothermal surface," the actual homogeneity of the cold plate surface depends on multiple parameters, such as the thermal conductivity of the plate material and the routing of the fluid channels in it. Even supposing a perfectly flat temperature distribution on the plate surface, due to the finite thermal conductivity of the applied thermal interface material, a typical bell-shaped temperature distribution develops on the package case.

The problems associated with the classical measurement of the junction-to-case thermal resistance R_{thJC} based on Eq. (3.3) are discussed further in details in the technical literature [88]. These are overcome by a relatively new thermal testing standard of JEDEC [40], using junction temperature measurements only, followed by structure function analysis.

3.1.5   Spatial Temperature Difference Replaced by the Temporal Difference of the Junction Temperature

Most of the problems associated with the measurement of temperature differences along the junction-to-case, or the complete junction-to-ambient heat conduction path suggested by Eq. (3.1), can be overcome as follows.

Let us apply, for example, Eq. (3.1) for the junction-to-ambient thermal resistance, R_{thJA}:

$$R_{thJA} = \frac{T_J - T_A}{P_H}. \tag{3.4}$$

Let us rearrange the equation to express the junction temperature from any heating power, P_H, applied at the junction and from the known reference temperature of the ambient, T_A:

$$T_J = R_{thJA} \cdot P_H + T_A. \tag{3.5}$$

If we apply two different, particular levels of the heating power, P_{H1} and P_{H2}, respectively, we can express the two different steady-state junction temperature values that develop in response to these powers:

$$T_{J1} = R_{thJA} \cdot P_{H1} + T_A, \tag{3.6a}$$

$$T_{J2} = R_{thJA} \cdot P_{H2} + T_A. \tag{3.6b}$$

Introducing the notations $\Delta P_H = P_{H2} - P_{H1}$ and $\Delta T_J = T_{j2} - T_{J1}$ and subtracting Eq. (3.6a) from (3.6b), we get:

$$\Delta T_J = R_{thJA} \cdot \Delta P_H. \tag{3.7}$$

thus:

$$R_{thJA} = \frac{\Delta T_J}{\Delta P_H}. \tag{3.8}$$

In general, the two power levels, P_{H1} and P_{H2}, are achieved if one follows the procedure mentioned as *static test method* in the JEDEC JESD51-1 thermal testing standard: first we apply the heating power P_{H1}, and when the junction temperature of the device under test got stabilized, we record the corresponding junction temperature T_{J1}. Then, we switch the heating power to P_{H2} and record the T_{J2} junction temperature when the final steady state is reached. The entire transient process is illustrated in Fig. 3.11.

Regarding the power step applied, the JEDEC JESD51-2A standard [31] recommends such levels that assure at least 20 °C junction temperature rise, but junction

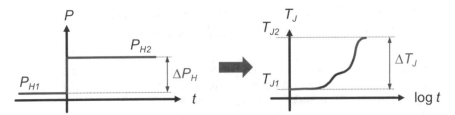

Fig. 3.11 Using temporal difference of the junction temperature for the determination of the thermal resistance-type thermal metrics

temperature elevations in the range of 30 °C–60 °C are recommended that better resemble chip temperatures under actual application conditions.

Note that due to the differential approach, the steady value of the reference temperature is cancelled out from (3.8). Therefore, there is no need to measure the reference temperature for deriving the thermal resistance. As the reference temperature disappeared from the equation, it is valid for any R_{thJX}-type thermal metric, both for R_{thJA} and R_{thJC}, provided that the measurements were carried out in the given standard thermal test environment.

We can further generalize our approach from thermal resistances to thermal impedances. Let us consider the actual $T_J(t)$ function. For the initial steady state, at time instance t_1 when the heating power is switched abruptly from P_{H1} to P_{H2}, we can write that $T_{J1} = T_J(t_1)$. Let us denote the time instance when we assume that the final steady state is reached by t_2; thus, $T_{J2} = T_J(t_2)$. With these notations we can express the junction temperature transient also as $\Delta T_J(t) = T_J(t) - T_{J1}$. For $t = t_1$, we get $\Delta T_J = 0$, and for $t = t_2$, the value of ΔT_J will be the same as used in Eq. (3.8). If we recall the definition of the $Z_{th}(t)$ thermal impedance function given by Eq. (2.17), we can write:

$$Z_{th}(t) = \frac{\Delta T_J(t)}{\Delta P_H}. \tag{3.9}$$

For convenience, the abrupt switching of the power may take place at $t_1 = 0$, and in theory one needs to wait until infinity for the steady state to occur: $t_2 = \infty$. In practice the final steady state can be assumed when the change of the temperature is smaller than the resolution of our electronic thermometer, i.e., the small temperature change cannot be detected any longer. With the above notations

$$Z_{th}(t_2 = \infty) = \frac{\Delta T_J(t_2 = \infty)}{\Delta P_H} = \frac{T_{J2} - T_{J1}}{\Delta P_H} = R_{thJA}, \tag{3.10}$$

that is, the *thermal resistance is the steady-state value of the thermal impedance*. If measured, e.g., in the standard still-air chamber such as shown in Fig. 3.2, Eq. (3.10) gives the *junction-to-ambient thermal impedance*.

If $P_{H1} = 0$ (no heating) is applied at $t_1 = 0$, then (3.6a) reduces to $T_J = T_A$, and $\Delta P_H = P_{H2}$ is the heating power switched on at the junction; thus, Eq. (3.10) becomes

$$Z_{th}(t_2) = \frac{T_J(t_2) - T_{J1}}{\Delta P_H} = \frac{T_J(t_2) - T_A}{P_{H2}} = R_{thJA}. \qquad (3.11)$$

It is easy to realize that (3.11) is exactly the standard definition of the junction-to-ambient thermal resistance given by Eq. (3.4).

As a summary, one can state that with measuring the entire $\Delta T_J(t)$ transient of the junction temperature change from the initial steady state when the heating power P_H is abruptly switched until the final steady state of the device under test is reached, there is *alternate way to measure the standard steady-state thermal metrics of semiconductor device packages*.

Calculating the ratio of the achieved total *change of the junction temperature*, ΔT_J, and the applied *change of the heating power*, ΔP_H, Eq. (3.9) provides *an equivalent definition* of the JEDEC standard steady-state thermal defined by Eq. (3.1).

The measurement of the $T_J - T_A$ *spatial difference* of the temperature along the heat conduction path requires the calibration of *two different types* of temperature sensors. On the contrary, measuring the $\Delta T_J(t)$ *temporal difference* of the junction temperature requires only the careful calibration of the semiconductor junction used as the sole thermometer in the test setup.

The advantages of this transient approach based on the ΔT_J and ΔP_H differences are obvious. For example, with properly chosen instrumentation during the calibration process of the diode sensor at the junction and the actual $\Delta T_J(t)$ measurement, some scale errors of the used electrical meters also cancel out.

Also, offset errors in temperature measurement cancel out.

Let us suppose that we measure T_J with a constant offset error of T_{JO} and T_A with an error of T_{AO}. Then, instead of a correct junction-to-ambient thermal resistance an inaccurate R_{AO} value is measured; from (3.4) R_{AO} would take the form:

$$R_{AO} = \frac{T_J - T_A + T_{JO} - T_{AO}}{P_H}. \qquad (3.12)$$

With the temporal difference approach in (3.12) all offset errors cancel out. The accuracy of the measurement is influenced only by the drift in time of the T_{JO} offset error on the single data acquisition channel used for junction temperature measurement.

In the physical realization of the JEDEC JESD51-1 electrical test method [30], the temperature measuring channel of the test equipment has a V_{JO} voltage offset; the T_{JO} temperature offset is provided by the calibration process as $T_{JO} = V_{JO}/S_{VF}$ where S_{VF} is the temperature sensitivity of the temperature-dependent parameter that is measured. For details on actual realization principles of the electrical test method, refer to Chap. 5. Calibration issues are further treated in Chap. 8.

It can be proven in a similar way that in case of measuring the R_{thJC} metric with the transient method, the offset errors also cancel out. The problems associated with the classical steady-state measurement of the R_{thJC} metric based on (3.3) are overcome by a relatively new thermal testing standard of JEDEC [40], using $\Delta T_J(t)$ junction measurements only, followed by structure function analysis.

This transient R_{thJC} measurement technique is known as the *transient dual interface method* (TDIM) first proposed in the papers [88] and [89]. The refinements of the originally proposed method were published in [90] and [91], forming the bases of the corresponding JEDEC standard [40]. This test method is further discussed in Sect. 7.1 of Chap. 7.

3.2 Thermal Metrics of Multi-Die Packages

In case of device packages in which there are multiple heat sources, the single die or, more precisely, single heat source thermal metrics (R_{thJA}, R_{thJC}) cannot be applied properly. Therefore, extended thermal metrics are needed that provide some insight into how such a package would behave if different combinations of powers are generated at the different locations. In this subsection the basic concepts will be briefly presented.

Semiconductor device packages with multiple heat sources may take different forms. Typical devices are packages in power electronics containing *laterally arranged chips within the package*, such as the one shown in Fig. 2.40 in Sect. 2.7 of Chap. 2 where the concept of driving point and transfer impedances was introduced. Similar, lateral arrangements can be seen in device examples presented in Chap. 6: in IGBT modules comprised of multiple chips (Fig. 6.54) or multicore IC chips (Fig. 6.94) where the individual cores are acting as individual heat sources. Multiple chips in a single package can be also arranged vertically; examples are the stacked die packages or real, 3D packaged ICs.

Nevertheless, in all of these cases, *the major question is: "what are the temperatures of the junctions when different levels of heating powers are provided by the different heat sources?"* Maintaining the generality of our discussion, from now on we shall refer to the multiple heat sources and temperature monitoring points as chips or dice, with all the statements made also valid for different heat generating and temperature sensing locations within a multicore IC package. Unfortunately, currently available thermal testing standards for multi-die packages [41, 42] provide little guidance. Despite its title, [41] recommends to measure the junction temperature elevations with respect to the ambient temperature of all chips, for different combinations of powers applied at the chips, e.g., for the i-th chip out of the n chips within the package: $\Delta T_i(P_{H1}, P_{H2}, \ldots, P_{Hi}, \ldots, P_{Hn})$, where P_{Hi} is the heating power applied at the junction of the i-th chip.

The [41] and [42] standards recommend to use the same test environments and testing procedures that were recommended to measure the standard thermal metrics of single die packages with a few modifications that are basically due to the fact that thermal measurements of multi-die packages require significantly more electrical connections than measurements of single die ones and the classical test boards were not designed for this. The [41] document recommends to measure and report the ΔT_i junction temperature elevations together with the power combinations (power maps) applied.

Assuming that linearity holds for the multi-heat source system that needs to be characterized (see Sect. 2.12.3 in Chap. 2), we can apply the superposition principle that would allow a systematic test procedure: we apply heating power only at one single die at a time, but we measure the temperature responses at all junctions. This way for a series of vectors of powers such as:

$$P_{H1} = 1 \text{ W}, \quad P_{H2} = 0, \quad \dots, \quad P_{Hi} = 0, \quad \dots, \quad P_{Hn} = 0$$
$$P_{H1} = 0, \quad P_{H2} = 1 \text{ W}, \quad \dots, \quad P_{Hi} = 0, \quad \dots, \quad P_{Hn} = 0$$
$$\dots$$
$$P_{H1} = 0, \quad P_{H2} = 0, \quad \dots, \quad P_{Hi} = 1 \text{ W}, \quad \dots, \quad P_{Hn} = 0$$
$$\dots$$
$$P_{H1} = 0, \quad P_{H2} = 0, \quad \dots, \quad P_{Hi} = 0, \quad \dots, \quad P_{Hn} = 1 \text{ W}$$

we obtain the vectors of junction temperature elevations, n^2 values for the n chips.

The resulting junction temperature elevations can be arranged into a matrix as follows:

$$\begin{bmatrix} \Delta T_{11} & \cdots & \Delta T_{n1} \\ \vdots & \ddots & \vdots \\ \Delta T_{1n} & \cdots & \Delta T_{nn} \end{bmatrix}$$

where ΔT_{ki} represents the temperature elevation of the i-th chip when all chips were unheated but $P_{Hk} = 1$ W power was applied at the k-th chip. Note that a $\Delta T_{ii} |_{P_{Hi} = 1 \text{ W}}$ value in this case represents the junction-to-environment X thermal resistance seen from junction i, that we shall denote by R_{ii}^*. The $\Delta T_{ki} |_{P_{Hk} = 1 \text{ W}}$ value, i.e., the temperature elevation of junction i when chip k was powered, represents the Ψ_{ki}^* thermal characterization parameter. The concept of the characterization process is illustrated in Fig. 3.12. With * in the upper index, we indicate that these thermal resistance and thermal characterization parameter values are not single die thermal metrics but are elements of the $\mathbf{R_{th}^*}$ *steady-state thermal characterization matrix*.

The characterization approach illustrated in Fig. 3.12 can be performed at any heating power level, provided that the linearity assumption is practically not violated.

Thus, as suggested by the [41] document, the power levels recommended by the JEDEC JESD51-2A standard [31] can be applied to achieve junction temperature elevations in the range of 30 °C–60 °C. According to Sect. 2.12 of Chap. 2 in case of such moderate junction temperature elevations, the assumption of linearity still

Fig. 3.12 The process of the steady-state thermal characterization of a multi-heat-source thermal system

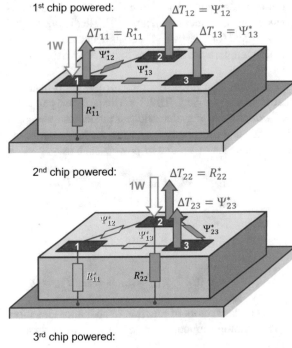

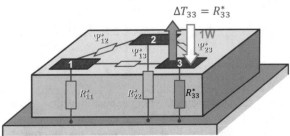

holds. Using the actual power levels, the ΔT_{ki} junction temperature elevations can be normalized in order to obtain the elements of the $\mathbf{R}^*_{\mathbf{th}}$ *steady-state thermal characterization matrix*:

$$R^*_{ii} = \Delta T_{ii}/P_{Hi}, \tag{3.13a}$$

$$\Psi^*_{ik} = \Delta T_{ik}/P_{Hi}. \tag{3.13b}$$

With this notation the $\mathbf{R}^*_{\mathbf{th}}$ matrix will look as follows:

$$\mathbf{R}_{th}^{*} = \begin{bmatrix} R_{11}^{*} & \cdots & \Psi_{n1}^{*} \\ \vdots & \ddots & \vdots \\ \Psi_{1n}^{*} & \cdots & R_{nn}^{*} \end{bmatrix}. \qquad (3.14)$$

Ideal linear RC systems are *reciprocal*. In this case this means that if chips i and k swap their roles as heat sources and temperature monitoring points, then for the same heating power, their ΔT_{ik} and ΔT_{ki} temperature rises will be the same, i.e., $\Psi_{ik}^{*} = \Psi_{ki}^{*}$. This reciprocity means that for the complete thermal characterization of a multi-die package with n chips inside, n multichannel measurements are needed. During each measurement the temperature responses of all the chips are recorded, as all the chips are individually powered in a sequence (from $i = 1$ up to n).

The R_{ii}^{*} thermal resistances in the main diagonal of the $\mathbf{R}_{th}^{*}$ matrix describe the effect of the self-heating of the chips, while the Ψ_{ik}^{*} thermal characterization parameters in the off-diagonal positions of the matrix represent the effect of the mutual heating between chips i and k.

Figure 3.12 suggests that the elements of the $\mathbf{R}_{th}^{*}$ matrix can be considered elements of a so-called fully connected thermal resistance model of a multi-heat source system. Actually, the $\mathbf{R}_{th}^{*}$ matrix includes all the information, but a few matrix transformations are needed to obtain the

$$\mathbf{R} = \begin{bmatrix} R_{11} & \cdots & R_{n1} \\ \vdots & \ddots & \vdots \\ R_{1n} & \cdots & R_{nn} \end{bmatrix}. \qquad (3.15)$$

matrix, the elements of which already represent real thermal resistances, constituting a real thermal network model suitable for simulation by a network solver such as shown in Fig. 3.13. The elements of $\mathbf{R}$ can be extracted from the elements of $\left[\mathbf{R}_{th}^{*}\right]^{-1}$, the inverse of the $\mathbf{R}_{th}^{*}$ steady-state thermal characterization matrix.

Fig. 3.13 Topology of a fully connected thermal resistance network for a multi-die package with three chips (such as shown in Fig. 3.12), suitable for simulation by a Spice-like circuit simulator

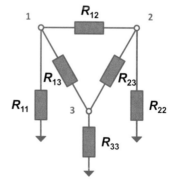

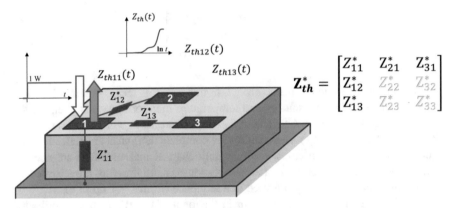

Fig. 3.14 The first step in the process of the dynamic thermal characterization of a multi-heat source thermal system

Regarding the steps of conversion, refer to paper [92], which also describes an example for system-level modeling of multi-heat source thermal systems[3].

The classical steady-state approach for measuring the thermal metrics can also be extended to the transient approach for the multi-die thermal metrics in a similar way as discussed in Sect. 3.1.5 for single die packages. The process is also similar, as it is illustrated in Fig. 3.14. The thermal metric we obtain is the *dynamic thermal characterization matrix* denoted by $\mathbf{Z_{th}^*}$:

$$\mathbf{Z_{th}^*} = \begin{bmatrix} Z_{11}^* & \cdots & Z_{n1}^* \\ \vdots & \ddots & \vdots \\ Z_{1n}^* & \cdots & Z_{nn}^* \end{bmatrix}. \tag{3.16}$$

The complete physical characterization is typically performed by a series of thermal transient measurements as illustrated in Fig. 3.14, but the measured thermal impedances can be converted into any equivalent representation; thus, they can be the originally measured time domain $Z_{th}(t)$ functions or their frequency domain counterparts, the $Z_{th}(\omega)$ functions as introduced by Eq. (2.44) in Sect. 2.8.1 of Chap. 2. Therefore, for the sake of generality for the elements of the $\mathbf{Z_{th}^*}$ matrix, we use the notation:

Z_{ii}^* for the driving point thermal impedances
Z_{jk}^* for the thermal transfer impedances

Here again, the prerequisite is that for the actual multi-heat source thermal system, the linearity assumption holds. With the transient approach here, the

---

[3] System-level thermal modeling using RC networks is beyond the scope of this book, though application of thermal transient testing to support compact thermal modeling is also discussed in Chap. 7.

superposition principle is applied not only for the calculation of the spatial distribution of the temperature in the system, but the superposition assures that by applying the convolution integral introduced by Eq. (2.12) in Sect. 2.3.3, the $T_{Ji}(t)$ dynamic temperature response to any form of $P_{Hi}(t)$ dynamic power profiles can be properly calculated. A rigorous description of the $\mathbf{Z}_{th}^*$ dynamic thermal characterization matrices is provided in [69].

Similarly to the $\mathbf{R}_{th}^*$ steady-state thermal characterization matrix, the $\mathbf{Z}_{th}^*$ matrix is also symmetrical if the heat sources and the temperatures are also monitored in a single point; thus, the thermal system is reciprocal.

In practice, however, *a slight violation of the reciprocity* is seen in the obtained transfer impedances. This is usually due to the differences in the area of the physical junctions used as heat sources and temperature sensors [93]. Such nonreciprocal behavior can be considered as a measurement artifact but can also indicate that the linearity assumption is violated (see also Sect. 2.12.3 in Chap. 2 on criteria of linearity of thermal systems).

There is currently active work, e.g., in the JEDEC JC15 standardization committee on Thermal Characterization Techniques for Semiconductor Packages [26] regarding the thermal metrics of multi-die packages. The $\mathbf{R}_{th}^*$ and $\mathbf{Z}_{th}^*$ thermal characterization matrices have been proposed [69] and are the best candidates to get accepted as such standard metrics.

3.3 Other Standards for Deriving Simple Thermal Metrics

In light of the former sections, those tests which rely on the temperature measurement by two sensors at two points seem to be simple. For example, for establishing the R_{thJC} junction-to-case thermal resistance, one has to determine the T_J junction temperature and the temperature reading of one of the sensors attached to the appropriate cooling surface as T_C. From Eq. (3.3) it can be deduced that $R_{thJC} = (T_J - T_C)/P$, where P is the applied power.

Regarding T_C, it is really just a simple reading of a sensor; it is more complex to determine T_J, which is an average value of the actual temperature distribution on the semiconductor surface. Moreover, there are only indirect ways to gain information on the chip temperature; for this reason, several standards call this quantity "virtual junction temperature" and denote it as T_{VJ}.

Taking a closer look at the measurement schemes in [19, 20, 25, 30, 45], we find that T_{VJ} is determined by:

- Putting a low I_2 current on the device under test in a thermostat and composing a calibration chart in one of the arrangements proposed in Chap. 5 Sect. 5.6
- Adding a high I_1 current to the device bias and heating it up by $I_H = I_1 + I_2$ as proposed in Sect. 5.4
- Periodically switching off I_1 and measuring the voltage on the device at low $I_M = I_2$ at a "proper" time

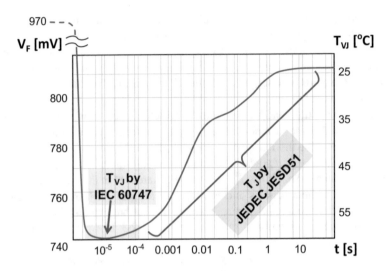

Fig. 3.15 Recorded electric and calibrated cooling thermal transient of a diode after switching from a higher heating current to low measurement current. Interpretations of T_J in the JEDEC JESD standards and of T_{VJ} virtual junction temperature in other standards are shown

Proper time is not clearly defined in the standards; there is some hint that the measurement should take place after an eventual electric transient but before considerable cooling of the chip.

We can recognize that determining T_{VJ} is a transient test, at least a shortened one. In Fig. 3.15, "proper time" would be somewhere between 100 μs and a few milliseconds. The transient measurement can be aborted after that time, but there is no statement in the standards for when it should be stopped, if at all. The voltage meter used typically has some integration time for suppressing noise; this way, actually, an average of the transient signal is recorded.

All standards prescribe an iterative process for the "virtual" junction temperature measurement but in a different way. The JEDEC JESD51 standards [29, 30, 40] aim at thermal characterization only. They assume that the cold plate in the measurement is kept at stable T_{cp} temperature, and a few trials are needed to find a proper I_H current which induces a "high enough" ΔT_J temperature elevation to keep low the influence of the limited accuracy of the test equipment, such as the offset errors mentioned previously.

The guidelines in the CIE Technical Report 225:2017 [45] comprise measurement of thermal and optical parameters of solid-state light sources. The light output of these devices strongly depends on the current and temperature, accordingly; the optical parameters have to be measured at a constant (T_J, I_F) pair. For this reason, the T_{cp} cold plate temperature is regulated at forced $I_H = I_1 + I_2$ driving current, until the pulsed voltage measurement at low $I_M = I_2$ corresponds to the target temperature determined in the calibration curve.

A comparative study on the T_J regulation defined in the JEDEC standards and CIE guidelines is presented in [94].

The IEC 60747 standards [19, 20] and the MIL-STD-750 standard [22] aim at measuring many various semiconductor parameters such as breakdown voltage, recovery time, etc. For all of these measurements, the T_{VJ} value, at which the measurement is carried out, has to be specified. The measurement of the virtual T_{VJ} is carried out mainly in the same way as in the CIE guidelines [45]. Still, the depicted measurement sequence in IEC 60747 is a bit obscure; it is not clear whether the iterative regulation of the cold plate temperature targets a predefined T_{VJ} or rather two different predefined T_{cp1} and T_{cp2} values at freely selected I_1 and I_2 currents.

Although the measurement of T_J does not conceptually differ in transient and static (that is truncated transient) measurements, the static approach needs simpler instrumentation, because the noise on the signal can be suppressed with integration along a short time period.

3.4   Brief Overview of Thermal Measurement Standards

We referred to several measurement standards in the previous sections; now we give a short but more systematic overview of them.

When the purpose of the measurements is building a properly accurate package model, there are no specific prescriptions on the number and style of the measurements needed. However, there exist guidelines for successful combination of measurement and simulation at various boundary conditions which yield a two-resistor model [27] or a compact thermal model consisting of a net of thermal resistances connecting simplified geometrical faces of a package [28].

On the other hand, when the purpose of the measurement is to produce *comparable* thermal data on packaged devices, a meticulous procedure has to be followed as listed in the appropriate standards.

Many relevant semiconductor test procedures, such as measurement of isolation voltages, parasitic inductances, capacitances, etc., are defined in the set of IEC 60747 (EN 60747) standards (e.g., [19, 20]).

In [20], several aspects of the thermal measurement of power modules are discussed. The measurement of the virtual junction temperature and for static methods also the position of thermocouples is specified. The transient methods are restricted to a short mentioning of Z_{th} curves as "transient thermal impedance."

The set of IEC 60747 standards differentiates between type tests and routine tests. Type tests are carried out on selected samples of new products in order to determine the electrical and thermal ratings of a type and for establishing test limits for further tests. The type tests are repeated regularly on a given number of samples taken from manufacturing batches at the manufacturer or from delivery batches at the end user in order to confirm the quality of the product. Routine tests are carried out on each sample of the production or delivery.

Thermal tests as routine tests are carried out only in mission critical industries (e.g., military, space).

The MIL standards [22] give some hint on the powering of the device for reaching a required temperature elevation in thermal tests, but the actual selection of voltages and currents for different semiconductor device categories seems to be ad hoc and sometimes poorly defined. A detailed review on the powering options is given in [95].

The most developed set for thermal testing is at present the JEDEC JESD51 family [29, 30, 40]. Especially, the JEDEC JESD51-14 standard [40] treats many aspects of the transient testing including the problem of removing eventual short-time electric perturbations from the thermal signal. Moreover, it introduces the concept of structure functions and the transient dual interface method (TDIM) introduced in Sect. 2.4.2.

Recent efforts of the European Center for Power Electronics (ECPE) have resulted in new guidelines for the automotive industry. This AQG 324 document serves validation purposes for different parameters of automotive power modules [25]. It restricts the thermal qualification to two-point methods, but, besides the junction-to-case thermal resistance of a module, junction to heat sink and junction to fluid thermal resistances are also defined for devices with an integrated cooling mount.

It has to be noted, however, that although thermal testing becomes more and more important in order to achieve reliable operation over a long lifetime, still, the construction of complete appliances often overlooks thermal testability aspects. Consequently, even the tests specified in various standards often need a work-around for accessing devices that are relevant for their power consumption or can be used as sensing points.

3.5 Problems of Using Standardized Thermal Metrics as Simple Thermal Models

In engineering practice, it is often the case that at the beginning of the design of an appliance, only a basic set of thermal parameters is available. This set is mostly limited to the R_{thJC} junction-to-case thermal resistance values of the packaged semiconductors or modules, in data sheets.

The published data can be suitable for finding promising types and for a prelim-inary back of the envelope calculation; still they should not be used for completing the final design. A prudent approach can be built on multiple subsequent steps, such as:

- Preselection of types based on data sheet values
- Measurement of the R_{thJC} junction-to-case thermal resistance on incoming samples
- Thermal simulation of several mechanical designs in FEM/FVM/CFD tool

- Thermal transient measurements on mock-ups or prototypes for design verification

This workflow can be highly improved if at the time of the thermal simulation *compact thermal models* are available from the manufacturer. Alternatively, these compact models can be built at the equipment designer if they possess the appropriate thermal transient measurement and post-processing tools presented in Sect. 7.7.

When the R_{thJC} junction-to-case thermal resistance is used in one of the design phases, it is necessary to know the limitations of its use.

As presented throughout this book, each measurement is a sort of a modeling process. For example, with a series of voltage measurements on the leads of a module at a several currents, a $V(I)$ model of the device is created. This model remains valid for a long time, until external factors cause the degradation of the device.

This longevity of the model is due to a number of factors. The physically realized current source, which represents the environment of the device, is very close to its abstract model. Moreover, the electric conductivity of the metal of the leads and that of the insulator material around differs in a ratio of 10^{12}. Keeping a few rules (e.g., four-wire measurements for applying the current and measuring the voltage), the measurement results remain highly repeatable.

The junction-to-case thermal resistance as characteristic thermal descriptor was originally developed for relatively small discrete packages and low heat flux densities. At these geometries it was justified to assume that the exposed cooling tab of the package is approximately at the same temperature. With the advent of large modules and high heat flux densities, it became an obvious oversimplification to distill a complex 3D heat spreading pattern into a single descriptive number.

At present, two alternative approaches exist to compose an R_{thJC} value for a device, the traditional two-point measurement method and the transient dual interface method (TDIM) defined in previous sections.

3.5.1 Two-Point R_{thJC} Measurement in Steady State

The fundamental problem of all two-point measurements is that R_{thJC} is not a single, unambiguous characteristic number describing the thermal performance of a package or module. Even if measured with great care, it is a function of applied fixing force, package surface roughness, and especially the heat transfer coefficient on the package surface in the measurement arrangement. This latter is influenced by the quality of the applied TIM, cold plate construction, velocity of the liquid flow in the cold plate, and similar other factors.

Figure 3.16 presents the 3D thermal simulation results of an assembly composed of a silicon chip in a TO220 package (similar to Fig. 2.59), mounted on cold plate. The figure represents the steady state after prolonged powering; the temperature

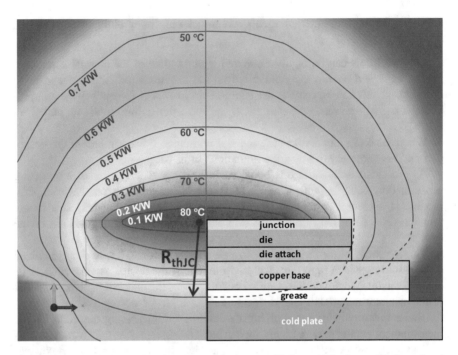

Fig. 3.16 3D simulation results of a packaged device on cold plate with isotherms. Horizontal and vertical scale is of 1:5 ratio. Two-point measurements would attribute R_{thJC} approximately to the 60 °C isotherm, corresponding to 0.5 K/W

distribution is visualized by colors and isotherms. For better discernibility of the details, the vertical scale of the picture is five times larger than the horizontal scale.

The isotherms are scaled both in temperature [°C] and in the thermal resistance measured from the junction [K/W]. It is easy to deduce the applied power from the distance of the isotherms. For example, the inner and the outmost isotherms in the figure are at a distance of 30 °C or 0.6 K/W; accordingly the chip was powered by 50 W.

It is a tempting approach to calculate the thermal capacitance of the matter enclosed between two isothermal surfaces and associate the thermal resistance and thermal capacitance growth with a section of the structure function. In a more precise assessment, the structure function is more of a kind of "frozen time"; as time rolls forward and the amount of matter enclosed between the reshaping isotherms changes, so thermal capacity develops.

The figure discloses that no uniform temperature exists on the physical copper base–grease interface. Cutting the isotherms along the physical package surface, the typical bell-shaped curve on the case can be produced.

In Fig. 3.17 the conceptual scheme of the two-point R_{thJC} thermal resistance measurement is shown. Regarding the chip temperature, a single T_{VJ} average value is known which is derived from a calibration and measurement process, described in Chap. 5. In this method R_{thJC} is defined as the difference between the average chip temperature and the measured temperature at the probe position, divided by the power on the chip.

Fig. 3.17 Conceptual scheme of the two-point R_{thJC} thermal resistance measurement. Temperature distribution on the chip and on the case surface along the symmetry plane is shown. R_{thJC} is derived from the difference between the T_{VJ} virtual chip temperature and the $T_C(x)$ temperature at the probe position

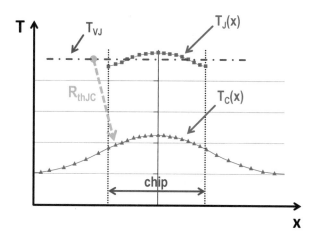

Several standards prescribe that the probe has to be placed on the case surface below the chip center. However, both Figs. 3.16 and 3.17 hint that even small lateral displacements can cause large change in the measured R_{thJC} value; the repeatability of the measured value depends largely on operator's experience.

Chapter 8 discusses further possible reasons of measurement uncertainties and repeatability issues in case of thermal transient measurements. Some of these inaccuracies have been classified as measurement errors; others were associated with the initial transient behavior of the devices under test and possible distortion in the data acquisition of the tester equipment. Besides other origins, a set of measurement uncertainties is related to the test environment in which the transient test is carried out. These can be best investigated in a simulation experiment, where the device and instrument induced errors play no role.

Example 3.2: The Effect of TIM Quality and Monitoring Point Selection on the Measurement Results in a Power Assembly
In this example simulated two-point measurements were investigated in a model composed of an IGBT module and various thermal interface material (TIM) layers under the base plate and a cold plate.

A simplified sketch of the arrangement is shown in Fig. 3.18.

The IGBT chips were of 11.2 mm × 11.2 mm size; the layers of the assembly are shown in Table 3.1. Under the silicon chips, a laminate of solder, copper, and ceramics layers is attached to an aluminum base plate.

The cold plate is modeled with a constant heat transfer coefficient of 3000 W/ m^2K which is a realistic value for an aluminum surface with internal water cooling. In order to examine the influence of the "base plate to cold plate" thermal interface, a 50 μm TIM layer is inserted between the module and the cold plate.

In this assembly, the transients were simulated in the SIMCENTER Flotherm tool [56] at a 100 W power step (heating), uniformly distributed on the die surface.

(continued)

Example 3.2 (continued)

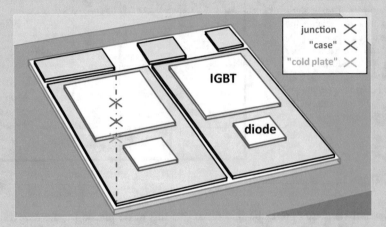

Fig. 3.18 IGBT module on cold plate; the left IGBT is powered

Table 3.1 Stack composition and the size of elements in the arrangement of the IGBT mounted on cold plate: x, z lateral size; y thickness in the stack; V volume of the element; λ thermal conductivity; c_V volumetric specific heat; C_{th} thermal capacitance of the element; ΣC_{th} cumulated thermal capacitance from the chip top

	x size [mm]	z size [mm]	y size [mm]	V [mm^3]	λ [W/mK]	c_V [kJ/m^{3}K]	C_{th} [J/K]	ΣC_{th} [J/K]
Silicon	11.2	11.2	0.3	37.6	450	1750	0.066	0.066
Die attach	11.2	11.2	0.1	12.5	100	1600	0.020	0.086
Copper	15	23.4	0.3	105	385	3440	0.36	0.45
AlN ceramics	32	30	0.3	288	250	2590	0.75	1.19
Copper	32	30	0.3	288	100	3440	0.99	2.18
Solder compound	32	30	0.3	288	100	1600	0.46	2.65
Aluminum	105	42	2.9	12790	150	2160	27.6	30.3
Thermal grease	105	42	0.05	221	**0.2, 1, 4**	3000	0.00066	30.3
Cold plate	$h = 3000$ W/m^{2}K							

The monitoring points for the simulated transients were selected as follows:

J: "Junction," center on the top of the powered semiconductor die,
   in the dissipating layer
C: "Case," center on the top of the TIM, below the semiconductor die
P: "Cold plate," center on the bottom of the TIM, adjoining the cold plate

(continued)

Example 3.2 (continued)

The temperature of J was considered to be the T_J junction temperature; the actual temperature distribution on the chip surface was not investigated.

The central monitoring point C mimics the prescribed placement of the probe as defined in [20, 25].

The monitoring point P corresponds to the situation when the probe does not (completely) penetrate the TIM layer. It corresponds approximately to the definition of the "junction to heat sink" thermal metrics defined in [20, 25].

For illustrating different measurement methodologies, the TIM layer was represented by different thermal conductivities, such as *dry surface* (0.2 W/mK) and *different interface materials* (1 W/mK, 4 W/mK). The two latter conductivity values corresponded to different qualities of thermal grease compounds.

Figure 3.19 shows the temperature change of the monitoring points at the different TIM conductivities. The improved thermal interface reduces the temperature elevation from 50 °C to 26 °C; the figure also proves that with a high-quality interface material it is less critical whether the reference probe really touches the module base plate or it is just "near" to it (J–C versus J–P distance).

The external monitor points reacted on the power change with a 0.5 s delay; accordingly, also in a live system, a slower data acquisition of the reference temperatures at low sampling rate (few samples/second) can be of appropriate time resolution. It can be observed that with more intensive cooling, the temperature stabilizes earlier. Steady state is approximately reached at 140 s, 50 s, and 30 s for the interface layers of 0.2 W/mK, 1 W/mK, and 4 W/mK thermal conductivity, respectively.

It would be hard to provide the full three-dimensional temperature distribution in the assembly as it develops in time; Fig. 3.19 shows only the temperature at a few characteristic points.

The steady-state temperature distribution on the case_bottom/TIM_top interface is shown in Fig. 3.20. The highest temperature under the chip center corresponds to the final transient value at C, shown as a blue "**x**" marker for the "dry" assembly in Fig. 3.19a and as black "**x**" and red "**x**" markers in Fig. 3.19b,c, respectively, for different TIM qualities. Note the large temperature difference even under the chip area.

The temperature record in Fig. 3.19 depicts only the outcome of one certain powering at three given boundaries. The normalized Z_{th} curves are presented in Fig. 3.21.

(continued)

Example 3.2 (continued)

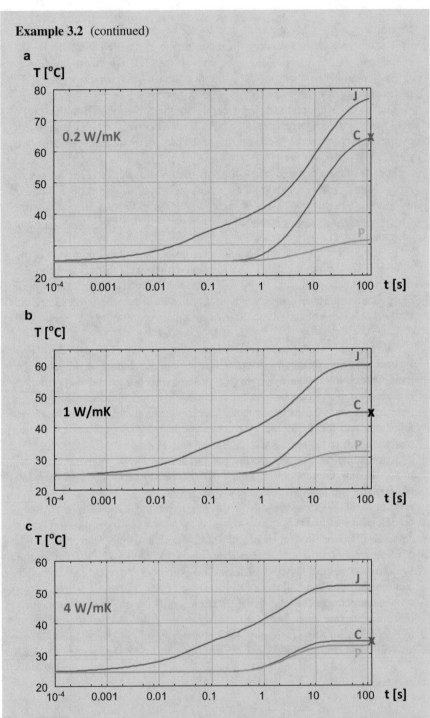

Fig. 3.19 Simulated temperature change at 50 W, thermal conductivity of the TIM: (**a**) 0.2 W/mK, (**b**) 1 W/mK, (**c**) 4 W/mK. J junction; C case center; P cold plate top position

(continued)

Example 3.2 (continued)

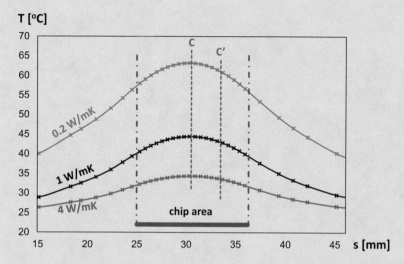

Fig. 3.20 Steady-state temperature distribution on the case_bottom/TIM_top interface. The peak temperature under the chip center corresponds to the final transient value at C, shown as "**x**" markers in Fig. 3.19. Location C' is displaced by 3 mm from the center

In case of a TIM layer of $\lambda = 0.2$ W/mK, first one can observe that the R_{thJA} total junction-to-ambient thermal resistance of the assembly is 0.52 K/W. This is the only true physical quantity in such a thermal measurement, based on the objective measured data without further assumptions on locations, divergence threshold, and other artificial elements introduced later on for other thermal metrics. The only approximation is assuming a uniform T_J junction temperature.

Composing the junction-to-case thermal resistance from the separate temperatures of the junction and the case yields $R_{thJC} = 0.13$ K/W if the probe penetrates the TIM, and $R_{thJC} = 0.45$ K/W if the probe just touches the lower surface of it ("**B = J–C**" and "**A = J–P**" in Fig. 3.21a, respectively). In the "**B = J–C**" curve, a characteristic overshoot explained in Fig. 3.27 can be observed.

Setting the TIM layer thermal conductivity to $\lambda = 1$ W/mK, separate measurements at the junction and at the case yield $R_{thJC} = 0.16$ K/W if the probe penetrates the TIM, and $R_{thJC} = 0.28$ K/W with the probe positioned on the lower surface of it ("**B = J–C**" and "**A = J–P**" curves in Fig. 3.21b, respectively). At this TIM quality for the whole assembly, R_{thJA} is 0.36 K/W.

With a TIM layer of $\lambda = 4$ W/mK, the two-point method yields $R_{thJC} = 0.18$ K/W junction-to-case thermal resistance with a penetrating probe and $R_{thJC} = 0.19$ K/W with the probe touching the lower surface of the TIM ("**B**" and "**A**" curves in Fig. 3.21c, respectively); R_{thJA} is now 0.28 K/W.

(continued)

Example 3.2 (continued)

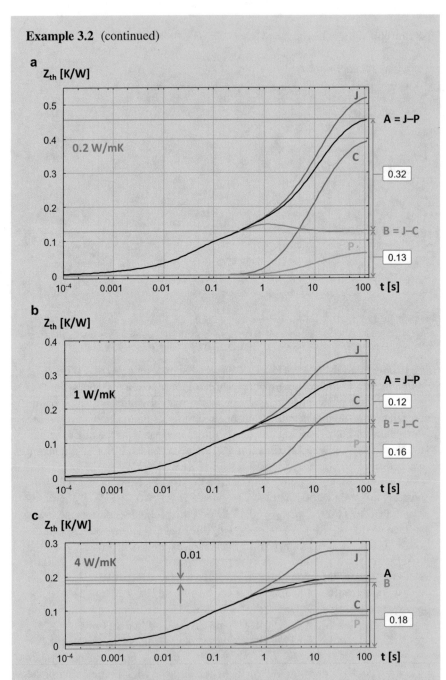

Fig. 3.21 Z_{th} curves, at junction and sensor locations, with TIM thermal conductivity: (**a**) 0.2 W/mK, (**b**) 1 W/mK, (**c**) 4 W/mK. J junction; C case center; P cold plate top position

It can be noted in Fig. 3.20 that with better TIM and cold plate qualities, the measured junction-to-case thermal resistance grows as the heat flow is more attracted to the center of the die–die attach–insulator–base plate sandwich, and the base plate temperature is more uniform.

In actual cold plate constructions, thermocouples or other sensors can be positioned only at a few fixed locations taking the mechanical design into consideration. The hole drilled into the plate for inserting the sensors distorts the temperature field, illustrated in Fig. 3.16 in its intact shape. Real thermocouples produce a further distortion of the measured case temperature because the probe tip is coated with an insulator layer and the wires draw some of the heat from the tip; this effect can be estimated based on Eq. (2.50). The measured thermal resistance can be well 100% larger than the ideal value obtained in a simulation[4].

In the previous simulation, the silicon material was modeled with the characteristic strong decrease of its λ_{Si} thermal conductivity, stored in the simulator tool. Numeric data on the change of λ_{Si} can be found at [17]; at 25 °C typically a value around 150 W/mK is assumed.

An important additional factor influencing the result of the two-point method is revealed in [96]. In the referenced document, a simulation experiment is carried out on a small silicon chip of 3 mm × 3 mm surface and 0.3 mm thickness, encapsulated in a package similar to Fig. 2.59 and mounted on cold plate.

The result of the simulations with 5 W homogeneous power on the chip surface and with different h heat transfer coefficients in the 100 W/mK–100 000 W/mK range is shown in Fig. 3.22. Simulating with constant λ_{Si}, the main effect influencing R_{thJC} is the flattening of the bell-shaped temperature distribution indicated in Fig. 3.20. Accordingly, the distance expressed in R_{thJC} increases monotonically with increasing heat transfer coefficient and is largest for the constant temperature boundary condition ($h = \infty$).

When the temperature dependence of λ_{Si} is taken into account, R_{thJC} increases also at low heat transfer coefficients, where the device heats up and the high thermal resistance of the silicon chip dominates the junction-to-case value. The maximum measured R_{thJC} is in the 2000 W/mK–10000 W/mK region, which corresponds to typical cold plate measurements.

---

[4]The additional heat loss through the thermocouple can be measured experimentally in two subsequent measurements. In the first measurement, the target arrangement is to be measured with a single thermocouple1. In the second measurement, an additional thermoucouple2 is to be mounted adjacent to thermocouple1, not connected to the measurement equipment. The measurement error due to the heat loss of the thermocouple is T1–T2, where T1 is the result of the first and T2 of the second measurement.

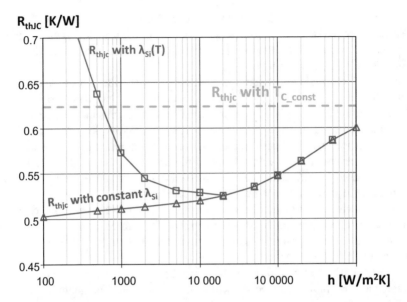

Fig. 3.22 The change of the R_{thJC} junction-to-case thermal resistance of a packaged silicon device, mounted on a cold plate surface of different heat transfer coefficient values and of fixed case temperature

3.5.2 Determination of $\mathbf{R}_{thJC}$ with the Transient Dual Thermal Interface Method

In the previous subsection, the influence of various surface qualities on the measured thermal resistance values was deemed to be a drawback. This effect is unavoidable, and for a long time, most effort was devoted to providing highly reproducible measurement conditions when using a two-point method.

In the revolutionary new approach introduced by the team of V. Székely, the drawback was converted into a real virtue. It was recognized that the structure function representation enhances the differences between results measured at different boundary conditions, but the portion belonging to the identical part, that is, to the device under test, remains unaffected. The following example illustrates the benefits of the methodology; a systematic elaboration of the concept is presented in Sect. 7.1.

Example 3.3: Determination of the R_{thJC} Junction-to-Case Thermal Resistance by the TDIM Method

The simulation experiments in Example 3.2 automatically yield several transients belonging to different boundary conditions. Figure 3.23 compares the Z_{th} curves at the J junction with various TIM qualities. One can observe that the heat flow arrived at the base plate at 1.7 s, and the curves deviated a bit below 0.2 K/W.

(continued)

Example 3.3 (continued)

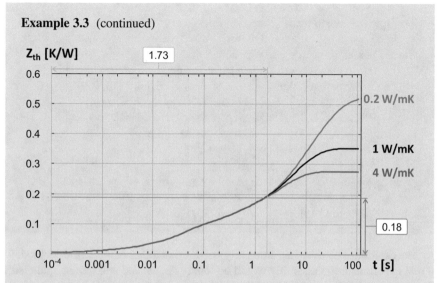

Fig. 3.23 Simulated Z_{th} curves, thermal conductivities of the TIM at 0.2 W/mK, 1 W/mK, and 4 W/mK

This difference is much more expressed in the structure functions, presented in Fig. 3.24. The curves belonging to different thermal conductivities start to diverge after 0.17 K/W. Until this point, we see the characteristic steps corresponding to the sandwich-like internal structure of the module composed of materials of diverse thermal conductivities.

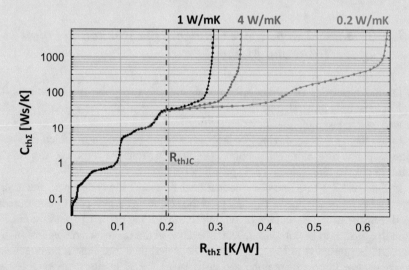

Fig. 3.24 Structure functions with thermal conductivities of the TIM at 0.2 W/mK, 1 W/mK, and 4 W/mK

(continued)

Example 3.3 (continued)

One can note that the figure describes well, besides the device under test, also the test fixture and the external cooler.

The approximate thermal capacitances of the components of the material stack are listed in Table 3.1. Figure 3.24 is cropped above 3000 J/K, as all the lines turn vertical, and no further change in the thermal resistance can be observed. In case of a real measurement on real cold plate, this capacitance would correspond to 700 liters of water, driven through the cold plate of the tester for more than 10 min at typical pump rates; one can rightfully assign this thermal capacitance to the "ambient."

At this high thermal capacitance, the structure functions end at the R_{thJA} junction to the ambient values (i.e., 0.52 K/W, 0.36 K/W, and 0.28 K/W) established previously.

The example underlines the extremely good repeatability of the TDIM method. In a real experiment, the assembly would be totally dismantled and rebuilt again; still the section belonging to inner parts would remain intact.

Focusing on the section of divergence, it can be observed that at high heat transfer coefficient, the structure function reaches higher capacitance at lower thermal resistance; the curves retreat towards the origin, due to different "spreading cone" in the package base before entering the TIM.

A more detailed use case on the comparison of two point methods and the TDIM method is presented in [97].

3.5.3 Dynamic Temperature Change at the Monitoring Points of the Two-Point Method

In the daily practice of laboratories which execute two-point measurements, the typical overshoot illustrated by the "$B = J–C$" difference curve in Fig. 3.21a is often viewed as a measurement error or an artifact.

In the reality it is an obvious consequence of the procedure in which a temperature response of a monitoring point is subtracted from the response of a driving point, in a three-dimensional heat spreading pattern.

In Sect. 2.4.2 a true one-dimensional heat flow was illustrated with the simple equivalent circuit scheme of Fig. 2.21. In the scheme of a two-point measurement, the difference of the Z_{th} curves belonging to the driving point Tjw and monitoring point Tc_wet is to be composed. The difference of the curves which were shown in Fig. 2.43 is presented in Fig. 3.25. It can be stated that "tapping" a homogeneous heat flow at an intermediate point and designating the difference of thermal impedance curves as a transient thermal descriptor of the structure between the heat source and the intermediate point is an acceptable approach.

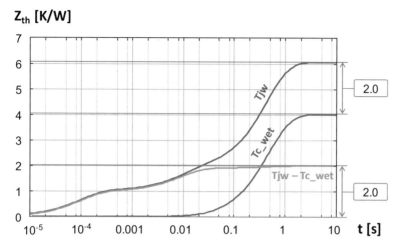

Fig. 3.25 The differential Z_{th} curve composed by subtracting the normalized temperature change of the monitoring point Tc_wet from the change at driving point Tjw

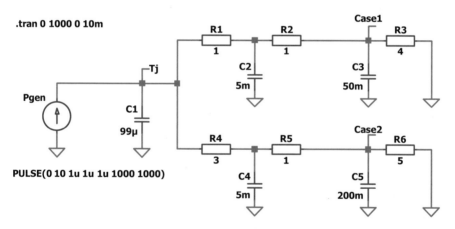

Fig. 3.26 Equivalent RC scheme of two heat-flow paths in a three-dimensional heat spreading pattern

In thermal systems with three-dimensional heat-flow pattern, different heat-flow trajectories progress through material sections of different length and composition. Figure 3.26 shows the simplest equivalent scheme, just two RC ladders of elementary thermal components of different values. Such a scheme can correspond to two arbitrary trajectories in Fig. 3.16, or to two essentially 1D heat-flow paths of a package with two side cooling.

In Fig. 3.27 the thermal impedance curves at the driving point Tj, at two monitoring points T_Case1 and T_Case1, and the difference of the driving point and monitoring point curves are presented. Although the transfer curves are

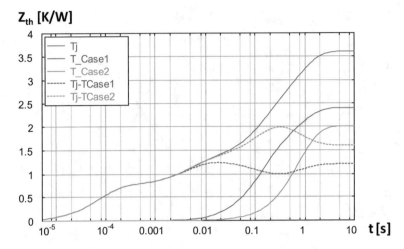

Fig. 3.27 Simulated Z_{th} curves at the driving point and at two monitoring points (solid curves); the difference of the driving point and monitoring point (dashed curves)

monotonous, the temperature change occurs at different pace when the heat progresses through diverse portions of the heat-conducting path. This is reflected in the "waves" of the subtracted curves, to which no plausible physical content can be assigned.

3.5.4 Interpretation of Thermal Metrics on Short Transients

The available time for thermal transient measurements is often limited, for example, in a production line. The possibility of using "incomplete" or short pulse transients, especially for the characterization of the die attach quality, has been experimentally studied quite early. Reports on experimental studies of LEDs were published as early as 2006–2007 [106, 107].

Figures 3.28 and 3.29 illustrate the findings of these studies, using LED packages assembled to a star-shaped MCPCB, attached and measured on a cold plate. As typical with such setups, the thermal response reaches steady-state cca. 100 s after the heating power is abruptly changed at the junction (curve 100 s in the diagrams). Taking this relatively short measurement as a "steady-state to steady-state" reference, characteristic structural elements can be identified in Fig. 3.29. From the cumulative thermal resistance and thermal capacitance values in such a structure function, thermal time constants can be calculated backwards, and such the length of the transient needed for characterization can be approximated.

The fundamental questions of short pulse thermal transient testing were discussed by V. Székely [108, 109]. The main message of these papers is summarized as follows: "*If the transient response is measured until a given t_x time instant, then one*

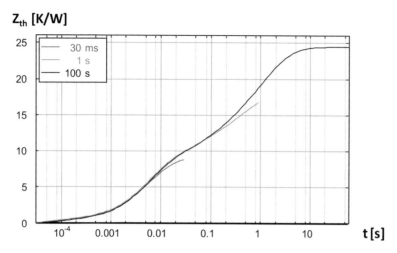

Fig. 3.28 Z_{th} curves measured on a LED sample, from steady-state to steady-state (curve 100 s), heated for 1 s and measured in cooling for 2 s (curve 1 s) and heated for 30 ms and measured in cooling for 60 ms (curve 30 ms)

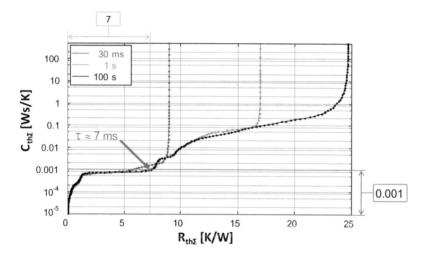

Fig. 3.29 Structure functions of the LED sample, calculated of steady-state to steady-state measurement and shorter pulses, from the Z_{th} curves of Fig. 3.28

is able to reconstruct the exact time constant spectrum till $0.09 \cdot t_x$. Thus, in order to investigate the time-constants of the $0.1\ldots10$ ms range (which is characteristic for the die attach quality) it is enough to measure the transient till 120–150 ms. This time requirement could be acceptable for in-line testing as well" [108]. Inspired by the earlier experimental studies [106, 107], the following example illustrates this for an LED package.

Example 3.4: Comparison of Complete and Short Pulse Thermal Transient Measurement Results of an LED Package
In Fig. 3.28 the thermal impedance curves of an MCPCB-assembled 1 W LEDs are shown.

In Fig. 3.29 the structure functions of the LED sample, calculated from the Z_{th} curves of Fig. 2.38, which are responses on heating pulses of different length, are shown. The initial flat plateau until approximately 7 K/W corresponds to the die attachment. When derived from a short transient of 30 ms, the die attach characteristics are slightly distorted in the structure function, but it is still suitable for comparative measurements. The short measurement of 1 s depicts the die attach and glue layers at the same precision as the full time scale (100 s) measurement. The thermal time constant corresponding to the end of the die attach region is estimated at 7 ms.

The practical consequence is that if thermal transient testing is aimed at qualifying heat-flow path features with small thermal time constants such as the die attach thermal interface resistance, then one does not need to spend time on capturing complete steady-state to steady-state thermal transients (including the stabilization time needed for the initial steady state to occur), but it is sufficient to capture the thermal transient response of a short pulse of heating power. With such a scheme, large numbers of packaged devices can be characterized under production line conditions, outside thermal laboratories. For example, in case of in-line testing of LED devices for each single LED package, a time-frame in the order of magnitude of 100 ms is available [48]. This time-frame is sufficient both for optical testing and thermal qualification of the LEDs' die attach layers with an associated thermal time constant of cca. 5–10 ms.

The estimated characteristic thermal time constant of 7 ms in the above example indicates that a transient response to a cca. 80 ms heating pulse would be sufficient to an uncompromised thermal testing of the die attach layer in the device, as suggested by the quote from [108]. The shorter transient captured after a pulse of 30 ms transient yields only the initial part of the structure function detail corresponding to the die attach layer.

These early experiments and a detailed theoretical analysis gave inspiration to a recent study [110] where a combination of thermal simulations and physical tests was carried out. As a result, a modified version of the TDIM method defined in the JEDEC JESD51-14 standard [40] has been suggested. It has been confirmed that two orders of magnitude reduction in testing time can be achieved without a significant sacrifice in the accuracy of the measured R_{thJC} values of power semiconductor device packages, carrying out the TDIM method on short pulse transient measurements instead of the classical full, steady-state to steady-state ones.

This drastic reduction of the required testing time (e.g., 1 s instead of 100 s) allows the modified test method to be implemented in an industrial environment as a standard in-line test method. Of course, this would require a systematic revision and possible extension of the JEDEC JESD51-14 standard, as outlined in [111]. The possible extensions and suggested revisions of the standard TDIM test method include:

- Use the local *thermal resistance diagrams* (see Sect. 2.4.3) to find better separation between the test results obtained for the two different qualities of the thermal interface at the package "case" surface.
- Use other new combinations of thermal interface qualities at the package "case" surface other than the present *dry* and *wet* conditions. For example, two different conditions with TIM applied (*wet1*, *wet2*) would allow testing of components operated near to their high power limits, when continuous proper cooling is indispensable.

In case of testing LED packages where the contamination of the LEDs' optics must be avoided, the best practice is to avoid the application of any thermal grease or paste. Therefore, two different dry conditions (*dry1*, *dry2*) can cause the measured structure functions to diverge.

- Use the concept of the TDIM method for the identification of standard thermal metrics other than R_{thJC}, such as the measurement of the junction-to-lead thermal resistance as shown in Example 3.1 of Sect. 3.1.2.

Chapter 4
Temperature-Dependent Electrical Characteristics of Semiconductor Devices

Gábor Farkas

Thermal transient testing is a true interdisciplinary segment of engineering. It has its roots in thermodynamics, but it also requires firm knowledge of electronics to ensure stable powering for the devices under test and to count with secondary effects caused by internal tester circuits and cabling.

The electrical characteristics of semiconductor devices, that is, what currents flow in various material regions at certain applied voltages and how these quantities depend on the temperature in these regions, are determined by the laws of solid-state physics.

In the typical academic curricula of engineers, the subjects are presented with focus on some narrow topics and wide gaps among them. For example, a student encounters electron shells and orbits in chemistry, then possibly learns about diode and transistor characteristics in electronics, jumping to logic gates in digital design, and suddenly switches to operating systems in computer science.

In this chapter, we do not want to convey a deep and continuous knowledge connecting all above. We also want to avoid just giving a reference to thick tomes of solid-state physics with many unrelated topics. Instead, in a short introduction (Sect. 4.1), we sketch the train of thought which leads from the physical background to the temperature-dependent electrical characteristics of semiconductor devices, and we refer to literature with exhaustive information for those who need a deeper insight. Those who focus on resulting device characteristics and their temperature dependence can directly refer to all subsequent sections following the introduction.

G. Farkas (✉)
Siemens Digital Industry Software STS, Budapest, Hungary
e-mail: Gabor.Farkas@siemens.com

© The Author(s), under exclusive license to Springer Nature Switzerland AG 2022   139
M. Rencz et al. (eds.), *Theory and Practice of Thermal Transient Testing
of Electronic Components*, https://doi.org/10.1007/978-3-030-86174-2_4

4.1 Basic Laws of Solid-State Physics

Solid-state physics is built on a few simple principles, but the elaboration to the level where the temperature-dependent equations governing the behavior of semiconductor devices can be obtained needs a difficult mathematical apparatus.

It is a common practice in this branch of physics to use theoretical models that yield analytic equations for physical processes and then to use empirical correction factors to fit the formulae to measured properties. In this section we use simple models that describe the basic physical effects, especially those related to the temperature dependence of semiconductor devices.

Those who need deeper knowledge of the physical background may refer to several classic books. General *semiconductor physics* is treated in depth in [10, 13]. A more focused work on the *physics of LED devices* is [11]; in an easier approach, also the first chapters of [7] can be useful.

There are also useful *online resources* on the topic in a well-structured linked arrangement; these are generally less bulky than printed books. Instead of reading them from begin to end, generic search for related keywords on the Internet often directly results in links to appropriate chapters of [12, 14].

Short and useful online summaries on the derivation of the equations which determine the characteristics of diodes or other devices built of pn junctions are given in [98–100].

A useful *Appendix to the solid state physics of heaters and sensors* is available online at [15] which embraces most of the solid-state physics background in moderate length.

4.1.1 Band Structure of Semiconductors, Electrons, and Holes

In atoms, the electrons can have only discretized energy levels assigned to electron shells. In crystals with a high number of atoms, the electron shells disperse into broader bands of allowed energy levels.

In semiconductors and insulators, the bands are separated with bandgaps, electrons cannot possess energies belonging to these forbidden energy ranges. In these materials there exists a last energy band which is nearly fully populated by electrons; this is called valence band, because the interaction of its electrons establishes the bonding between crystal atoms. The next higher energy band, called conduction band, is nearly empty and is separated from the valence band by a bandgap of W_g energy distance.

The distribution of electrons in a space with potential energy is governed by the *Fermi-Dirac* (F-D) function. It specifies the probability that an electron occupies an existing state at energy level W:

$$f(W) = \frac{1}{e^{(W-W_F)/kT} + 1} \tag{4.1}$$

The average energy of all electrons is the W_F Fermi energy. In general, particles of average energy do not necessarily exist.

When the particle concentration is low, the Fermi-Dirac distribution does not significantly differ from the *Maxwell-Boltzmann* (M-B) distribution which says that at higher energy, the concentration of particles diminishes exponentially:

$$f(W) \approx e^{-(W-W_F)/kT} \tag{4.2}$$

The F-D and M-B distributions are temperature dependent; at higher temperatures, more electrons appear around the same energy level.

Those materials in which at room temperature (or at the temperature of use) a small but observable number of electrons, corresponding to the F-D distribution, have a high enough energy to escape from the valence band into the conduction band are called *semiconductors*. In a pure (intrinsic) semiconductor, the number of empty energy levels left in the valence band is necessarily equal to the number of electrons elevated into the conduction band.

As it is rather hard to track the behavior of the large number of electrons in the valence band, the concept of *holes*, quasiparticles corresponding to the missing electrons, is introduced. Instead of following the position of empty places in the abundance of electrons, holes are handled as particles of positive charge, residing at the highest energies of the valence band. The elevation of an electron into the conduction band is interpreted as the generation of an electron-hole pair; the return to the valence band is handled as a recombination of two particles of different energy.

The equal number of electrons in the conduction band of the intrinsic semiconductor and the holes in the valence band inflicts that the average W_F energy is near to the middle of the bandgap. This energy level is called the W_i intrinsic Fermi energy.

The temperature dependence of the F-D and M-B distributions determines the $n_i = p_i$ concentration of electrons and holes in intrinsic semiconductors, as proved in [15]

$$\begin{aligned} n_i &= G \cdot T^{\frac{3}{2}} \cdot e^{-\frac{W_g}{2kT}} \\ n_i^2 &= G^2 \cdot T^3 \cdot e^{-\frac{W_g}{kT}}, \end{aligned} \tag{4.3}$$

where T is the absolute temperature, k is the Boltzmann constant connecting thermal energy to temperature and the G factor cumulates structural parameters of the crystal lattice with no significant temperature dependence.

When an external E electric field is applied on the crystal, all mobile electrons start moving in it against the direction of the electric field, due to their $-q$ elementary negative charge. This movement is known as electric current. Again, it is easier to describe the flow of the high number of electrons in the valence band as the movement of a few holes of positive charge toward the direction of the external field.

In free space (vacuum), an electron can "fly" freely. It may have any energy and momentum, both of them determined by its m mass and v velocity. The momentum is proportional the velocity, $p_w = m \cdot v$; the kinetic energy is quadratically dependent as $W_{kin} = \frac{1}{2} m \cdot v^2$. A deeper interpretation of p_w is presented in [15].

At absolute zero temperature, the lattice atoms, which are of positive charge, form a perfect periodic potential in space, repeated in all directions by the lattice constant of the crystal. In this undisturbed periodic potential, braking and accelerating forces influence the movement of an electron. As a consequence, electrons and holes act as having an m_n^* and m_p^* effective mass, not equal to the mass of the free electron. As a further consequence, the laws of quantum mechanics imply that in a periodic potential space due to the braking and accelerating forces, the minimum energy of the moving electron does not necessarily coincide with the minimum momentum. (Momentum has some more generalized meaning in quantum mechanics; a deeper definition is presented in [15].) In the perfect periodic potential space, no obstacle perturbs the movement of electrons; they can accelerate freely as determined by their m^* effective mass.

Those semiconductors in which the minimum energy of electrons and the maximum energy of holes belongs to the same (typically zero) momentum are called semiconductors of *direct bandgap*. In other semiconductors, as silicon or silicon carbide, electrons of lowest energy have nonzero momentum; these are materials with *indirect bandgap*.

4.1.2 The Concentration of Charge Carriers and Their Motion in Semiconductors

At nonzero temperatures, the lattice atoms oscillate around their position in the crystal lattice. Moving electrons interact with the vibrating atoms, "collide" in a broader sense, and exchange energy with them. These interactions ensure that the temperature of the electrons remains the same as the temperature of the crystal atoms until no extreme external energy is applied on the material. The electrons at T temperature move stochastically at v_{th} average velocity; as a consequence of the M-B distribution, their kinetic energy is

$$W_{kin} = \frac{1}{2} m^* \cdot v_{th}^2 = \frac{3}{2} kT \tag{4.4}$$

Equation (4.4) is the definition of the T absolute temperature, and it presents how the k Boltzmann constant connects the concept of kinetic energy to temperature.

In real crystals also other, less temperature-dependent imperfections of the lattice arise. These crystal defects can be atoms of other elements occupying regular positions in the lattice, or in interstitial position, and dislocations. A large imperfection is the surface of the bulk semiconductor itself.

Applying E electric field on the semiconductor, the persistent energy exchange with the lattice prevents unlimited acceleration of the charge carriers. The braking effect of the interactions with lattice atoms and crystal defects can be expressed in the empirical quantity of the μ mobility. The electric field causes a drift of carriers of v velocity: $v_n = \mu_n \cdot E$ for electrons and $v_p = \mu_p \cdot E$ for holes. The movement of charge carriers provides the electric current:

$$J_{ndrift} = q \cdot \mu_n \cdot E, \quad J_{pdrift} = q \cdot \mu_p \cdot E, \qquad (4.5)$$

J is the current density and q is the elementary charge. The electrons have negative charge and move opposite to the E field; both current components have the same direction. Equation (4.5) is the differential form of the Ohm law, $J = \sigma \cdot E$, where σ is the electrical conductivity of the material and J is the sum of the currents of electrons and holes. The thermal power generated in unit volume is $P_V = J \cdot E$.

In electronics generally current is imagined as a continuous flow of charge carriers, and a small perturbation, *noise*, is superposed on this continuous flow. A signal-to-noise ratio is an eminent descriptor of a measurement. In reality, the electrons at T temperature stochastically move at high v_{th} thermal velocity, and a small v drift velocity is superposed on their random movement. Better to say, because their temperature does not change independently from the temperature of the crystal, the stochastic movement is turned slightly in the direction of the E field by a v drift velocity.

The v_{th} thermal velocity in bulk semiconductor is about 10^7 cm/s. For comparison, the μ_n electron mobility in intrinsic silicon is about 1400 cm^2/Vs, and the μ_p mobility of holes is around 480 cm^2/Vs at room temperature; and both quickly diminish at higher temperature. The field strength is limited by the presence of mobile carriers; the charge carriers' movement is really best described as a small superposed drift.

The density of electrons and holes in a semiconductor can be influenced by doping, by substituting, for example, silicon atoms in the lattice with trivalent atoms (acceptors) or pentavalent atoms (donors). The substituent atoms can be considered dopants, if the 3/2 kT thermal energy at room temperature (or at the temperature of actual use) is sufficient to ionize almost all the dopant atoms. In a neutral material section of doped semiconductor, $n \approx N_D$ in an *n-type* material of N_D donor concentration and $p \approx N_A$ in a *p-type* material of N_A acceptor concentration.

In an *n-type* semiconductor, for example, the large N_D concentration of freely moving electrons enhances the probability of the recombination of holes, the amount of holes will drastically sink in donor-doped regions. In such a material, the electrons are considered majority charge carriers and the holes are minority charge carriers.

In most cases even the concentration of the majority charge carriers remains low in the sense that only a low number of available energy levels of the conduction band are occupied. The majority and minority carriers are in dynamic balance through generation and recombination. It is a general law of reaction kinetics that as long as the concentrations of two reagents are "thin," the elevation of the concentration of one component proportionally lowers the concentration of the other. This is the mass action law, in the actual case resulting in $np = n_i^2$.

The mass action law is maintained through different recombination mechanisms. In direct bandgap materials, electrons and holes can directly recombine in a single recombination event, emitting a photon of zero mass and approximately W_g energy. Such materials are intermetallic compounds like GaAs or GaN. The rate of recombination depends on both n and p concentrations, such it is related to n_i^2 and on its temperature dependence described in (4.3). In the literature these recombination mechanisms related to the concentration of two particles are called *bimolecular recombination*, a term borrowed from the description of chemical reactions.

In materials with indirect bandgap, the carriers have to lose their excess momentum during the travel from the conduction band to the valence band; such recombination has to occur through interaction with particles of high mass. These can be primarily vibrating lattice atoms. In order to formalize this interaction, the thermal energy of the vibration is attributed to virtual particles called phonons.

The energy and momentum can also be dispersed in an interaction with dopant atoms and crystal defects. Moreover, special intermediate energy levels, "traps" can be established in the bandgap with intentional placement of appropriate impurity atoms of other elements into the crystal lattice. With offering multiple successive jumps through the intermediate levels, the presence of traps increases the recombination probability. This trap-assisted recombination mechanism is generally known as Shockley-Read-Hall (SRH) recombination.

The concentration of majority carriers is equal to the dopant concentration, a temperature-independent parameter permanently determined by the semiconductor technology. Similarly, the number of crystal defects and traps is less dependent of the temperature. This way the rate of indirect recombination predominantly depends on the concentration of minority carriers, which is governed through the mass action law on the temperature dependence of the n_i intrinsic carrier concentration. These recombination mechanisms are frequently referred to as *monomolecular recombination*.

SRH recombination can be seen as being in a halfway between different mechanisms; the number of traps which are used as intermediate levels for the travel of electrons is present in unvarying number, but at the end electrons recombine holes.

The high number of majority carriers and the reduced number of minority carriers in doped semiconductors also shift the average W_F Fermi level from the W_i halfway position of the bandgap toward the band edges where majority carriers reside. From the M-B distribution approximation

$$n = n_i e^{\frac{W_F - W_i}{kT}}, \qquad p = n_i e^{\frac{W_i - W_F}{kT}}, \tag{4.6}$$

and n or p are tied to the N_A or N_D concentration of dopants.

The particle concentration of charge carriers can be kept at steady high n_1 or p_1 level with an appropriate charge transfer mechanism into a region of the semiconductor. If other neighboring regions are of lower carrier concentration, the random thermal movement of particles results in a net particle diffusion toward those regions. The movement of charged particles represents a J_{diff} diffusion current.

Formerly it was observed that the E electric field represents a force driving the charged particles, while their motion is hindered by "collisions," interaction with obstacles in the lattice. The obstacles were condensed into a μ mobility (or rather into $1/\mu$).

It is less obvious but still true that concentration difference of randomly moving particles on two sides of a surface corresponds to a force.[1] The velocity of the particles is v_{th}, and the obstacles against their movement are the same in the drift and diffusion current mechanisms.

Introducing now an empirical D diffusion constant to consider the obstacles against the particle stream induced by the concentration gradient, we get that

$$J_{ndiff} = q \cdot D_n \cdot dn/dx, \qquad J_{pdiff} = q \cdot D_p \cdot dp/dx, \qquad (4.7)$$

again the obstacles rather correspond to $1/D$. In (4.7) it is exploited that the particle flow streams toward lower concentrations, but the electron has $-q$ charge.

The relation between μ and D can be easily constructed in a simple "thought experiment" in which a dn/dx concentration gradient is caused by an E field which is maintained until the drift and diffusion current components equalize. The result is the well-known Einstein relation, $D_n = \mu_n \cdot V_T$, $D_p = \mu_n \cdot V_T$.

The $V_T = kT/q$ temperature-dependent factor is known as thermal voltage. This quantity is of principal importance in the temperature dependence of the characteristics of semiconductor devices. Around room temperature (300 K), the thermal voltage can be calculated, with $k = 1.38 \cdot 10^{-23}$ J/K and $q = 1.6 \cdot 10^{-19}$ As, one gets $V_T = 26$ mV.

4.1.3   The pn Junction

When in a single crystal a p-type and an n-type semiconductor are brought together, the large difference in electron concentration of the two sides causes a diffusion current of electrons from the n-type material across the metallurgical interface into the p-type material. Similarly, the difference in hole concentration causes a diffusion current of holes from the p-type to the n-type material. Due to this diffusion process, the region at the interface becomes almost completely depleted of mobile charge carriers. The gradual depletion of the charge carriers gives rise to a space charge created by the charge of the ionized donor and acceptor atoms that is not

---

[1]This force drives, for example, the steam engine.

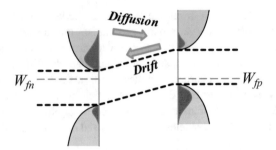

Fig. 4.1 A pn junction in equilibrium. The diffusion and drift components of the current are in balance in the depletion region. The V_D diffusion voltage corresponds to the energy difference either in the conduction or the valence band

compensated by the mobile charges any more. This region of the space charge is called the *space-charge region* or *depleted region* and is schematically illustrated in Fig. 4.1. Regions outside the depletion region, in which the charge neutrality is conserved, are denoted as the quasi-neutral regions.

The space charge in the depleted region results in the formation of an internal electric field which forces the charge carriers to drift in the opposite direction than the concentration gradient. The diffusion currents continue to flow until the forces acting on the charge carriers, namely, the concentration gradient and the internal electrical field, compensate each other. The driving force for the charge transport does not exist any longer, and no net current flows through the pn junction.

Integrating the electric field along the depleted region yields the *contact potential* between the two material types, which is called at semiconductors, referring to the process which established it, the V_D *diffusion potential* (also called built-in potential). The integral can be calculated only knowing the actual doping profile, the concentration of donors and acceptors around the metallurgical interface. However, V_D can be also established directly from the energy levels in equilibrium.

Each electron which stepped from the n region into the p region caused a sinking in the energy on the p side and an elevation of it on the n side; the process ended when the average electron energy, the W_F Fermi level, equalized over the whole structure. From the Maxwell-Boltzmann approximation of the Fermi-Dirac particle distribution, it directly follows through (4.6) that

$$V_D = [(W_F - W_i)|_{\text{n side}} - (W_F - W_i)|_{\text{p side}}] = \frac{kT}{q} \cdot \ln\left(\frac{N_A \cdot N_D}{n_i^2}\right) \qquad (4.8)$$

Equation (4.8) yields the temperature-dependent diffusion voltage of the junction calculated from the N_A and N_D doping concentrations and the temperature-dependent intrinsic carrier concentration.

In Fig. 4.1 the charge carrier densities on the two sides of the depletion region and the band edges are shown, in a pn junction in equilibrium. The densities of available states for electrons and holes are shown as light red and light blue shapes; the dark blue and red "bubbles" correspond to the actual particle densities at a given energy

Fig. 4.2 A forward-biased pn junction

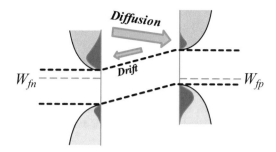

level. The n and p total concentrations can be calculated integrating the "bubbles" along the conduction and valence band. The V_D diffusion voltage corresponds to the energy difference either in the conduction or the valence band.

When an external V_F voltage is applied to a pn junction, the potential difference between the n-type and p-type regions will change, and the electrostatic potential across the space-charge region will become $V_D - V_F$ (Fig. 4.2).

The height of the energy barrier which hinders the flow of the majority carriers of one side toward the other side is diminished by V_F; the tail of higher-energy electrons (high end of the dark blue bubble) extends over the barrier with exponentially growing number of electrons. Assuming a few conditions, such as the number of injected carriers remains significantly below the native majority carriers at both sides and no significant recombination occurs in the short depletion layer, the I–V characteristics of a pn junction can be easily derived.

This is a train of thought of simple elementary steps, but because of its length, we have to refer to [15]. The calculation yields the expected exponential growth of the current in a forward-biased pn junction, the ideal Shockley equation:

$$I_F = I_0 \cdot \left(e^{\frac{V_F}{V_T}} - 1 \right). \tag{4.9}$$

4.2   Resistive Heaters and Sensors, Active Devices

The previous section focused on the temperature-dependent quantities in semiconductor materials and on the characteristics of the pn junction composed of semiconductor regions of different doping. Still, this overview already contributes to understanding the operation of many actual devices used as heaters and sensors in thermal transient testing, such as the following:

Resistive Sensors
Equation (4.5) describes the drift current in semiconductors. It has to be noted that metals do not differ from highly doped semiconductors in that aspect that a high number of electrons ("majority carriers") are available in them next to an abundance

of available empty states which can be occupied. The equation reveals the origin of the Ohm law in both metals and semiconductors and indicates that the probability of interaction of the charge carriers with lattice atoms grows at higher temperature. This explains the positive thermal coefficient of series resistance in diodes and MOS devices, resistive heaters, and resistor sensors.

The temperature dependence of $1/\mu$ can be well approximated by an exponential function. For example, the popular PT100 platinum sensor has $100 \; \Omega$ resistance at $0 \; °C$, and it grows by 385 ppm/K with the change of temperature.

Resistive Heaters

Equation (4.5) also indicates that in a unit volume in which an E electric field causes a flow of J current density, a thermal power of $P_V = E \cdot J$ is generated. This results in the whole volume of a resistive material region $P = V \cdot I = I^2 \cdot R$ heat generation (V is the voltage drop across the region and I is the current flowing through it, $R = V/I$.)

It has to be noted that recently novel *active devices* have been introduced based on sudden drastic, thermally influenced change of the electric conductivity. For example, vanadium dioxide (VO_2) transits from semiconductor phase into metal phase at around 67 °C, changing its electrical conductivity from $\sigma = 10^{-1} - 10^{-2}$ S/cm to $\sigma = 10^3 - 10^4$ S/cm [102].

Thermistors with Negative Thermal Coefficient

These temperature sensors are simply a piece of semiconductor (sintered metal oxide) in which the number of charge carriers quickly grows with the temperature, as indicated by (4.3).

Thermocouples and Thermoelectric Cooler (TEC, Peltier) Devices

In (4.8), we introduced the V_D contact potential (diffusion potential) as the difference of the Fermi levels in materials when they are in contact and equilibrium. The formula is equally valid for two metals in contact; the contact potential is the difference of the Fermi levels; just the "depleted region" is infinitesimally thin, or not that thin at all when their connection is oxidized or contaminated.

This contact potential cannot be measured by a voltage meter, or used as voltage source in a circuit in which all components are at the same temperature, because the other components are also constructed of metals and semiconductors and the sum of V_D values in a closed circuit equals to zero. However, as (4.8) indicates, keeping different components at diverse temperatures, the different $V_D(T)$ temperature-dependent values in series result in a net temperature-dependent voltage (Seebeck effect).

In a reverse effect, a flow of current generates temperature difference at the junction of two materials (Peltier effect). In TEC devices the net resulting temperature can be calculated from a superposition of the Ohm, Seebeck, and Peltier effects.

Active Devices

The previous section expounded that when there is a potential barrier in the way of the flow of charge carriers (current), then a tiny shift in the barrier height causes a major alteration in the current. Beyond the barrier height, the current also depends on the presence of charge carriers, and such depends heavily on the temperature.

In the last 120 years of electronics, the operation of all active devices has been based on the shifting of a W_b barrier by an action needing low energy. The shift regulates the stream of (charged) particles, resulting in high energy change.

This operation is analogous to the function of a valve which can modify a flow of a substance with minor energy investment. This analogy is reflected in the names of active devices from the beginning, for example, vacuum tubes are referred to as *valve*, and analogous solid-state active devices are *TRAN*-sfer re-*SISTORS*, such as BJT, FET, and HEMT.

4.3   Diodes

As a consequence of the previous assumptions in Sect. 4.1, the forward current (I_F) – forward voltage (V_{Fpn}) characteristics of a pn junction follows the Shockley equation:

$$I_F = I_0 \cdot \left(e^{\frac{V_{Fpn}}{mV_T}} - 1 \right). \qquad (4.10)$$

The V_T and I_0 parameters are temperature dependent:

$$V_T = kT/q, \qquad (4.11)$$

and

$$
\begin{aligned}
I_{01} &\sim n_i = G \cdot T^{3/2} \cdot e^{\frac{-W_g}{2kT}} \\
I_{02} &\sim n_i^2 = G^2 \cdot T^3 \cdot e^{\frac{-W_g}{kT}},
\end{aligned}
\qquad (4.12)
$$

where I_{01} and I_{02} are the saturation current constituents maintained by monomolecular and bimolecular recombination mechanisms, respectively.

The two recombination components are typically reflected, *more or less justified* in a single m parameter:

$$I_0 = G_{sum} \cdot T^{3/m} \cdot e^{\frac{-W_g}{mkT}}. \qquad (4.13)$$

In practical measurements, as justified in Sects. 5.4 and 6.1, diodes are typically driven by a controlled I_F forward current, for electric and thermal stability reasons. The V_{Fpn} forward voltage at controlled I_F forward current will be

$$V_{Fpn} = m \cdot V_T \cdot \ln \frac{I_F}{I_0} \tag{4.14}$$

Real diodes have characteristics similar to (4.14) at lower current. At higher current, additional recombination and diffusion mechanisms modify the basic effect expressed by the Shockley equation.

Moreover, the ohmic regions in the semiconductor at the two sides of the junction and the wiring in the package add an additional voltage drop to the total V_F forward voltage on the device. Cumulating these secondary effects into an R_S internal electrical series resistance (and neglecting the very small $-I_0$ term), we get

$$V_F = V_{Fpn} + V_{FRs} = m \cdot V_T \cdot \ln \frac{I_F}{I_0} + I_F \cdot R_S, \tag{4.15}$$

m is the device specific constant from (4.13) called *ideality factor*. An ideal diode is supposed to have $m = 1$.

Real silicon diodes may have various m values. Signal diodes are heavily doped with impurities establishing trap levels in order to promote SRH recombination; their ideality factor is nearer to $m = 2$. The shorter lifetime of charge carrier results in faster switching operation. In power devices SRH recombination enhances current leakage at high reverse voltages and reduces injected current density, it is advantageous to keep the trap density low, and their ideality factor is nearer to $m = 1$.

The number of available energy states at the edge of the bandgap in semiconductors differs from the number of states in the partially filled conduction band of metals. For this reason the $3/m$ exponent in (4.13) changes to $2/m$ in Schottky diodes, where the potential barrier is formed at the interface of a metal and a semiconductor.

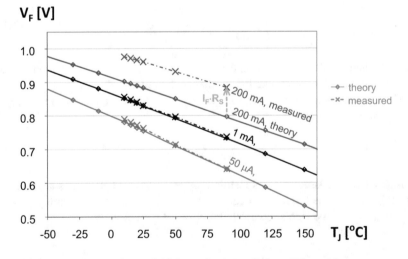

Fig. 4.3 Temperature dependence of the forward voltage, diode at different bias currents

Equations (4.13) and (4.15) offer a good analytical approach for the nonlinear diode characteristics in a broad current and temperature range.

Figure 4.3 shows the V_F forward voltage of a silicon diode at different bias currents (50 µA, 1 mA, 200 mA). The x-shaped marks in the plot show measured V_F values of the actual diode at various temperatures from 10 °C to 90 °C. Using the measured values at 1 mA bias, we calculated the I_0 and m values and produced analytic curves (dash-dotted lines) for other bias than 1 mA and in an extrapolated temperature range (-50 °C to 150 °C).

The curves, corresponding to the analytic equations above, are obviously nonlinear, but show only very small nonlinearity over a broad temperature range. The R_S series resistance can be ascertained from the difference of the modeled and measured values at higher currents, as shown by the green $I_F \cdot R_S$ vector in the chart.

4.3.1 Differential Properties of the Diode Characteristics

In many cases we study the change of the I–V characteristics at small current or temperature changes. The quantities depicting this partial variations around an operating point are the $R_D(I_F, T_J) = dV_F/dI_F$ differential (electric) resistance and the $S_{VF}(I_F, T_J) = dV_F/dT$ temperature sensitivity. Sometimes in engineering the so-called K factor is used, $K = 1/S_{VF} = dT/dV_F$.

Composing the dV_F/dI_F derivative of (4.15) one obtains that the differential resistance of a diode is approximately inversely proportional to the I_F forward current:

$$R_D = m \cdot V_T/I_F + R_S \qquad (4.16)$$

In most cases R_D is rather low, 26 Ω at 1 mA forward current, or 26 mΩ at 1 A forward current for an "ideal" diode ($m = 1$) at room temperature. This ensures high noise tolerance of thermal measurements as it will be shown in Sect. 5.7.

In a few steps, compiling the temperature derivative of (4.14), at a proper point inserting the dI_0/dT derivative from (4.13), it can be found that

$$S_{VFpn} = \frac{dV_{Fpn}}{dT} = \frac{V_F - 3V_T - W_g/q}{T} \qquad (4.17)$$

It is important that m which was present in both (4.13) and (4.14) has not disappeared; it is just incorporated into the V_F term.

A first estimation on the value S_{VF} can be done easily. Around room temperature (300 K), as a consequence of (4.8) and (4.15), the forward voltage of realistic *silicon* diodes in their typical operating points is around $V_F = 500$–800 mV. Inserting this V_F into (4.17), Table 4.1 lists a few calculated S_{VF} values. For the homogeneity of the measures, the bandgap is given as the $V_g = W_g/q$ bandgap voltage; its change with the temperature is estimated based on the empirical Varshni formula.

Table 4.1 Calculated S_{VF} sensitivity values at some temperatures

T [K]	V_{Fpn} [mV]	V_T [mV]	$3V_T$ [mV]	V_g [mV]	S_{VF} [mV/K]
300	500	26	78	1120	−2.33
300	800	26	78	1120	−1.33
400	500	34.5	103.5	1097	−1.75
400	800	34.5	103.5	1097	−1.01

The table can be easily extended to other temperature and forward voltage ranges, and other semiconductor materials of different W_g bandgap. With the actual values it is proven that silicon diodes expose around room temperature and in a broad current range an S_{VF} temperature sensitivity factor between −1 and −2.5 mV/K, a negative temperature coefficient. In wide bandgap materials, V_F and S_{VF} can be significantly higher.

One can express the dependence of S_{VF} on the current density through the Shockley equation. It is generally known that at *higher current density*, the S_{VF} *sensitivity diminishes*, but only modestly through many orders of magnitude as the logarithmic nature of (4.15) indicates.

From a more detailed calculation:

$$\Delta S_{VF} = \frac{V_{F2} - V_{F1}}{T} = \frac{m \cdot V_T (\ln \frac{I_{F2}}{I_{F1}})}{T} = \frac{m \cdot kT (\ln \frac{I_{F2}}{I_{F1}})}{q \cdot T} \tag{4.18}$$

Surprisingly, ΔS_{VF} does not seem to depend on temperature. If still it does, it is just because the Shockley equation is merely an approximation of physical reality. I_F is typically the forced quantity; in this way, it is an external parameter not dependent on temperature. Thus

$$\Delta S_{VF} = \frac{m \cdot k}{q} \cdot \ln \frac{I_{F2}}{I_{F1}} \tag{4.19}$$

The k/q constant is 86 μV/K. According to this; and considering the logarithm of the current ratio, 10% decrease in I_F current will result in $8.22 \cdot 10^{-3}$ mV/K growth in the temperature dependence (with $m = 1$ ideality factor, and any temperature).

It has to be noted that S_{VF} is a negative number, and so is ΔS_{VF} because $W_g/q > V_F$. If I_F grows a full decade, ΔS_{VF} changes by $k/q \cdot \ln(10) = -0.2$ mV/K. At a growth by 5 decades, $\Delta S_{VF} = k/q \cdot \ln(100000) = -0.99$ mV/K.

So far we have proven that V_F is of negative S_{VF} temperature coefficient which slightly diminishes at high current densities. However, the R_S component can also have negative temperature coefficient until secondary diffusion effects around the depleted region and included into the series resistance dominate. R_S turns to positive temperature coefficient at high currents due to the ohmic nature of the semiconductor regions adjacent to the pn junction.

Table 4.2 Calculation of the Z temperature-induced relative growth of the forward current at various constant V_F forward voltages, expressed in percentage

V_F [V]	$Z_{@T\,=\,300\ K}$ (%)	$Z_{@T\,=\,400\ K}$ (%)
0.4	10.27	5.80
0.5	8.98	5.07
0.6	7.70	4.35
0.7	6.41	3.63
0.8	5.12	2.90

The change of V_{Fpn} can also be calculated analytically for larger I_F current changes. Equation (4.15) yields for large current differences in an ideal diode:

$$V_{F1} - V_{F2} = V_T \cdot \ln\left(I_1/I_2\right) + R_S \cdot \left(I_2 - I_1\right) \tag{4.20}$$

For calculating the change of the voltage caused by the first, ideal term one can use the approximate values $V_T = 26$ mV (at room temperature) and $\ln(10) \approx 2.3$. At ten times higher current, the V_{Fpn} forward voltage on the junction grows by approximately $V_T \cdot \ln(10) = 60$ mV. Similarly, at hundred times higher current, the forward voltage grows by 120 mV, and at ten thousand times higher current, it grows by 240 mV.

When for some reason a pn junction is driven by steady V_{Fpn} forward voltage, the current grows extremely fast with increasing temperature. In order to obtain a manageable expression for this change, it is reasonable to relate the current change to the current itself. From the Shockley equation, the Z ratio of the temperature-induced current change and the forward current itself can be expressed as

$$Z = \frac{1}{I}\frac{dI}{dT} = \frac{1}{mT}\left(3 + \frac{V_g - V_F}{V_T}\right) \tag{4.21}$$

The Z ratio of the current growth at fixed V_F forward voltage is calculated in Table 4.2 with V_T and V_g values taken from Table 4.1, at 300 and 400 K absolute temperature and $m = 1$ ideality factor.

A further consequence of the high dependence of the forward current on temperature at fixed forward voltage is the *current crowding* effect. This may undermine the accuracy of the forward voltage to temperature calibration and of the thermal transient measurement process because it causes an uncertainty in the interpretation of the results.

A diode can be imagined as many elementary pn structures in parallel. As it was presented in Sect. 3.5.1 when uniform power is applied on the chip surface, then a bell-shaped nonuniform temperature distribution develops on it. At high currents uniform powering can be achieved if the series R_S resistance has positive thermal coefficient. Switching to a low measurement current, the elementary junctions have the same V_F forward voltage as they are connected in the chip. This causes the current threads to cumulate into the hotter elementary structures.

Electrothermal simulation tools can properly model the temperature and current distribution on the chip surface, but they yield the result only for one certain

Table 4.3 Calculation of the current crowding of a low I_F forward current at temperature inhomogeneity on the diode surface

ΔT [°C]	$K_{@Z\,=\,4\%}$	$K_{@Z\,=\,5\%}$	$K_{@Z\,=\,6\%}$
1	1.04	1.05	1.06
2	1.08	1.10	1.12
3	1.12	1.16	1.19
4	1.17	1.22	1.26
6	1.27	1.34	1.42
8	2.36	2.46	2.56
10	3.45	3.58	3.72

geometry and one powering. An estimation on the current crowding can be done simply, supposing that the temperature difference on the surface is typically a few centigrade between the hottest and coldest elementary diodes.

Table 4.2 indicates that 1 °C temperature difference causes a few percent change in the forward current.

Table 4.3 presents the typical K ratio of the current density between the coldest elementary junction and an other junction in parallel, which is hotter by ΔT. Supposing 4–6% relative Z growth of the current density, $K = (1+Z)$ for 1 °C difference. More centigrade difference multiplies this change; the K current ratio between elementary diodes follows a "compound interest" scheme.

The table proves that a temperature gradient of a few degrees makes the current cumulate into the hottest surface region of a diode.

4.3.2 Negative and Positive Thermal Coefficients in the Diode Characteristics

In the next example, the sections of negative and positive thermal coefficient in the diode characteristics are analyzed in a simulation experiment.

Example 4.1: Negative and Positive Thermal Coefficients in the Diode Characteristics

In the example we use the basic diode model of the popular LTSpice software for modeling the internal pn junction. Realistic devices have positive thermal coefficient at high currents, which can be attributed to the decrease of mobility in bulk semiconductor and to the wiring within the package. These effects are modeled by external resistors of 0.15 Ω resistance and a positive thermal coefficient of 0.2%/K (shown as 0.15 tc $=$ 2 m in the figure).

Separate diodes and resistors are used to represent the devices kept at $T_J = -25$ °C, 25 °C, 75 °C, and 125 °C temperatures (Fig. 4.4).

The resulting temperature-dependent diode characteristics are shown in Fig. 4.5; the negative and positive thermal coefficient regions and the zero thermal coefficient point ZTC can be observed.

(continued)

Example 4.1 (continued)

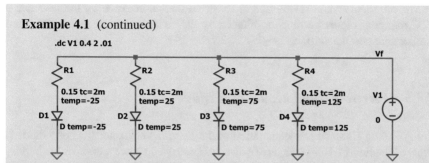

.dc V1 0.4 2 .01

Fig. 4.4 Circuit scheme for analyzing the temperature and current dependence of the diode characteristics in an LTSpice simulation. The resistor added to the standard diode model represents the external portion of the series resistance

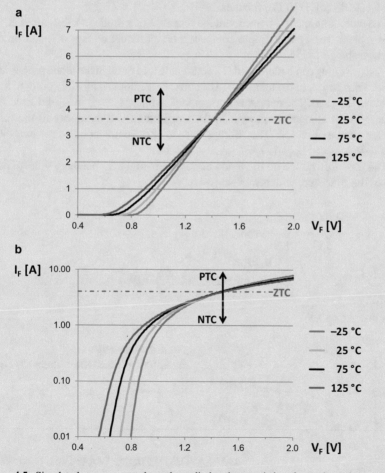

Fig. 4.5 Simulated temperature-dependent diode characteristics from the scheme of Fig. 4.4, plotted with linear (**a**) and logarithmic (**b**) current scale

We can identify an NTC region at lower currents dominated by the junction, a PTC region at higher currents dominated by the series resistance and a ZTC zero temperature coefficient point at ~1.4 V, 3.7A.

4.3.3 Electrothermal Model of a Diode

In Fig. 4.6 an LTSpice implementation of the electrical characteristics of a diode is shown. The Eqs. (4.10)–(4.13) are realized as controlled sources. In order to simplify the formulae in the source definitions, the conversion between centigrade and kelvin is realized as a constant "voltage" source and V_T is provided by a "temperature" controlled voltage source. R_S is constant in this simple model, a more complex representation can be a current controlled voltage source, with, e.g, a V = I (IF) * ({R0}+{dR}*V(Tj)) formula.

The temperature and forward voltage response of a diode in the thermal environment represented by the simple compact thermal model of Fig. 4.7, can be monitored, as shown in Fig. 4.8.

First, a heating can be followed. The current forced through the diode grows linearly from 10 mA to 1 A in the first 1 μs. (The current is multiplied by 10 in order to make it visible in the chart. The curvature corresponds to linear growth in log-lin scale.)

In the next second the growth of the temperature and the decay of the forward voltage can be observed. The "bumps" of the curve correspond to the two thermal time constants in Fig. 4.7, τ_1=1 ms, τ_2=100 ms.

The forward voltage on the diode corresponds to the V_g bandgap voltage parameter of the I_0 source, which was set to 1.1 V.

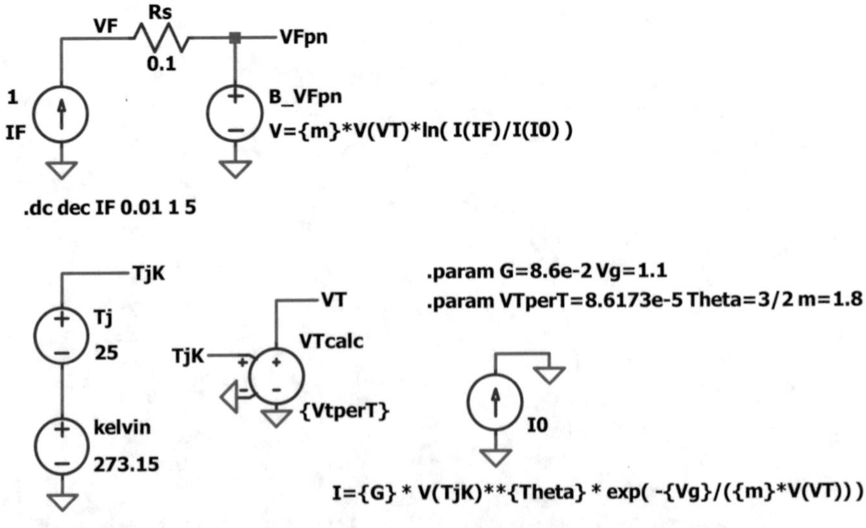

Fig. 4.6 LTSpice implementation of the electrical characteristics of a diode with the help of controlled sources

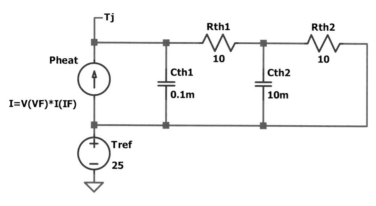

Fig. 4.7 Thermal subcircuit of the diode; heat source and a simplified equivalent Foster chain

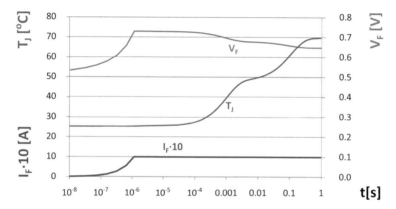

Fig. 4.8 Transient simulation of the equivalent LTSpice circuit. The diode is switched on to $I_{heat} = 1$ A heating current. Values of I_F, T_J and V_F are shown

4.4   MOS Transistors

A correct analysis of thermal effects in field effect devices can be carried out using simulator programs following the flow of particles in the crystal, charge carriers, and phonons representing the thermal behavior of the lattice atoms. An example of such tools is presented in [101].

The characteristic quantity of insulated gate devices is the V_{th} threshold voltage, above which the device current grows in a quadratic way. Because of the quick growth, the definition of the threshold shall be not very sharp. Practical guides define the presence of an I_D drain current of 1 μA or 1 mA or similar for an actual device as current limit, and all these yield very similar V_{th} threshold voltages.

A more physical approach can be based on the operation of the device. Figure 4.9 presents the cross section of a (lateral) MOS structure. The n^+ doping profiles at the source and drain diffusion (S and D in the figure) induce the appearance of depleted regions around the metallurgical interface of p- and n-type semiconductor (Fig. 4.9b).

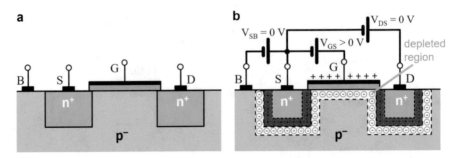

Fig. 4.9 Schematic drawing of a MOSFET device, (**a**) cross section with doping profiles (**b**) depleted area around pn junctions of the source and gate diffusion and under the gate

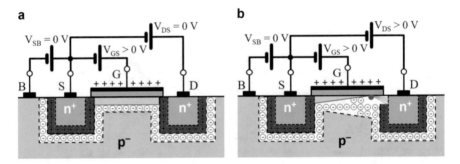

Fig. 4.10 Schematic drawing of a MOSFET device, (**a**) inverted channel under the gate biased by higher V_{GS} voltage (**b**) increased depleted area around the drain diffusion and pinch-off at elevated V_{DS}

An external voltage applied *between* the gate (G in the figure) and the bulk semiconductor, which is most frequently connected to the source, bends the band structure on the semiconductor-oxide interface. At a high enough V_{GS} voltage, the bending of the potential causes the W_i intrinsic level to turn to the opposite side of the W_F Fermi level, as explained more in detail in [15]. The surface will be *inverted*; in the actual case, it starts behaving as p-type semiconductor instead of the bulk n type (Fig. 4.10a).

The accepted interpretation of inversion is that a semiconductor is *fully inverted* when the n concentration below the gate reaches the original p concentration of the bulk material; the conducting *channel* below the oxide reaches the electric conductivity of the bulk.

For composing the device characteristics, it is easier to describe the equations as potentials, rather than energies. Using the $\Phi = q \cdot W$ notation for potentials, the threshold voltage of an n-channel device can be written as

$$V_{th} = \Phi_{GC} - 2\Phi_{FS} - \frac{Q_{B0}}{C_{ox}} - \frac{Q_{ox}}{C_{ox}} \qquad (4.22)$$

the inversion channel appears at $2\Phi_{FS}$ bending of the Fermi level, that is, the electron concentration in the channel is the same as the hole concentration of the p-body. Φ_{GC} is the contact potential between the gate material and the p-body, $\Phi_{GC} = \Phi_{FS} - \Phi_{FG}$, with a $\Phi_F = q \cdot W_F$ notation.

Φ_{FS} is dependent on T temperature. From Sect. 4.1, at N_A acceptor dopant concentration of the body:

$$\Phi_{FS} = (kT/q) \cdot \ln(n_i/N_A), \quad \text{and} \quad n_i^2 = G \cdot T^{3/2} \cdot e^{-W_g/kT} \tag{4.23}$$

(as before, W_g is the bandgap in the semiconductor, k is the Boltzmann constant and q is the elementary charge).

In (4.22) C_{ox} is the gate oxide capacitance per unit area, and Q_{ox} is the charge per unit area on the surface due to oxide traps, interface states, etc. At least these two latter are not strongly temperature dependent. However, the charge of ionized acceptors in the depleted layer under the channel is, as calculated from the length of the depletion region at V_F voltage, from [15]:

$$-\frac{Q_{B0}}{C_{ox}} = \frac{\sqrt{2qN_A\varepsilon_{Si} \cdot |-2\Phi_{FS}|}}{C_{ox}} \tag{4.24}$$

where ε_{Si} is the permittivity of the silicon. All these temperature-dependent factors result in an approximately -4 mV/K temperature coefficient of V_{th}.

For calculating R_{DSON} (the resistance of the channel when the device is switched on) we introduce the $V_{OV} = V_{GS} - V_{th}$ "overdrive" voltage, which produces electrons in the channel. If C is the capacitance of the gate/body structure,

$$C = (\varepsilon_{ox}/d_{ox}) \cdot W \cdot L = C_{ox} \cdot W \cdot L \tag{4.25}$$

the total charge in the channel of W width and L length is $Q = C \cdot V_{OV}$. Applying a small V_{DS} voltage to the device; an electric field of $E = V_{DS}/L$ appears along the channel. Electrons in the channel move from the source to drain with the drift velocity:

$$v_{drift} = \mu_n E = \mu_n(V_{DS}/L) \tag{4.26}$$

In a typical power MOS, we can count with the following numeric data when it is switched "on": $\mu_n = 1000$ cm^2/Vs, $V_{DS} = 100$ mV, $L = 1$ μm $= 10^{-4}$ cm.

With these data from (4.26), we get $E = 1$ kV/cm and $v_{drift} = 10^6$ cm/s. This superposed v_{drift} is low; the speed of the random thermal movement of electrons at room temperature from the $\frac{1}{2} m^* \cdot v^2 = 3/2 \, kT$ formula is over 10^7 cm/s.

In case of a saturated MOSFET, most dissipation occurs between the pinch region and drain (right side of Fig. 4.10b), and here a few "hot" electrons, that is, electrons of high speed, carry the same current which is forwarded by plenty of carriers of slow v_{drift} in the open channel.

From the above equations, the so-called modified Shichman-Hodges equation for the MOS characteristics can be derived [12, 14]. It distinguishes between *subthreshold* mode, *triode* mode, and *saturation* mode.

The separation curve between the triode and saturation range is $V_{DS\_sat} = V_{GS} - V_{th}$. At this gate voltage, the channel thickness goes to zero; this is the pinch-off situation.

The characteristics in different operation ranges can be formulated as:

- $V_{GS} \leq V_{th}$, $I_F = 0$, cutoff/subthreshold operation range with low current;
- $V_{GS} > V_{th}$, $V_{DS} \leq V_{DS\_sat}$, linear/ohmic/triode operation range:

$$I_D = \mu C_{ox} \cdot \frac{W}{L} \cdot \left[(V_{GS} - V_{th}) \cdot V_{DS} - \frac{V_{DS}^2}{2} \right] \tag{4.27}$$

- $V_{GS} > V_{th}$, $V_{DS} > V_{GS} - V_{th}$, saturation operation range:

$$I_D = \mu C_{ox} \cdot \frac{W}{2L} \cdot \left[(V_{GS} - V_{th})^2 \right] \cdot \left[1 + \lambda_L \cdot (V_{DS} - V_{DS\_sat}) \right] \tag{4.28}$$

Condensing $\mu \cdot C_{ox}$ into a K_p parameter, and denoting $V_{GS} - V_{th}$ as V_{OV}, one gets the more convenient form:

for $V_{OV} > 0$, $V_{DS} \leq V_{DS\_sat}$, linear/ohmic/triode operation range

$$I_D = K_p \cdot \left[V_{OV} \cdot V_{DS} - \frac{V_{DS}^2}{2} \right] \tag{4.29}$$

for $V_{OV} > 0$, $V_{DS} > V_{DS\_sat}$, saturation operation range

$$I_D = \frac{K_p}{2} \cdot V_{OV}^2 \cdot [1 + \lambda_L \cdot (V_{DS} - V_{DS\_sat})] \tag{4.30}$$

The λ_L empirical parameter describes the shortening of the MOS channel due to the voltage drop between the pinch and the drain.

Figure 4.11 presents an LTSpice simulation with a generic MOSFET model. The main parameters are set to $K_p = 60$ A/V^2, $V_{th} = 2$ V with a β negative thermal coefficient of $S = 6$ mV/K, and channel-length modulation parameter of $\lambda = 0.1$ 1/V.

Figure 4.12 is the output characteristics of the MOSFET device when simulated in the circuit scheme of Fig. 4.11.

Each curve in the plot shows the I_D drain current value when V_{DS} drain to source voltage is applied. Applying higher V_{GS} voltage on the control pin which is now the gate, one will experience higher current at the same drain-source voltage.

Figure 4.13 is the logarithmic version of the plot.

Focusing now our attention on the conducting open channel, we get for the initial section of Figs. 4.12 and 4.13 where V_{DS} is low:

.dc V1 0 3 .1 V2 2.0 4 .5

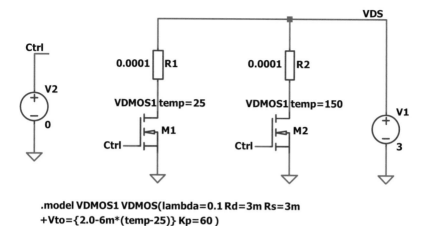

.model VDMOS1 VDMOS(lambda=0.1 Rd=3m Rs=3m
+Vto={2.0-6m*(temp-25)} Kp=60)

Fig. 4.11 LTSpice analysis of two identical MOS transistors at 25 and 150 °C device temperature. The equation of the MOS model is shown. The λ channel shortening parameter is set to 0.1, the threshold voltage V_{th} to 2 V, its S thermal coefficient to -6 mV/K

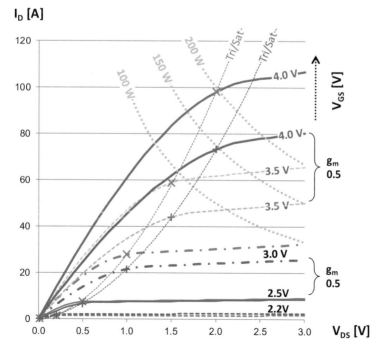

Fig. 4.12 Output characteristics of a MOSFET, cold (blue) hot (red), lin-lin scale. At fixed high current V_{DS} is of positive thermal coefficient, at low current of negative thermal coefficient. The zero thermal coefficient point is around $V_{GS} = 2.5$ V. Triode/ohmic and saturation regions are shown

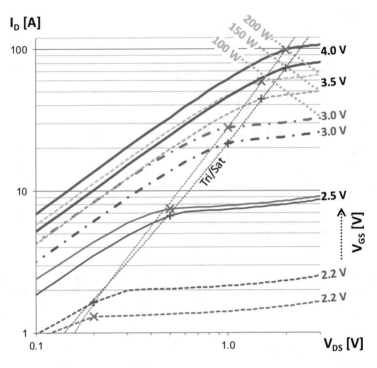

Fig. 4.13 Output characteristics of a MOSFET, cold (blue) hot (red), log-log scale. At fixed high current, V_{DS} is of positive thermal coefficient, at low current of negative thermal coefficient. The zero thermal coefficient point is around $V_{GS} = 2.5$ V. Triode/ohmic and saturation regions are shown

$$I_D = \mu_n C_{ox}(W/L) \cdot V_{OV} \cdot V_{DS}$$

$$(4.31)$$

$$G_{DSON} = \mu_n C_{ox}(W/L) \cdot V_{OV} = \mu_n K \cdot V_{OV}$$

the device acts as a resistor; its G_{DS} conductance is proportional to V_{OV}. In a smaller range, we can assume a linear dependence of V_{th} and μ_n on temperature with a gradient of α and β, respectively. Fixing the gate overdrive at $V_{OV} = V_{OV0}$, we get:

$$G_{DSON} = \mu_{n0}(1 - \alpha \cdot dT) \cdot K \cdot [V_{GS0} - V_{th0}(1 - \beta dT)];$$

$$G_{DSON} = \mu_{n0}(1 - \alpha \cdot dT) \cdot K \cdot (V_{GS0} - V_{th0} + V_{th0}\beta dT)$$

$$= K \cdot (\mu_{n0} - \mu_{n0}\alpha \cdot dT) \cdot (V_{OV0} + V_{th0}\beta dT)$$

$$= K \cdot (\mu_{n0}V_{OV0} - \mu_{n0}V_{OV0}\alpha dT + \mu_{n0}V_{th0}\beta dT - D\alpha\beta dT^2)$$

$$(4.32)$$

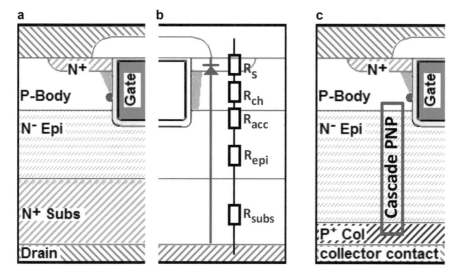

Fig. 4.14 (**a**) Elementary section of a vertical TrenchMOS device, (**b**) constituents of the R_{DSON} resistance (**c**) elementary section of a vertical IGBT device with the cascade pnp bipolar junction transistor shown

Hence, α and β are small; the quadratic term at the end of (4.32) can be neglected; G_{DSON} is determined by a mix of positive and negative temperature coefficients. This can be well observed in Figs. 4.12 and 4.13.

A typical construction of power MOSFET devices is the TrenchMOS structure. A dense net of elementary vertical MOSFETs ensures equal current distribution on the surface. Figure 4.14a, b shows an elementary section of a device.

One can identify the reverse body diode (in red) between the p-body and N^+ source. For illustration only, on the left side, a saturated channel with a pinch point is shown; on the right side, there is a continuous channel corresponding to low V_{DS}.

The figure indicates that R_{DSON} is influenced by the following components:

- R_s: source resistance, low due to high N^+ doping.
- R_{ch}: channel resistance, determined by gate voltage.
- R_{acc}: resistance from the accumulation region.
- R_{epi}: resistance from the epitaxial top layer of silicon; this layer controls the blocking voltage of the MOSFET.
- R_{subs}: resistance from the silicon substrate.

The technological parameters of different regions contributing to the R_{DSON} resistance can be adjusted so as to minimize the channel resistance. However, their setting compromises other device parameters, especially the blocking (breakdown) voltage. In low-voltage transistors for switching purposes, the p-type body can be doped more heavily in such a way substantially decreasing the channel resistance.

The critical technological parameter is the thickness and donor doping of the N^- doped epitaxial layer. Thin epitaxy of higher doping results in low blocking voltage

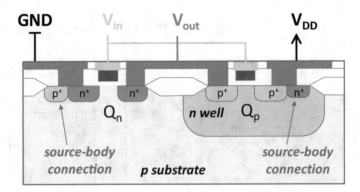

Fig. 4.15 CMOS inverter, cross section. A lateral NMOS transistor in the p substrate and a lateral PMOS transistor in the n well are shown

Fig. 4.16 CMOS inverter, schematic

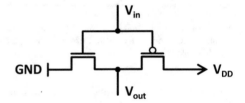

and low associated R_{epi} series resistance. In high-voltage transistors, this layer is thick and lightly doped; the associated resistance increases R_{DSON} critically.

Depending on gate material and oxide construction, the MOSFET can be of depletion or enhancement type, that is, it does or does not possess a conducting channel at $V_{GS} = 0$.

The construction of MOS transistors has also a primary role in the thermal transient testing of integrated circuits. Nowadays the majority of these circuits is manufactured in some version of the CMOS technology. Figure 4.15 shows the cross section of an inverter circuit. This is the basic building block for most functionalities in the device, such as logical gates, amplifiers, and output stages. Figure 4.16 presents the two complementary transistors in the equivalent circuit scheme of the inverter.

As the figures illustrate, there is a chain of p and n regions in series between the VDD and GND rails of the circuit. This large diode covers the active area of the whole chip. Reverse biasing this large diode, we get a "dull" but robust and nondestructive way for powering and sensing during a thermal transient test.

4.5 Insulated Gate Bipolar Transistor (IGBT) Devices

In the previous subsection, we stated that high-voltage MOS devices perform poorly because of the long and lightly doped epitaxial layer which is the drain of the transistor. The voltage drop at high current in high-voltage insulated gate devices can be significantly lowered if a large number of charge carriers is injected into the epitaxial layer externally, for example, from an additional pn junction.

IGBTs have many construction variants alike MOS transistors. A frequent form is presented in Fig. 4.14c. This structure is identical to the vertical MOSFET of Fig. 4.14a, except that the N^+ drain region responsible for lowering the series resistance is replaced with a P^+ collector layer, thus forming a vertical pnp bipolar junction transistor. The additional P^+ region forms a cascade scheme of the pnp transistor and the surface n-channel MOSFET (Fig. 4.17).

IGBTs combine the simple control of power MOSFETs due to their insulated gate with the high-current and low-saturation-voltage capability of bipolar transistors.

At high currents IGBTs feature a low V_F forward voltage drop on the base-emitter junction of the pnp device, compared to high-voltage MOSFETs, while those are more advantageous at lower current where a proportional voltage drop occurs just across R_{DSON}.

In IGBTs the injection of minority carriers (holes) from the collector P^+ region into the N^- drift region considerably reduces the resistance of this latter. Moreover, at higher current this injection increases, resulting in the known merely logarithmic increase of V_F.

However, this resultant reduction in on-state voltage drop brings about several disadvantages:

- The series pn junction blocks reverse current flow; IGBTs do not conduct in the reverse direction. In switching applications where reverse current flow is needed, a discrete freewheeling diode is to be placed antiparallel with the IGBT to ensure conduction in the opposite direction.
- The reverse bias rating of the N^- drift region to collector P^+ diode is low, only tens of volts. If in an application a reverse voltage appears on the device, an additional series diode must be used.

Fig. 4.17 Equivalent circuit scheme of an IGBT device

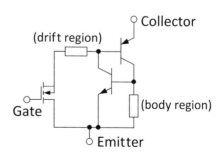

- The minority carriers injected into the N⁻ drift region take time to enter and exit or recombine at turn-on and turn-off. This results in longer switching times, and consequently in higher switching loss compared to MOSFETs.

All these drawbacks can be mitigated using MOSFET devices manufactured on wide bandgap materials, where high blocking voltage is granted as a result of the high W_g value.

4.6 Semiconductor Devices on Wide Bandgap Materials

During the previous decades, the devices of power electronics were mostly realized on silicon. This offered a very robust technology working fine even at 150 °C operating temperature. For this reason, the thermal testing standards (e.g., of JEDEC or AQG) have taken for given such features of silicon like

- pn junctions available for powering and sensing
- Mostly linear temperature dependence of parameters
- Availability of normally-off devices for power electronics
- No slowly moving surface charges in MOS structures
- Rather long lifetime of charge carriers in depleted regions

Newly, compound semiconductor devices have been introduced into power applications where their excellent properties can justify their higher cost. They work well at extremes like at 77 K in liquid nitrogen for minimum noise, or at high temperatures up to 350 °C. They show low channel resistance due the high carrier mobility and high blocking voltages, thanks to their wide bandgap.

Silicon carbide (SiC) is a material of a wide and indirect bandgap; for this reason, it has always been intended for power and high-temperature devices.

Gallium nitride (GaN) has a direct bandgap; minority carriers are not present in it as they would recombine after an extremely short lifetime, they have no chance to reach an opposite material layer (i.e., collector). Accordingly, only unipolar transistors can be made of it. On the other hand, on an AlGaN/GaN interface a

Table 4.4 Thermal testing methods for device types realized in certain materials

Material	Si	SiC	GaN	Note
Device				
BJT	St	E	×	
MiSFET	St	Td	Td	Also as MIS gate HEMT
JFET	St	E	Td	Also as Schottky gate HEMT
HEMT	×	×	Td	
IGBT	St	Td	×	
pn diode	St	E	×	
Schottky diode	St	Td	E	

two-dimensional (2DEG) electron gas forms easily and offers low sheet resistances in a high electron mobility transistor (HEMT) device.

Optoelectronics used to be the main market for the direct III–V-semiconductor GaN. In this role it became better and a cheaper and entered the power market.

Table 4.4 lists the existing device types and also hints on the existing knowledge base available for their thermal testing. (In the table the abbreviations mean: St: standard testing method, E: the device exists, ×: the device does not exist, Td: thermal transient is defined and explained in this book.)

The preeminent material properties of these materials are given in [17, 103]. In most parameters, GaN is slightly superior to SiC and provides a three to five times better ON resistance in insulated gate devices. The thermal conductivity of GaN is low that would impede its use in power applications; for this, it is always used as a thin epitaxial layer on a cheap substrate material.

Generally, Eqs. (4.22)–(4.32) also apply for MOSFET devices built on wide bandgap material, resulting in device characteristics similar to Figs. 4.12 and 4.13.

Equation (4.32) predicts that the $R_{DSON}(T)$ channel resistance has a temperature dependence with sections of positive and negative thermal coefficients, governed by the shrinking of the V_T threshold voltage at lower and the increase of the μ mobility at higher temperatures. At practical current levels, the negative thermal coefficient can be observed only at extreme low temperature in silicon devices. A simple test proving this effect in the $-196\ °C$ to $25\ °C$ range is presented in [83].

In SiC devices, the turning point between negative and positive temperature coefficient sections can be around room temperature; its actual position depends on the operation point, that is, the V_{DS} drain-source voltage and I_D drain current.

Figure 4.18 investigates the temperature dependence of the channel resistance in two MOSFET devices. R_{DSON} is interpreted in a "static" manner, that is, as V_{DS}/I_{DS} at low drain-source voltage (see also Chap. 6, Sect. 6.2.1). The chart compares the

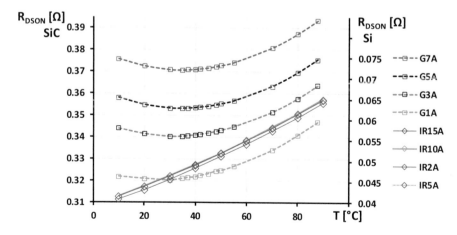

Fig. 4.18 Measured temperature-dependent "static" $R_{DSON} = V_{DS}/I_{DS}$ channel resistance of a SiC MOSFET and a Si MOSFET. The nonmonotonous temperature dependence of R_{DSON} in the SiC device can be observed

popular IRFP260N silicon MOSFET with 40 mΩ nominal channel resistance to the commercially available silicon carbide G3R350MT12D MOSFET, with a claimed channel resistance of 350 mΩ at room temperature and 499 mΩ at 175 °C in its datasheet.

The chart presents the change of the channel resistance of the SiC MOSFET at $I_D = 1$ A, 3 A, 5 A, and 7 A current values (key G1A to G7A, curves scaled on the left vertical axis). The channel resistance of the larger Si MOSFET, which has a ten times lower channel resistance, is shown at $I_D = 2$ A, 5 A, 10 A, and 15 A (key IR2A to IR15A, curves scaled on the right vertical axis).

The higher current dependence and the nonmonotonous temperature dependence of R_{DSON} in the SiC device can be well observed. The turning point from negative to positive temperature coefficient is around 30 °C and shifts toward higher temperatures at higher current.

Conversely, for the Si device, the temperature coefficient of R_{DSON} is positive in the examined 10 °C–90 °C range, and so it remains from deep sub-zero to the maximum rated temperature of 175 °C. The temperature dependence follows an $R_{DSON}(T) = R_{DSON0} \cdot \exp(\alpha_\lambda(T - T_0))$ formula, as a consequence of (4.29) and (2.53).

The R_{DSON} channel resistance is one of the most often used temperature sensitive parameters (TSPs) in thermal transient measurements, described in detail in Chaps. 5 and 6, Sect. 6.2. The nonmonotonous nature of the channel resistance in SiC MOSFETs limits its use as TSP for this device category.

4.7 High Electron Mobility Transistor (HEMT) Devices

HEMT is a type of field effect transistor, where the channel is formed at the interface of two layers (usually AlGaN and GaN) of different bandgap (heterojunction). The layers are grown upon each other on top of a carrier substrate. As the consequence of the abrupt change of the different electric fields at the material interface (polarization fields), a two-dimensional electron gas (2DEG) layer is formed at this which acts as the channel of the transistor. In this 2DEG layer, the electron mobility is much higher than in the bulk material, which makes the device very efficient in high-frequency applications [104, 105]. The substrate material has no direct effect on the electric behavior; however, it affects significantly the price, yield, and the thermal performance. Often used materials are silicon, sapphire, SiC, or GaN. As the channel forms naturally without external biasing, the classic HEMT devices have a depletion mode (normally ON) characteristics; a negative gate voltage needs to be applied to turn the device off. The classic GaN HEMTs have a Schottky barrier gate contact.

HEMT devices are widely used as millimeter wave RF amplifiers; newly they are conquering high-frequency switching mode power conversion applications as well. In order to meet industry requirements, for safer operation and simpler gate drive circuits, there is extensive development of modified structures to achieve optimal performance with enhancement mode (normally normally off) operation [129].

Using band engineering techniques like "recessed gate" [130], "fluorine gate" [131], and thin AlGaN barrier layer [132], the threshold voltage can be increased above 0 V realizing the enhancement mode operation. Recent advancements combining these techniques with additional gate dielectric layer (MIS HEMT) can reduce gate leakage current and decrease switching and conduction losses [129, 133].

In GaN devices, the R_{DSON} channel resistance, related to the sheet resistance of the 2DEG electron gas, has positive thermal coefficient in the temperature range of interest for power devices. However, many constructions suffer from a time-variant behavior, due to the varying amount of trapped charge at off-state at the AlGaN barrier surface (current collapse). Recent constructions aim at reducing this effect which restricts the use of GaN devices in all power switching applications; this also improves the usability of channel resistance as TSP in thermal transient measurements.

Chapter 5
Fundamentals of Thermal Transient Measurements

Gábor Farkas

As discussed deeply in Chap. 2, a change in the power applied on a system results in a transient change of the temperature. The growth of the powering from a lower P_L level to higher P_H initiates a heating transient. When the power diminishes from P_H to P_L, we observe the cooling of the system (Fig. 5.1).

The transient change is influenced by the material properties and geometry of the structural elements in the heat-conducting path. Consequently, thermal transients carry information on the composition of the structure and eventually on its health and location of potential failures. Also, with an appropriate number of recorded transients, thermal models of the system of different complexity can be built.

Thermal system descriptors such as Z_{th} curves and structure functions can be derived, and thermal models of the system can be constructed with a relatively simple mathematical apparatus if the change in powering occurs as sharp switching between two stable power levels.

Looking into the physical background, we obtained that the origin of thermal transients in a system is a lasting imbalance between the energy *generated* or *applied externally* at specific system locations and the energy which leaves the system.

In this book thermal transients of electronic systems are discussed, where the applied energy is of electrical nature, converted to heat in the tested devices. The energy removal occurs mostly in the form of heat flux, with the important exception of solid-state lighting devices where a large part of the energy is emitted in the form of light [7].

G. Farkas (✉)
Siemens Digital Industry Software STS, Budapest, Hungary
e-mail: Gabor.Farkas@siemens.com

© The Author(s), under exclusive license to Springer Nature Switzerland AG 2022   171
M. Rencz et al. (eds.), *Theory and Practice of Thermal Transient Testing of Electronic Components*, https://doi.org/10.1007/978-3-030-86174-2_5

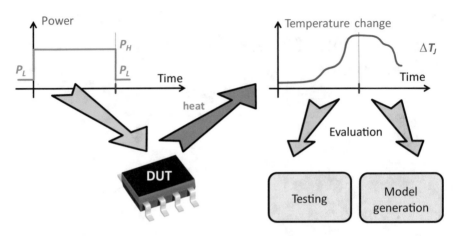

Fig. 5.1 Thermal transients in a system and their use for testing and modeling

In Chap. 2 the energy imbalance was formulated between the momentary power values, the power generated (P_{gen}, P_{in}) and removed (P_{diss}, P_{out}).

The heat removal, more often called dissipation, is governed by the structure in which the powered locations reside. The actual heat flux and the shape of its trajectories depend on the material properties and the geometry of the structure when *conduction* is the mechanism of the heat removal. The heat flow in a continuous, thermally conductive body is determined by a set of partial differential equations, the Fourier law, formulated for an elementary structural detail in Eqs. (2.1), (2.2), (2.3), (2.4), (2.5), (2.6), (2.7) and (2.8). Heat conduction is the most important form of heat transport in the internal parts of electronic systems.

In case when *convective* heat removal is present in some system portions, a different set of equations is to be applied. *Radiation* may play a role at systems operated at high temperature.

The thermal behavior of the system can be investigated based on the *continuous model* of the system if the full geometry and properties of the material components are known [56].

It was presented in Chap. 2 that the conductive heat removal can be modeled by a net of discretized thermal resistance and capacitance elements. An equivalent RC model can be also constructed for convection. In case of radiation, such models are of poor accuracy because of the strong nonlinearity of the effect.

In the practical task of *thermal testing*, it is obvious to partition the above theoretical "structure" into the *device under test* (DUT), the thermal properties of which are to be determined and the *test environment*. The two are separated by a *thermal interface*.

The DUT in the transient test can be a packaged semiconductor device or a larger subsystem in a system-level structural analysis.

Figure 5.1 hints that the thermal transient tests always include a heating and a cooling phase. Placing the obviously non-energized DUT into the test system, these two phases are typically realized as consecutive processes.

In order to provide repeatable testing results, the procedure of thermal testing is regulated in *thermal testing standards* built on the above considerations. Different standards exist for the accomplishment of the task; these will be presented in later chapters in details.

In most standards first the *device categories* to be tested are defined, classified upon their electrical and mechanical construction. The *thermal test equipment* comprises the electronics for powering (excitation), and the electronics for data acquisition. Some standards also specify the conductive or convective thermal test environment around the device.

Today the data acquisition is always accomplished by recording electrical signals. This implicates that thermal sensors (transducers) are needed which convert the transient temperature of the locations of interest into an electrical signal.

The most critical factor in the reliability and lifetime of power applications is the temperature of the semiconductor power devices themselves. As discussed in Chap. 4, practically all parameters of these devices are strongly affected by the device temperature. These temperature-dependent parameters include voltage and current at input and output pins in an actual operating point, the timing of their change, etc. In such a way, semiconductor devices are perfect thermometers; besides they amplify, oscillate, switch current, or emit light. In the literature and in standards, the temperature-dependent electrical parameters that are used for measuring the temperature are referred to as TSP (temperature-sensitive parameter) or TSEP (temperature-sensitive electrical parameter).

The temperature of other accessible points on the device package or in the test environment can be transduced by external *dedicated temperature sensors*, based on the various temperature-dependent effects presented in Chap. 4, e.g., change of resistance or Seebeck effect. In large modules for current switching purposes, it is not uncommon to mount internal resistive temperature sensors on the baseplate hosting the power semiconductors.

5.1 Tester Instrumentation and the Environment

The major elements needed for thermal transient testing are as follows:

- A power switching unit to provide the exact powering, typically switching between two power levels
- A measurement or data acquisition unit to detect the temperature-dependent electric signal
- A temperature-controlled environment to maintain a prescribed ambient into which the DUT is placed (e.g., cold plate, or still-air chamber, etc.)
- A data processing unit, as the measured temperature response has to be further processed

All the above functions have to be controlled by a computer.

The general scheme of a thermal transient tester is shown in Fig. 5.2. The DUT in the figure is a thermal test vehicle with separated powering and sensing, a device in which the functionalities of the thermal testing are least intermingled.

Thermal transient tester

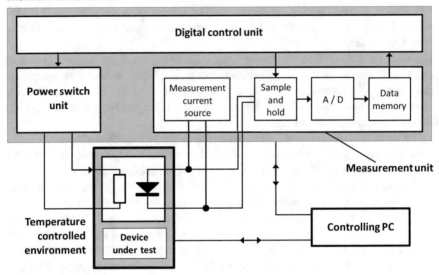

Fig. 5.2 The general scheme of the tester and test environment

The major functionalities of the tester are discussed below.

Powering

In Chap. 2 it was proved that the mathematical procedure of converting the transient temperature data into relevant thermal descriptors such as Z_{th} curves, time constants, and structure functions is the easiest if a power step is applied on the investigated device, and the power is maintained until a steady state is reached again. In the case when a sharp, steplike power change cannot be produced; this conversion is still feasible, but it needs recording of the variation of the power profile and then a complex calculation of convolution.

It has to be noted that the powering units of testers have to serve drastically different needs and have to be of different power and timing specification. Tiny microelectromechanical systems (MEMS) or laser diodes respond in the microsecond range when milliwatts are applied on them; large press-pack IGBT modules stabilize in a few minutes when powered by appropriate kiloamperes. Large street luminaires with their normal air cooling reach steady state in a daylong transient.

In typical testers the diverse demands are served by a base instrument fulfilling the common voltage and current requirements and further accessories, amplifiers, and boosters which extend the voltage and current range.

In many cases applying a long power step until steady state is reached is not feasible.

For example, in a production line, the measurement time is inherently limited. Considering the number of devices to be measured, the targeted throughput restricts the time slot for thermal testing sometimes just to milliseconds [106, 107].

In a type of lifetime testing (active power cycling), pulses at an appropriate duty cycle are applied on the device until its deterioration (Chap. 7, Sect. 7.4). The wear mechanism typically differs when the pulses are short, below a few seconds, and when they are longer, lasting for minutes [21]. Keeping the power for a prolonged time would cause an abrupt breakdown of the device.

However, also in these cases, a sharp swing between the power levels, in the form of a single pulse or multiple pulses, facilitates the calculation of a limited set of the thermal descriptors.

For satisfying all these demands, the sophisticated hardware of transient testers has to incorporate a powering unit with controlled voltage and current sources, which switch between preprogrammed levels fast and reach the accurate "set value" with short settling time.

The settling time of a current, or power in general, is also related to the targeted power magnitude, partly because of limitations in the electric circuitry, partly because of the inductive nature of the cabling towards the DUT.

Regarding the accuracy of power sources, there is a general contradiction in electronics between the expected speed of change and precision. High accuracy of 4–7 digits of the target value can be achieved with long integration times. Typical power supply units in tester appliances have approximately 0.5% accuracy and a typical relaxation (regulation) time of a few milliseconds at the sudden drastic load change associated with applying or revoking power.

As explained later in this chapter, in a properly constructed measurement arrangement, testers typically switch from high power level to low in a very fast process. In this concept, a higher heating current is switched off by a fast switching circuitry, while a constant sense current from a separate source is maintained on the DUT. The relaxation of the powering source continues in the millisecond order of magnitude, but this wobbling occurs behind the switch, separated from the DUT.

This way, a switching time of a few microseconds can be achieved at the start of a *cooling transient*. In high power testers at hundred and thousand amperes, careful design can achieve a settling time below 100 μs for cooling transients. Power edges of this slew rate are in most cases unfeasible for heating transients due to the relaxation time of power supplies.

As an exception, some advanced testers can turn on several amperes in 10 μs. This can be advantageous when beyond thermal transients also transients in other physical domain (optical, magnetic) are to be investigated.

It also has to be noted that the excitation for thermal testing is not necessarily provided by a dedicated powering unit. In many cases the inherent power belonging to the normal operation of the investigated system can be used for powering. In order to meet the sharp timing constraints between the applied power and the captured temperature response, either the recording starts on a trigger event generated from the power change or the operation of the tested system is to be synchronized with trigger signals from the tester.

This external powering approach also opens the way for thermal investigation of live functional systems during their normal operation.

Sensor Biasing

As discussed in Chap. 4, the temperature-to-voltage transducers are dedicated sensors or semiconductor devices operating at some bias. In case of resistive or diode sensors, a separate source of programmable but constant value, referred to as "measurement current" or "sensor current" maintains a low power level on the sensor element.

Measuring thermal transfer effects needs multiple sensors at various locations. For example, when the thermal coupling between different chips in a power module is investigated, each chip has to be biased. Accordingly, testers typically offer multiple independent sensor sources.

The good quality of sensor biasing sources is indispensable in a thermal tester; it highly affects the usability of the recorded thermal signal. Keeping their noise low is essential for identifying tiny variations in the temperature change. They are exposed to sudden harsh jumps in voltage and current levels on the DUT during the switching process; the distortion of the temperature related signal is influenced by the time and amplitude of their relaxation after the perturbation. The contradicting requirements of low noise, precise value, and short relaxation time have to be fulfilled at the same time.

In some practical realization of thermal transient testers, some bias sources are integrated into the powering units in order to simplify the wiring of transient measurements at the driving point, where power is applied.

Several standards define requirements on the expected ratio of the power level in the heating phase of the transient and in the cooling phase when normally only this sensor bias is present. The validity of these expectations is investigated in Chap. 6.

Data Acquisition

The data acquisition part of a thermal tester is also demanding. A thermal tester equipment seems to be an "inefficient" construction in terms of electronics, as hundred watts or kilowatts are fed into the device under test as excitation, while the resulting electrical signals delivered by semiconductor devices and dedicated sensors are in the order of just a few millivolts. The tiny signal has to be formed by amplifiers, which have to fulfill contradictory requirements again: low noise, high speed, and high accuracy. Overdrive conditions frequently occur at switching; in this case fast recovery to normal operation is indispensable.

A tester typically comprises a number of identical data acquisition functionalities, commonly referred to as measurement channels.

As the subsequent sections confirm, a minimum data acquisition rate of 1 megasample/second and a voltage resolution finer than 100 μV are needed to adequately separate the electric and thermal constituents of the recorded signal and to capture the smallest thermal time constants.

Measurement channels designed with this specification can ensure proper suppressing of the inherent noise of the device and such eliminate the need for complex noise attenuation procedures, e.g., capturing the signal multiple times and averaging it.

Sample and hold circuits and high-resolution analog-digital converters in the channels accomplish the transformation of the analog signal into digital data which are stored in the tester and transferred to the external measurement control for further processing.

Temperature-Controlled Environment

An obvious precondition for repeatable measurements is a proper definition of *repeatable boundary conditions*. The related standards [24, 29, 30], define *convective* and *conductive* environments.

Typical convective boundaries are still-air chambers, wind tunnels, and liquid baths. Conductive environment is realized by different cold plate structures.

Plates and baths can be programmed to force different temperature values to the device under test during the transient measurement. In a similar way, they can be used in the calibration process, in which the devices are kept at preprogrammed temperatures for prolonged time and their temperature-sensitive parameter (TSP) value is recorded.

The temperature range and the heat sinking capability of the test environment have to be selected carefully.

The calibration of the TSP typically occurs on a cold plate or in a liquid bath at low power but in a broader range than the expected temperature transient.

During the heating phase of the transient test, the environment may need to keep its own temperature stable while balancing several kilowatts of dissipated power. However, the actual "set temperature" range can be smaller than that of the DUT; it has to match the intended temperature range on the *accessible package surface* only.

As an example, a typical automotive power module built on silicon chips is often calibrated between 25 °C and 150 °C maximum temperature. Then, transient tests for providing thermal data of the module for data sheet are carried out at room temperature. After finishing these tests, for lifetime analysis, the power cycling occurs at an elevated 80 °C baseplate temperature to mimic the typical operating temperature under the hood. In such cases the selected thermostat has to be able to work in wider temperature range and needs to have a high enough heat sinking capability to balance the power dissipation of the component.

Calibration thermostats are often based on thermoelectric coolers (TEC). These fast thermostats may use a combination of resistive heaters for heating and Peltier elements for cooling with a few watts of heat sinking capability at room temperature.

The performance of cold plates is determined by the programmable liquid circulator equipment and the properties of the coolant material. These thermostats also have asymmetric characteristics, with high heating and lower cooling capability, this latter being limited by the cooling aggregate.

In all thermostat constructions, the heat sinking capability diminishes at lower set temperature; the minimum temperature of a construction is defined at zero applied power.

In many branches of electronics, critical subassemblies rely on air cooling (e.g., automotive applications), or complete appliances are cooled by air exclusively (mobile phones). In these constructions the integrated circuits and discrete

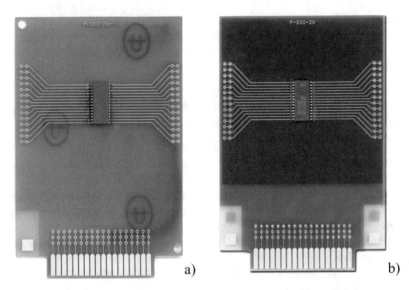

Fig. 5.3 Integrated circuit in DIP package on standard printed boards as defined in the JEDEC JESD51 set of standards, (**a**) worst-case cooling with one patterned copper layer (**b**) with two solid internal copper layers for enhanced cooling. Separated force and sense traces can be observed

components dissipate towards the printed boards hosting them, mostly towards the copper layers.

In order to qualify the thermal performance of packaged semiconductors designed for heat removal towards the board, the whole thermal environment around them must be highly standardized. This uniformity facilitates clear distinction of the share of heat flow in the packaged device and in the environment and helps gaining a valid comparison between different package constructions.

The first elements in the heat-conducting path which belong to the standardized thermal environment are the printed boards. For comparability of measurement results, these have to follow the related standards regarding their geometry and layer structure. Figure 5.3 demonstrates a pair of standard constructions, a worst-case solution, in which cooling occurs towards a single patterned copper layer on board surface with thin traces and a version where two solid internal copper layers ensure enhanced cooling. The JEDEC JESD51 set of thermal measurement standards [29] defines several standard printed board constructions, adjusted to the style of package pins or solder balls, and the presence of cooling tabs or fillers, etc. The two formations in Fig. 5.3 host leaded surface mount packages. An example with their comparison is shown in Chap. 3, Sect. 3.1.2.

The standardization concept is also extended to the farther environment; the JEDEC JESD51 set of standards includes a worst-case still-air conductive environment (Fig. 5.4, JESD51-2A standard) and wind tunnels (JESD51-6 standard).

The construction of the conductive or convective test environments also determines the time needed for a complete thermal test. At aggressive cooling on cold

Fig. 5.4 One cubic feet still-air chamber and the location of a standard printed board as defined in the JEDEC JESD51-2A standard

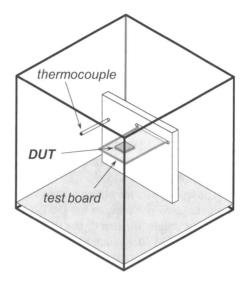

plate, steady state is reached within a couple of minutes; in still air tens of minutes or hours are needed for equalization time.

In case of solid-state lighting devices, additional radiometric and photometric standards (listed in Chap. 6) prescribe an appropriate temperature-stabilized environment for LED measurements, e.g., an integrating sphere with temperature-stabilized fixture. In Sect. 6.10 we present a combined methodology for radiometric/photometric and thermal testing.

5.2 The Interaction of the Components in a Complete Test System

At macroscopic scale the measurement instrument and the entity to be measured have typically minor interaction in a measurement. For example, a tape measure has almost no influence on the size of a measured object; neither do scales impact the weight of it. On the contrary, at atomic magnitudes a measurement severely influences the observation of the phenomena which were discussed in Chap. 4.

In case of thermal testing, the components of the test system and the DUT have much stronger influence on each other. Moreover, this interaction of tester, test environment, and DUT takes place in the electric and the thermal domain in a simultaneous and strongly coupled way. In the testing of power LED devices, this interdependence is extended to the optical domain as well [8] in [7].

In the electric domain, the testing scheme can be partitioned into subsystems such as the tester, the DUT, and the cabling between the two, but all these form a single electric circuitry.

One might think that with the proper construction of the tester, the effect of the attached cabling and DUT can be minimized; the internal sources can be considered as ideal voltage and current generators at a stable operating point. To some extent this can be achieved with careful design. However, the external circuitry can be considered as a load on the sources in the tester, not necessarily an exclusively resistive and time-invariant one. Conversely, the tester constitutes the load on the complex external circuitry composed of the DUT and the cabling. All these together may combine into a high gain amplifier with feedback through the cabling and the parasitic effects of the internal circuitry in the tester. Such a feedback can be present also through the temperature-triggered parameter shift of the DUT and can cause oscillations and runaway.

The testing scheme allows a bunch of distinguished operating points, belonging to the targeted powering of the DUT. Circuit theory offers stability criteria for the operating points in general, but these criteria are always to be interpreted on the complete set of interacting subsystems in the scheme.

The transit between dedicated operating points occurs through transients of electric and thermal nature. After applying a certain power on the DUT, a transient starts, which ends, in best case, in another stable operating point of the DUT and also of the internal circuit elements in the tester.

As discussed throughout this book, the aim of thermal transient testing is to restrict all observed and recorded transients in the complex system to those electric changes, which can be unambiguously tied to the shift of temperature-sensitive parameters of the DUT. Time-variant changes of other root cause are classified as "electric transients" and are suppressed by various means.

Several time-variant effects are inseparable from the physics of the tested device itself. These can be the recombination of charge carriers in pn junctions, modulation of the base length of bipolar transistors, or channel length modulation in field effect transistors. These two latter are classified as "backlash effects."

As it will be presented in subsequent chapters, these effects cannot be completely eliminated, but they can be significantly restrained with diminishing the difference of the high P_H and low P_L power levels applied.

However, other electric changes are consequences of the transit of the of tester-cabling-DUT compound circuitry between operating points. For damping the transient or suppressing eventual oscillations, the cabling has to be arranged properly, and capacitive and inductive add-ons may be needed as illustrated in some examples in this chapter and in Chap. 6.

In standard electronics design, the decay of a transient is generally considered finished after a period of a length of four or five times longer than its characteristic time constant.

In thermal testing the electric transients have much higher amplitude (in the range of volts) than the thermally induced change (in the range of millivolts, or just a fraction of them). Still, the electric transients fall off in microseconds, while the thermally induced change is to be recorded from microseconds to minutes or hours continuously.

In thermal transient testing, the data acquisition magnifies the tiny change around the aimed operating point. This way, the falloff of the transient is also recorded in a "zoomed in" way; it perturbs the thermal signal until ten times longer than its characteristic time constant.

At higher currents also the resistive and inductive properties of the cabling may add significant distortion to the thermal signal.

Even with a perfect tester construction and proper routing of the cables, a serious disturbance can be superposed on the temperature-induced voltage signal due to the coupling of the powering and the data acquisition within the device under test.

Inductive and capacitive coupling through the cabling and the DUT may distort the initial, early section of the thermal signal. This effect can be partially eliminated with differential measurement techniques.

The voltage drop on the ill-separated, common resistive sections of the external cabling or the internal wiring of a power module adds a constant error to the measurement of the power level in the heating phase. During the cooling transient, this error typically vanishes; the voltage drop at low measurement current becomes negligible.

Considering the thermal domain, the device under test and the temperature-controlled environment constitute a coupled thermal system. The thermal transient test characterizes the whole heat-conducting path, first its sections within the device and then the thermal interface and the cooling mount. The structure function methodology helps in separating the DUT and the environment, either with repeated tests using different TIM layers or based on the characteristic capacitance of structural elements in the DUT.

The structure functions also portray the TIM layer and the cooling apparatus in the test bench and identify their properties and potential failures.

5.3   Device Under Test Categories and the Related Electrical Arrangement

Thermal testing of power devices is an intricate topic which can be pictured from the perspective of the test system, considering how current or voltage are applied on the DUT and what is the appropriate measurement range in data acquisition channels. Devices are represented by a generic model in this treatment.

In this chapter the test procedure is examined according to this concept.

A deeper analysis of the behavior of particular devices in a test system is given in Chap. 6, with a focus on the temperature-related change of the device characteristics, thermal coefficients, and stability of the operating point.

5.3.1 Devices with Separate Heaters and Sensors

The way of powering and sensing is quite different for different device categories.
The highest flexibility in powering can be achieved in devices with fully separated
heaters and sensors.

Such devices of monolithic realization and smaller size are called *test chips*.
These are used to verify packaging concepts in the design phase when the actual
semiconductor device is still not available, but many parameters as the chip size,
wafer thickness, and an approximate surface power pattern are already known. Test
chips are broadly used for comparing the quality of die attach technologies.

Larger and more sophisticated devices with separate heaters and sensors are
denoted as *thermal test vehicles*, TTVs. Using these the thermal management of a
complex assembly can be optimized in a mock-up which includes the packaged
device and its near environment, thermal interfaces, and cooling appliances.

Test chips and TTVs are typically realized with resistive heaters. The sensors can
be resistors or diodes (Fig. 5.5); in TTVs sometimes thermocouples are also used.

In Fig. 5.5 the power step is applied by switching the I_{drive} source on or off, or
between I_H and I_M levels in a more general case. A constant I_{sense} source provides the
required bias for the sensor diode (or a resistive sensor).

In TTV devices the power step does not severely interfere with the thermal
response transduced by the sensor. A short initial electric transient can appear on
the thermal signal because of capacitive or inductive cross talk within the device, or
in the cabling, the influence of this latter can be reduced with appropriate routing.

It can be observed that the data acquisition occurs on two separate channels. The
voltage on the heater (V_H and V_L at the two levels of the power step) has to be
measured in order to ascertain the $\Delta P = V_H \cdot I_F - V_L \cdot I_M$ power difference which is
used during the postprocessing of results for Z_{th} and structure function calculations.
V_H and V_L voltages are typically of several volts.

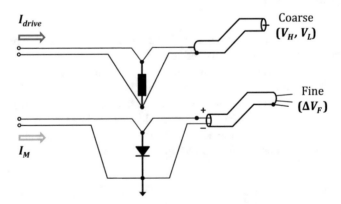

Fig. 5.5 Thermal transient measurement of a thermal test vehicle with separated powering and
sensing

Another acquisition channel serves the capture of the thermally induced voltage change on the sensor. For both resistive and diode-type sensors, this change is small; a ΔV_F shift of a few millivolts occurs on the top of a larger V_F bias of many hundred millivolts or volts.

A separated force and sense wiring (four-wire method, Kelvin contact) ensures that the voltage drop caused by I_{drive} or I_{sense} on the wires of the force side is not added to the true voltage on the heater and the sensor element.

It has to be noted that the shielding on the cables in Fig. 5.5 is only needed when the heater or sensor is of resistive type, in the range above a few hundred ohms. At higher impedance the external perturbation on the measurement occurs through electric, capacitive coupling, and shielding can effectively keep off the disturbance.

At low source impedance of the "signal source," which is in this case a low resistance or a diode in forward operation, the source of perturbation is of inductive, magnetic nature. Shielding of wires does not improve the signal quality; tight routing or twisting of the wires is the recommended measure.

The disturbance through common wiring sections and the magnetic interference through grounding loops can be minimized by an appropriate grounding scheme. According to general construction principles in electronics, a single grounding point (star point) minimizes both effects.

Although this clear scheme of powering and sensing cannot be achieved in other classes of thermal systems, the principles of separation, perturbation suppression, grounding, etc. can be used for improving the quality of their thermal testing.

5.3.2   Two-Pin Devices, Diodes

In the general testing scheme of Fig. 5.2 and also in the previous scheme of Fig. 5.5, it was supposed that the I_{drive} current which ensures the powering, the I_{sense} current which provides the bias for the sensor, and the measurement channels are well separated. In electronic systems where the heating source is a device with fully or partly separated input and output ports, this separation can be reached with more or less compromise.

In many cases the powered device is a diode, or another device with just two pins. Moreover, many tutorials and thermal measurement standards recommend a "simplified" test scheme for devices having three or more pins, in which some pins are shorted so that the scheme can be traced back to a two-pin measurement.

This reduction to two pins has some merits and some drawbacks. For example, when a large integrated circuit is to be tested with a high number of pins and limited knowledge on the functionality is available, then a transient measurement on the inherent reverse substrate diode of the chip (Fig. 4.15) can serve as a suitable backup methodology. Also, this approach is the only one which enables a thermal transient measurement in a test system equipped with a single driving current source and some sensor sources only. Up-to-date testers frequently offer more sophisticated powering

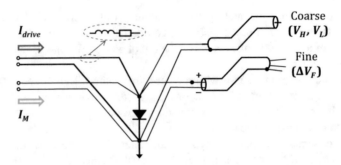

Fig. 5.6 Thermal transient measurement of a diode, powering, and sensing arrangement shown

by a variety of voltage and current sources, but these are not necessarily available at high power levels, for example, in the kiloampere range.

In Fig. 5.6 the typical connections for a thermal transient measurement of a diode are shown. One can recognize again the two current sources and the two data acquisition channels of the previous schemes, but now all these are inevitably connected at the two pins of the device.

The power step is accomplished again by applying and revoking a high I_{drive} current while keeping a constant I_{sense} current on the device. Although the clear separation between the functionalities is now lost, still, a huge change in the power at the input side (left side in the figure) does not rule out a fine detection of a temperature change-induced signal at the output (right side in the figure).

First, as proved in Chap. 4, a diode or other similar device under test has a logarithmic $I_F(V_F)$ characteristic, at least for its internal pn junction.

As many conclusions in this chapter will be built on this characteristic, it is worthwhile to summarize in Table 5.1 the basic equation and some of its consequences.

All equations referred to in this chapter can be looked up either at the place of their original definition or in the table.

The Shockley equation (4.15) implies that if the ratio of I_{drive} and I_{sense} is 10:1, then the V_{Fpn} forward voltage change on the internal junction is just 60 mV at room temperature (300 K) or 80 mV at 127 °C (400 K). Even at 1000:1 ratio, this elevation is just 180 mV on the top of a typical 0.5 V–0.8 V forward voltage of a silicon diode. The voltage growth on the series R_S resistance is higher, but R_S is kept low in power devices. The logarithmic nature of the characteristics ensures a strong *decoupling* between the input and the output sides.

The same statement can be formulated for the external disturbance, caused by the inherent noise from the tester, electromagnetic interference, or other effects. These effects can be modeled as an additional current source at the input side, superposing a further excitation on the device at various frequencies and amplitudes. The measured noise voltage amplitude will be the product of the noise current amplitude and the differential (electric) resistance of the diode.

Table 5.1 The diode equation and some affiliated relationships

Equation	Reference in Chap. 4	Notes, typical values
Diode equation, *Shockley equation* for V_{Fpn}; *voltage drop* on the series resistance		
$V_F = V_{Fpn} + V_{FRs} = mV_T \ln \frac{I_F}{I_0} + I_F R_S$	(4.15)	V_{Fpn} typical value 0.5 V to 0.8 V
Differential (electrical) resistance, $dV_F(I_F,T)/\,dI_F$		
$R_D = R_{Dpn} + R_S = m \cdot V_T \,/\, I_F + R_S$	(4.16)	V_T/I_F value at 300 K 28 Ω @ 1 mA, 28 mΩ @ 1A, etc.
Temperature sensitivity, often used as TSP, $dV_F(I_F,T)/\,dT$		
$S_{VF} = \frac{dV_F}{dT} = \frac{V_F - 3V_T - W_g/q}{T}$	(4.17)	S_{VF} typical value −1 mV/K to −2.5 mV/K
Change in forward voltage at constant temperature		
$V_{F1} - V_{F2} = V_T \cdot \ln(I_1/I_2) + R_S \cdot (I_2 - I_1)$	(4.20)	Difference of $V_T \cdot \ln(I_1/I_2)$ at ratio 10:1, 60 mV; at ratio 100:1, 120 mV; at ratio 1000: 1, 180 mV; etc.

As it was formulated in (4.16), the R_{Dpn} differential resistance of the internal junction is around 30 Ω at 1 mA sensor current and in the mΩ range at 1 A driving current (at room temperature). This low differential value represents the source resistance of the temperature-induced voltage signal towards the data acquisition part of the tester, and ensures the strong decoupling of possible perturbations from the input side. Routing and shielding rules declared in the previous section apply again.

A clear separation of the "force" wiring at the input and the "sense" wiring at the output side prevents adding the voltage drop on the cables to the measured voltage on the device. Due to the inductivity of the cabling, at switching off a high I_{drive} current also an additional voltage spike is superposed on the "force" wiring. These effects are indicated by the resistor and inductance symbols above the corresponding cable section in Fig. 5.6.

In many testers the two measurements at the "sense" side are integrated into the same unit; the scheme of Fig. 5.6 is reduced to six wires, four for applying the driving current and the sensor current and two towards the data acquisition. Separation of the wires of I_{drive} and I_{sense} is essential at higher currents. The resistive and inductive voltage jumps at switching I_{drive} may cause wobbling effects in the sensor current and may cause initial artifacts in the measured signal.

Even in an integrated data acquisition unit, two interrelated measurement functions are to be performed, still of different required accuracy.

The measurement of the full V_F forward voltage is performed as a "coarse" functionality; the V_H voltage on the device at I_{drive} and the V_L voltage at I_{sense} have to be measured on the diode for calculating the power step. In case of a standard silicon diode, the V_H and V_L values are in the 0.5 V–1 V range, and it is sufficient to measure these at the 0.5% accuracy mentioned previously. Transient testing of a complex subassembly in a two-pin arrangement may involve a high voltage measurement, e.g., the driving current may induce a cumulated forward voltage above 100 V on a screen backlight unit built of a chain of LED diodes.

The other function of the data acquisition unit is to record the tiny ΔV_F change of the forward voltage induced by the temperature change, typically in the range of a few millivolts or tens of millivolts.

5.3.3 Discrete Devices with Three or More Pins

Three-pin devices are typically discrete with a single active component. The three accessible leads offer a wide variety of powering and sensing options. Several pins can be assigned as "power ports" where the powering occurs, and one pair of pins can be used as "sense port" where the temperature-sensitive parameter is measured.

Connecting two of their pins, three-pin devices can be "degraded"; in this case the measurement falls back to one of the methodologies used for two-pin devices.

The general thermal transient testing methods of three-pin devices will be presented in Sect. 5.5; specific methods for different device types will be shown in Chap. 6 in a detailed way.

In some designs the powering and sensing of the active element are separated; such devices have four or more pins.

Certain amplifiers and voltage stabilizers are also encapsulated in a package with three (or a few) pins. It has to be noted that the active operation of such devices may impede the fast switching between power levels as it is targeted in thermal transient testing. For relevant transient measurements, the internal complex circuitry is to be thwarted, by controlling it out of its normal operation range. More details on the measurements of amplifiers and voltage stabilizers are presented in Sect. 6.11.2.

5.3.4 Modules with Multiple Active Devices

In modules and in other subassemblies containing more devices, the powering and sensing schemes may became very intricate.

The devices are always coupled in the thermal domain, various thermal transient measurements can be carried out, and different effects of self-heating and transfer heating between devices can be examined.

In the electric domain, the voltage and current constraints are determined by the circuit topology. Independent voltage and current values cannot always be maintained on electrically connected devices.

Figure 5.7 illustrates an arrangement in which the self-heating and transfer heating between transistors in a half bridge module can be measured. The I_{drive} current is always applied on one transistor only. Several sense currents of I_{sense} value enable the transient temperature measurement on both devices.

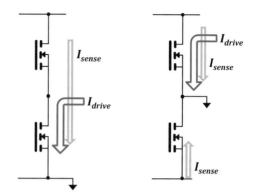

Fig. 5.7 Thermal transient measurement in a half bridge module, (**a**) low-side transistor heated, transient measured on both conducting channels, (**b**) high-side transistor heated, transient measured on the conducting channel of the high-side transistor and on the reverse diode of the low-side transistor

5.4   Thermal Transient Measurement of Two-Pin Devices

As discussed above, several device classes can be measured in the two-pin measurement scheme. In this subsection a simplified diode model will be used to outline the generic rules of this type of testing. A detailed examination of the behavior of real diodes in a thermal transient test will be given in Chap. 6, Sect. 6.1. Other examples of testing more complex semiconductor structures (bipolar transistors, MOSFETs, resistive structures) as two-pin devices will be also shown in that chapter.

The test techniques treated in this section differ in their powering and timing concept. Still, in all methodologies heating and cooling transients follow each other, caused by switching between different current levels. Some aspects on the selection of the current or power levels, especially its relation to reliability testing, will be expanded in Chap. 7, Sect. 7.4.

5.4.1   Continuous Cooling Measurements

Continuous cooling is the most frequently used thermal transient measurement technique. In this methodology the change between power levels occurs only once, as shown in Fig. 5.8.

Instead of using the notation I_{drive} and I_{sense} used throughout this book, several standards define the switching between an I_H *heating current* and an I_M *measurement current*.

Figure 5.9 shows the theoretical measurement arrangement for diodes as recommended by the classical technical literature [112] and the JEDEC thermal testing standard [30].

The scheme is clearly just of symbolic nature; the I_H or I_M current sources cannot be left open in the transient phases when they are not used.

In Fig. 5.10a, b, the practical realization of the switching is shown.

Fig. 5.8 Stepwise change
of the forward current on the
diode under test

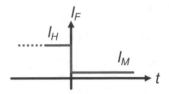

Fig. 5.9 Diode
measurement scheme as
defined by the JESD51-1
standard

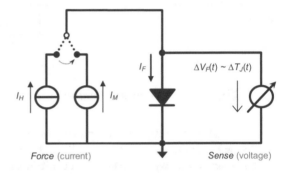

The I_H heating current is provided as the sum of the currents of the I_{drive} and I_{sense} current sources; thus, in the heating period, both currents are applied on the device for an equalization time of appropriate duration (Fig. 5.10a). When the device reaches the "hot" steady state, the drive current is switched off.

At the moment of the switching, the transient recording of the cooling starts at I_M current bias, provided by the I_{sense} current source (Fig. 5.10b).

In this book we shall use the I_{drive} and I_{sense} notation when we refer to the instrumentation of a test, and the I_H and I_M notation when the operating point of the DUT is considered. In most cases the heating current is simply $I_H = I_{drive} + I_{sense}$ and the measurement current is $I_M = I_{sense}$.

In some modules with complex current paths, the heating and sensing currents are composed as sums and differences of several currents from different sources. In modules with simple serial elements like in Fig. 5.7, the sum of I_{drive} and I_{sense} provides the powering on the heated device; heat transfer can be measured on the other devices biased by I_{sense}.

Equation (4.17) indicates that diodes have negative temperature sensitivity at low current level ($S_{VF} \approx -1$ to -2.5 mV/K). The V_F forward voltage at high I_F forward current may have positive or negative thermal coefficient depending on the extent of the effects cumulated into the R_S series resistance. Considering moderate heating currents, the negative thermal coefficient dominates; consequently during the heating period the temperature increases and the forward voltage decreases by a ΔV_{FH} value.

At constant I_H the decrease of ΔV_{FH} results in a proportional decrease in P_H during the heating (Fig. 5.11), $\Delta P_H = \Delta V_{FH} \cdot I_H$.

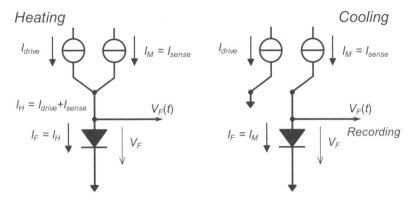

Fig. 5.10 Practical realization of the diode measurement scheme: (**a**) applying heating current; (**b**) cooling at constant measurement current

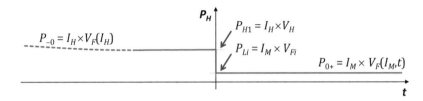

Fig. 5.11 Power change before and during the cooling measurement on a diode

The change of the heating power in time can be mathematically handled with the convolution apparatus, even if the change to lower power occurs before steady state is reached. Nevertheless, in thermal transient testing, the common practice is that I_H is maintained until a "hot" steady state is reached and P_H is considered to be P_{H1}, the value measured just before switching off.

In case of in-line measurements on the production lines, the high-powered steady state cannot be reached, because the production throughput allows only a short time heating and recording only a partial transient. The related problems are exposed in details in [106, 107].

When the test is carried out between two steady-state situations, the power levels can be calculated as

$$P_H = I_H \cdot V_H, \quad P_{Li} = I_M \cdot V_{Fi}, \tag{5.1}$$

and the power step is

$$\Delta P_H = P_H - P_{Li} = I_H \cdot V_H - I_M \cdot V_{Fi}. \tag{5.2}$$

where V_H is the forward voltage of the hot diode biased with the I_H heating current at the time instance of switching and V_{Fi} is the initial value of the forward voltage at the beginning of the cooling transient, when only the small I_M measurement current flows through the diode.

During the cooling period, the V_F forward voltage will grow; it nearly repeats the change that took place during the equalization, but now in the opposite direction. The slightly changing power during cooling can be expressed as

$$P_L = I_M \cdot V_{Fi} + I_M \cdot \Delta V_F(t) \tag{5.3}$$

If the I_M measurement current is significantly lower than the I_H heating current, then the *power change* during cooling can be neglected.

Example 5.1 Thermal Transient Testing of a Medium Power Rectifier Diode

In order to follow the course of a typical thermal transient measurement on a medium power rectifier diode, let us consider powering parameters as $I_H = 20$ A and $I_M = 100$ mA. Assuming a forward voltage of about 0.8 V at I_H and 0.6 V at I_M, the total power step will be (20 A × 0.8 V) − (100 mA × 0.6 V) = 15.94 W.

Suppose that the temperature sensitivity of the diode is −2 mV/°C at the measurement current and the temperature change during the transient is 50 °C in the actual cooling environment. Accordingly, the temperature-induced forward voltage change is 100 mV. The error term (power instability during the cooling) will be $I_M \cdot \Delta V_F(t) = 0.1$ A × 0.1 V = 10 mW which is about 0.06% of the total power change. Thus, at low ΔV_F and low I_F, the change of the power on the device is a secondary order effect only. In such a way, the cooling starts with a nearly perfect *power step*.

After switching off first a sudden large voltage jump occurs on the diode – the V_F forward voltage sinks from the V_H value belonging to the "hot" diode characteristics at I_H to the lower V_{Fi} value belonging to the "hot" diode characteristics at the smaller I_M (Fig. 5.12).

This change, referred to as electrical transient, can be many hundred millivolt for silicon diodes, and for diodes with higher series resistance even more than one volt. The voltage change is usually in the 10 μs–100 μs range, because a large amount of stored diffusion charge has to be removed from the forward-biased *pn* junction (see, e.g., [10]).

After this electric transient, the pure thermal part of the transient can be recorded. The V_F forward voltage slowly increases from its V_{Fi} initial value as the operating point moves from the "hot" to the "cold" diode characteristics, reaching its final, V_{Ff} value, always at I_M bias (Fig. 5.13).

Fig. 5.12 Electrical transient of the hot diode shown in its *I–V* characteristics, switching from the heating current to the measurement current. This chart is an enlarged portion of Fig. 6.1

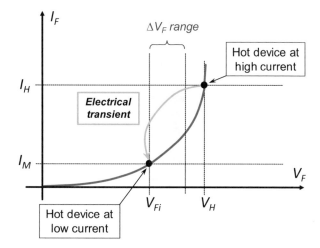

Fig. 5.13 Thermal (cooling) transient of the diode at constant measurement current shown in its *I–V* characteristics. This chart is an enlarged portion of Fig. 6.1

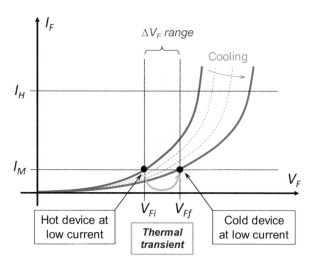

The measurement channel has to be capable to record the complete transit of the V_F voltage between its hot and cold value, in an appropriate "fine" measurement range covering ΔV_F. Still, the fast travel of the signal during the electric transient is not recorded always in its entirety; the signal may jump into the "fine" measurement range from an "out of range" condition. The necessary ΔV_F range of recording is highlighted in Figs. 5.12 and 5.13.

The V_H and V_{Fi} endpoints are to be recorded in the "coarse" range of the data acquisition unit.

For achieving proper signal to noise ratio in the record, it is often proposed to select the smallest "fine" measurement range in which the temperature-induced signal still fits. Selecting larger range enhances quantization noise on the measured signal.

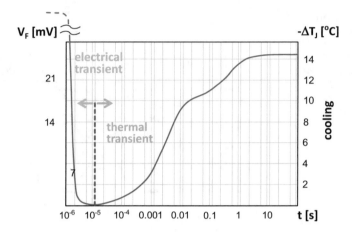

Fig. 5.14 Recorded transient of an actual diode, scaled in voltage (on the left) and in temperature (on the right)

In Fig. 5.14 a recorded transient of a diode is shown; the electric transient finishes at approximately 10 µs.

Various techniques are available to reconstruct the early part of the measured transient which is covered by the electric distortion. Some of these are based on the considerations formulated in Example 2.6 of Chap. 2. A detailed analysis on the initial transient correction techniques is given in Chap. 6 Sect. 6.1.4.

So far in the examples, the I_H and I_M values used in the measurements have been taken for granted. Even in the daily engineering practice, their selection is often based on data sheet values and local traditions.

As thermal testers are typically sensitive, for analyzing the structure integrity, a relatively low power level is sufficient to ensure a few centigrade temperature elevation. However, in case of nonlinearities in the material parameters of the devices, it is safer to accomplish the measurements with the power levels that are foreseen in the normal operation. For reliability testing always the normal operational power has to be ensured. For accelerated reliability testing, higher than normal power has to be applied unto the device. A comparison of results measured at various heating currents is presented in Chap. 6.

A common misconception broadly publicized in the literature is that the measurement current should be kept low in order to avoid the self-heating of the component. In reality, self-heating has only some minor influence in an *absolute* measurement technique, and can be nearly neglected when using a *differential* technique. That means that sensor current values can be freely selected in a very wide range.

One of the advantages of selecting higher measurement currents is that the initial electric transients are smaller and faster. Moreover, at higher measurement current, the noise on the measured signal is lower and external disturbances are better suppressed. The physical background of these effects is explained in depth in Sect.

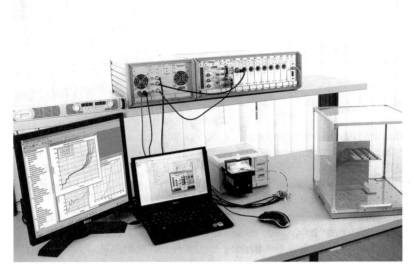

Fig. 5.15 Typical thermal transient measurement arrangement: test equipment, thermostat for TSP calibration or to be used as a cold plate, natural convection test environment, and PC with measurement control and data processing software

5.7. A comparison of results measured at various measurement currents is presented Chap. 6, Sect. 6.1.4.

An actual thermal measurement arrangement with the tester and the equipment realizing the thermal boundary is shown in Fig. 5.15.

5.4.2   Continuous Heating Measurements

In these measurements, only the sensor current is applied to the device for an equalization time of appropriate duration.[1] When the device reaches the "cold" steady state (the equilibrium at $P_{H1} = I_M \cdot V_{F1}$) the drive current will be switched on in addition to the sensor current, and the transient recording of the heating will start.

---

[1] In a strict sense, heating transients can be carried out without an applied measurement current. Still, because of the difficulties of calibration at high heating current, in practical cases heating measurements always are carried out with interlaced cooling transients. The calibration of the heating occurs with fitting towards calibrated cooling transients. In many tester solutions, "single power pulse" and "multiple power pulse" test modes are provided for various test purposes. These test modes play a primary role at thermal quality test applications in production where the measurement time is inherently limited.

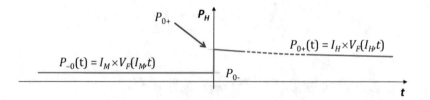

Fig. 5.16 Power change before and during the heating measurement on a diode

Most considerations introduced for the cooling operate in the same way. However, here we experience a larger power change on the hot device during the recording time; approximately the opposite forward voltage change of what we saw at cooling measurements occurs now at high current. The powering is far from an ideal step function, as shown in Fig. 5.16.

This problem can be again mathematically handled, but it is easier to use the cooling method for the thermal measurements of two-pin devices.

Both the cooling and the heating measurements described up to this point correspond to the transient extension of the JEDEC JESD51-1 "static" test method, defined in [30].

5.4.3 Pulsed Measurements

Older tester types use the so-called "dynamic" method, also defined in [30]. The measurement principle is illustrated in Fig. 5.17. In this method the measurement is based on a series of high current pulses for heating and switching back to low current for temperature recording.

The temperature value T_1 corresponding to t_1 time instance is measured such that a power pulse with a duration of t_1 is applied to the device under test. When t_1 time is elapsed, the power pulse is switched off (switching from I_H heating current to the I_M measurement current), and with a short delay t_{MD} (called the measurement delay), the value of the TSP (forward voltage) is measured and through the K-factor is converted to junction temperature. Then, the device under test needs to cool down; the cooling time must be at least as long as t_1. Then, the process is repeated for a longer pulse width t_2, etc.

The test result (referred to as the heating curve) is composed from these responses to individual heating pulses of different length. (Such heating curves obtained by two different commercial testers can be found in the technical literature, e.g. in [113] and [114].) This technique distorts each recorded point by an electric transient, and the data correction problem (back extrapolation of the measured i-th junction temperature at $t_i + t_{MD}$ time instance to the t_i time instance) is also present at every data point of the composed Z_{th} curve. Last but not least, the physical time needed for the measurement by the dynamic test method is by orders of magnitude longer than the length of the real cooling transient measured by the transient extension of the static test method, as demonstrated in [115].

a

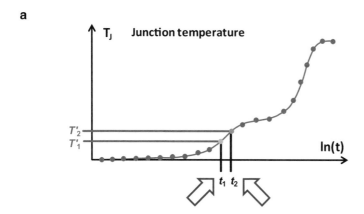

b

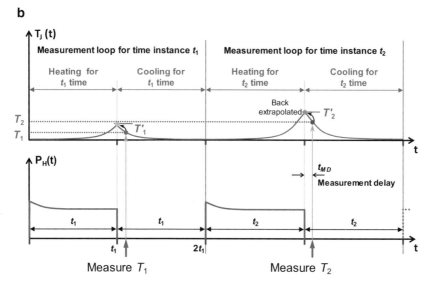

Fig. 5.17 The principle of the JEDEC JESD51-1 dynamic test method: (**b**) the series of heating pulses and corresponding temperature responses, (**a**) heating curve composed of individual temperature values measured at the end of the heating pulses

5.5 Measurement of Three-Pin Devices

Three-pin devices offer certain flexibility, as outlined in Sect. 5.3. Several pins can be assigned as "power ports" where the powering occurs and one pair of pins can be used as "sense port" where the temperature-sensitive parameter is measured.

We shall discuss below the measurement of bipolar junction transistors, but all considerations apply in the same way to MOSFETs, IGBTs, etc.

In the powering and sensing scheme of three-pin devices, one pin is shared between the input and output port. Figure 5.18 presents the common base

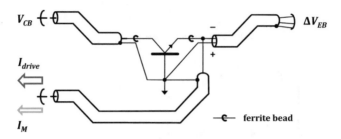

Fig. 5.18 Measurement of a transistor in common base configuration

arrangement, in which the collector-base "port" is of high impedance while the emitter-base "port" is of low impedance. This determines the type of the power sources, which can be used, a V_{CB} voltage is sustained on the collector-base "port" and an I_E current is applied on the emitter-base "port".

Figure 5.18 corresponds to the measurement of an npn bipolar transistor (positive V_{CB} voltage, negative I_E current), but all statements are valid for pnp devices as well, with appropriate signs.

Transistors in the common base setup can be characterized with high gain and high bandwidth; the device with the source units in the tester and the cabling forms a high gain amplifier with feedback, which tends to oscillate at a frequency of many megahertz. This also justifies the extensive use of cooling transients in device characterization. The characteristic amplification parameters of three-pin devices (current gain of bipolar devices, transconductance of MOSFETs) are lower at the operating point of the low measurement current; this makes the DUT-cabling-tester compound less susceptible to oscillations.

Ferrite beads and capacitors are often added to the circuit scheme in order to lower the cutoff frequency.

The measurement scheme in the figure implies that the power on the device can be altered in two obvious ways.

In one mode a fixed V_{CB} voltage is maintained between the collector and emitter, and the I_E emitter current jumps between a higher and a lower level. This mode resembles very much the previous measurement technique introduced for two-pin devices, and is also called *current jump mode*.

Another way for inducing a power step is a sudden change in the collector-emitter voltage, at steady emitter current. This mode is called *voltage jump mode*.

More details about measuring transistors and other three-pin devices will be presented in Sect. 6.1.

5.5.1 Current Jump Measurement of Three-Pin Devices

In *current jump mode*, the power can be calculated as the sum of a $P_{CB} = I_E \cdot V_{CB}$ and a $P_{EB} = I_E \cdot V_{EB}(I_E, T)$ constituent. I_E changes between an I_H and an I_M level.

The measured temperature-induced signal is the $V_{EB}(I_M,T)$ emitter-base voltage at I_M current.

This measurement mode has many advantages. For example, the highest power step can be achieved on a three-pole device in this mode, without special fast high voltage switches in the tester equipment. Due to low power at the low current state, the calibration of the devices can be easily carried out.

On the other hand, the charging or discharging of the diffusion capacitance of the forward-biased emitter-base *pn* junction can be slow; a long electric transient is to be expected at the beginning. Figures 5.12 and 5.13 are valid again for the electric and thermal transient sections.

A significant drawback of the measurements on diode-like devices is the limited power step, constrained by the I_H heating current and the V_F forward voltage. In case of three-pin devices, the high voltage on the collector or drain pin allows a significant increase of the applied power.

In Example 5.1 the diode was powered at $I_H = 20$ A, $I_M = 100$ mA, which resulted in $V_{FH} = 0.8$ V and $V_{FM} = 0.6$ V corresponding forward voltages. The power step was calculated as $\Delta P_D = (20$ A $\times 0.8$ V$) - (100$ mA $\times 0.6$ V$) = 15.94$ W.

Suppose a transistor is measured in the arrangement of Fig. 5.18, in current jump mode with identical I_H and I_M, at $V_{CB} = 20$ V collector-base voltage. The power generated on the collector-base junction also heats the device; further $\Delta P_{CB} = (20$ A $\times 20$ V$) - (100$ mA $\times 20$ V$) = 398$ W adds to the power step.

The error term (power instability during the cooling) calculated for the diode in Example 5.1 was $I_M \cdot \Delta V_F(t) = 0.1$ A $\times 0.1$ V $= 10$ mW which was about 0.06% of the total power change; now this 10 mW is to be related to the 413.94 W total dissipation.

Details of these testing schemes are presented in Sects. 6.2 and 6.3.

5.5.2   Voltage Jump Measurements of Three-Pin Devices

In *voltage jump mode*, a steady I_E current has to be maintained, and the voltage is to be switched between a high and a low V_{CB} value for generating a power step (Fig. 5.18). The $V_{EB}(I_M,T)$ voltage change is the temperature-dependent signal again which can be captured. Using different pins, now powering and sensing are well separated.

This measurement mode offers the best resolution when mapping the fine details of the thermal structure, belonging to shortest time constants.[2]

In voltage jump mode, the I_E emitter current does not change; such the diffusion charge profile in the emitter-base *pn* junction remains nearly unchanged. The electric

---

[2]It was presented in Chap. 4 that in bipolar devices, the major source of the electric transient is the diffusion capacitance.

Fig. 5.19 Electric transient of a transistor after switching V_{CB} from high to low voltage

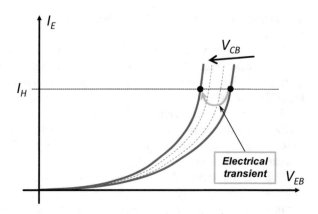

Fig. 5.20 Thermal transient of a transistor after switching V_{CB} from high to low voltage

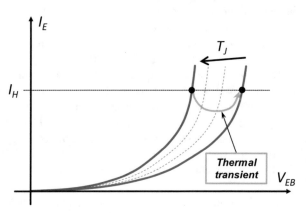

transient is very fast, just a few μs, which is near to the slew rate of the power sources in the best transient testers.

The electric transient is caused by a backlash effect introduced in Sect. 5.2.

The physical background of backlash is at bipolar transistors the Early effect: at higher V_{CB} the base length of the transistor shortens. This effect is small and very fast.

MOSFET devices have a much stronger backlash, especially at short channel devices. Here the cause of the backlash is the channel length modulation of the device, i.e., the increase of the depletion layer width at the drain as the drain voltage is increased.

At both device categories, an elevating V_{CB} voltage causes some decrease in V_{EB} while I_E stays constant (using the indices for bipolar device). Figure 5.19 illustrates the electric transient at a sudden change from higher to lower V_{CB} voltage.

After this electric transient, the cooling can be observed as growing V_{EB} forward voltage at steady I_E current (Fig. 5.20).

The powering in this mode is not limited by the forward voltage on the emitter-base junction either. In case of, e.g., constant $I_H = 2$ A and a V_{CB} jump from 20 V to

2 V, the power step on the collector-base junction is obviously $\Delta P_{CB} = 36$ W. The power change on the emitter-base junction is caused by the backlash effect and is typically insignificant compared to ΔP_{CB}.

The disadvantage of the method is that the calibration of the devices should occur at higher power.

Assuming $V_{EB} = 0.8$ V forward voltage at 2 A, similarly to the previous section, the calibration is to be carried out at $P_M = 2.8$ V $\cdot$ 2 A $= 5.6$ W power. This level already needs a precise calibration arrangement with negligible change in the R_{thJA} junction to ambient thermal resistance (see Sect. 5.6).

Details of measuring various three-pin devices will be presented in Chap. 6.

5.6   The Voltage to Temperature Calibration Process

The calibration process connects a measured temperature-sensitive parameter value to the actual device temperature. Its precision is of high importance because this step influences the overall accuracy of the measurement. All other steps in thermal measurements are practically voltage and current measurements for which instruments of high precision and high time resolution are available. On the contrary, it is easy to perform a bad calibration and undermine the validity of thermal data.

5.6.1   The Temperature-Sensitive Parameter

The most often used parameter for temperature sensing is the forward voltage of a diode-like structure. Previously we defined this parameter as $dV_F(I_F,T)/dT$ and deduced an appropriate formula for it in (4.17) and also in Table 5.1.

In the thermal transient testing, other thermally sensitive parameters can be equally used. In this section we denote the TSP as V_F, but substituting it by other electric parameters, all considerations apply in the same way.

Thermostats produce highly repeatable boundary conditions at various temperature levels. So they can be used for *temperature-voltage calibration*, which means *recording V_F forward voltage* (or other parameter) values *at different component temperatures*.

We have to make a difference between *absolute* and *relative* calibration. The former means mapping the $V_F(I_F,T)$ function, the latter deriving the S_{VF} sensitivity parameter only.

In case of relative calibration, the temperature-related change of the parameter is approximated by

$$V_F(I_M, T_J) \cong V_{F0}(I_M) + S_{VF} \cdot (T_J - T_{J0}). \qquad (5.4)$$

Absolute calibration has a power-dependent error, but is exempt from nonlinearity problems.

For operating a component, some P power needs to be applied on it. Supposing all paths in the heat conduction path arrive at the same temperature-controlled surface named X, one experiences a

$$T_J = P \cdot R_{thJX} + T_X \tag{5.5}$$

junction temperature value, the junction temperature differs from the controlled temperature.

For the differential measurement principle mentioned in Chap. 3 Sect. 3.1.5, only the S_{VF} sensitivity value (or its reciprocal, the *K-factor*) is to be determined in the calibration process; this approach remains valid as long as the temperature to voltage mapping is obviously linear.

The power-dependent error of the absolute calibration procedure expressed in (5.5) has no practical consequences when the calibration occurs at low power and the result is used in cooling measurements, as demonstrated below in Example 5.3. In fact, there is no reason to use the "sensitivity" or "K-factor" quantities in the measurement evaluation; up-to-date testers are able to provide an absolute temperature calibration. The only reason to use a sensitivity value instead is when many semiconductor devices of the same manufacturing batch are to be compared. Typically, the reproducibility of the samples of the same batch is better than the repeatability of the calibration procedure; in such a way, forcing the same sensitivity value to all samples ensures a comparison of higher accuracy.

It has to be noted that during the calibration, the chip surface and the junction are practically at homogeneous T_J temperature due to the low power applied. In the actual test process at the beginning of a cooling transient, the chip surface has a bell-shaped temperature distribution, as illustrated in Fig. 3.17. The recorded T_{VJ} virtual temperature is a weighted average of the temperature distribution over the chip surface. Still, as Eq. (4.21) and Table 4.3 in Sect. 4.3.1 indicate, the hot center is represented with an overwhelmingly high weight in the average; T_{VJ} corresponds to the hottest point. During the cooling transient, the temperature of the chip surface levels out; the measured junction temperature matches the calibrated value.

In this section we focus on the calibration of devices with single energy transport, where all input energy is converted into heat. Further important aspects of the TSP calibration will be expounded in Chap. 6 Sect. 6.10.5.

5.6.2 Calibration on a Cold Plate

Components having a large cooling surface (case, tab) where most heat flows through can be measured and calibrated when mounted on a cold plate (Fig. 5.21). This is actually an easy process:

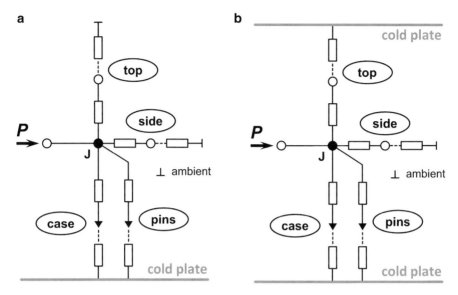

Fig. 5.21 Calibration of a component mounted on single and dual cold plate

- Set the cold plate temperature to several values.
- Record the corresponding V_F forward voltage.

Despite its simplicity we have to be aware of some important issues for doing the calibration correctly:

- As S_{VF} depends on the operating point, always *apply the voltage and/or current on the component to be calibrated*, which *corresponds exactly to the transient measurement* circumstances.
- For cooling measurements, this should be the lower, for heating the higher powering of the two levels used in the differential method.
- For diode-type components, this practically means applying the sensor current only if cooling (or pulsed method) is the selected transient type. The same is true for other devices in current jump mode.
- The power sinking capability of (liquid circulator driven) cold plates is high, even the calibration at high power needed for heating can be carried out easily.

Figure 5.21 shows the simplified model of a packaged device mounted on a single-side and a dual cold plate. The figure reveals that some surfaces of the package are terminated by the ambient (room temperature) rather than by the temperature-controlled plate. The junction temperature is "downscaled" by the appropriate thermal resistances in the thermal circuit; it does not follow exactly the set point of the plate.

Figure 5.22 shows the consequences. The dashed line labeled as "absolute" shows the $R_{thJX} = 0$ case ("chip not packaged, just attached to the cold plate"). The solid SC curve corresponds to the forward voltage of the packaged component

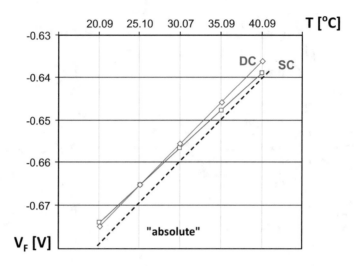

Fig. 5.22 Example on the absolute calibration error

on a single cold plate, while the solid DC curve corresponds to the dual cold plate case. The temperature axis is scaled in the measured *cold plate temperature*. V_F is negative (calibration of an anode-grounded device).

If the cold plate is set to room temperature (25 °C in the figure), then the SC and DC curves coincide. The junction-to-ambient thermal resistance can be derived from the plot, using the $T_J = P \cdot R_{thJA} + T_A$ equation. At all other temperatures, the junction is between the ambient and the cold plate temperature; we underestimate the actual S sensitivity.

Figure 5.21 also hints that if a large portion of the heat leaves through the pins (package with small tab and many pins), also a good thermal contact is needed between the wires feeding the package through the pins and the cold plate.

The calibration process can be manual or automatic, using calibration software. In both cases the following steps have to be carried out:

- Select four or five temperature set points, spanning the whole temperature range of the future measurement.
- Apply the appropriate power on the device.
- Program the lowest temperature value, and wait for t_1 time until the cold plate temperature stabilizes.
- Checking the component voltage wait for t$_2$ time until the voltage stabilizes.

t_1 and t_2 waiting times are needed because the "thermal resistance" elements shown in Fig. 5.21 are complex impedances; their capacitive part expresses heat storage in different material sections. If t_2 is too short, we face the problem shown in Fig. 5.23. The component voltage (dynamic curve labeled "fast" in the figure) follows the cold plate with some delay; at the lowest temperature point, we also see the previous cooling from room temperature.

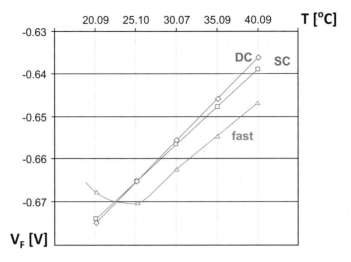

Fig. 5.23 Example on the dynamic calibration error

Even waiting for long t_2 times, we cannot get rid completely of the dynamic effect. On the other hand, even with long t_1 times, we experience small changes also in the cold plate temperature due to the control loop of the liquid circulator. The best practice to minimize these effects is:

- Record the actual temperature of the cold plate after a t_1 stabilization time instead of the programmed set point temperature (as in Fig. 5.23).
- Omit the first voltage reading at lowest temperature.
- Try to read all voltage values at equidistant $t_1 + t_2$ times (in such a way the dynamic curve runs parallel to the SC or DC curve).

Example 5.2: Calibration of a Package on Single and Dual Cold Plate
A flat power package with exposed cooling surface (tab) is calibrated in single-side cold plate setup.

Suppose that in the model of Fig. 5.21 the junction-to-top resistance is approximately 10 K/W and the still air cooling on the top surface corresponds to approximately 20 K/W.

The parallel path composed of the junction-to-case and junction-to-pin resistance is 2 K/W, the spreading in the cold plate is below 0.1 K/W. The wires towards the pins are in good thermal contact with the cold plate. The junction-to-side path is approximately 200 K/W.

Without detailed calculations we can see that nearly 10% of the heat leaves towards the ambient; we underestimate the sensitivity by almost 10% in this setup.

In the dual cold plate setup, we have some heat loss towards the sides only. However, as there is a thin air gap between the two temperature-stabilized metal plates, the air is practically of the same temperature as the cold plate. The side resistance is connected to the cold plate temperature rather than to room temperature.

5.6.3   Calibration in a Closed Chamber or Bath

In case all branches of the heat conductance path end at the same temperature, many problems of the previous section are automatically solved. This is the case when using a closed chamber with thermo-electric heating and cooling, or a liquid bath (Fig. 5.24).

Otherwise, the way of programming temperatures and reading voltages is much the same as in case of cold plate calibrations.

The t_1 and t_2 equalization periods are not fixed values. Instead, we should accept the state as thermal equilibrium if the changes of the chamber temperature remain below a given limit in a given time window, and after reaching this, also the measured voltage does not change more than a predefined limit for a similar time window.

For cold plates and chambers, the actual liquid or plate temperature can be different from that of the sensor used for controlling the system. In real life 3–5% repeatability can be expected for TSP calibration.

As mentioned before, all other steps in thermal tests need some sort of voltage and current measurement instruments which are of high accuracy and stability. Recalibration of these is also only very rarely needed. Calibration thermostats need regular recalibration using stable reference devices.

Fig. 5.24 Calibration of a component in closed chamber or bath

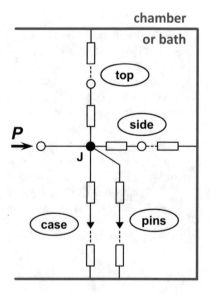

Example 5.3: Calibration of Devices Mounted on Printed Board in a Bath

The natural thermal environment for surface mount devices (SMD) is being mounted on a printed board. For this reason, their thermal transient test typically occurs on test boards similar to the design of Fig. 5.3. The boards represent a reproducible thermal boundary, and their edge connectors provide electrical access to the relevant pins of the device for powering and for measurement.

Several guidelines for thermal testing propose the calibration of board-mounted devices in a temperature-stabilized liquid bath filled with dielectric oil. Although this approach seems to be rational because the dielectric oil ensures homogeneous temperature on the whole board and package surface, the handling of the boards contaminated with sticky, greasy material during the test is rather inconvenient.

Many years of practice confirms that good thermal contact towards the liquid and the electric insulation of the board can be realized in an easy way, without compromise.

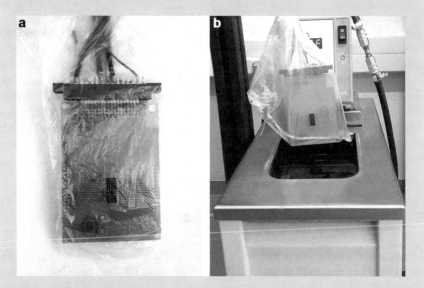

Fig. 5.25 Water-filled liquid circulator with open vessel. Printed board placed in three plastic bags prepared for TSP calibration

A simple test arrangement is presented in Fig. 5.25. After plugging the edge connector of the board into an appropriate socket with the powering and sensing wires, the board is placed into thin plastic bags of appropriate size. Two or three plastic bags layers are recommended in order to avoid leakage problems.

(continued)

Example 5.3 (continued)

Immersing the board in bags into the vessel of a liquid thermostat, an air bubble of continuously changing shape and volume builds around the board. Now, analyzing the influence of this fluctuation on the measured TSP value, we find that it is negligible.

Suppose the junction to ambient thermal resistance from the device to the ambient (which is in case the water in the vessel) changes massively, for example, between 10 K/W and 20 K/W.

If the forward voltage on the device is 0.7 V when it is driven by 10 mA sensor current during the calibration, the resulting power dissipation will be 7 mW. The difference of the junction temperature from the ambient will swing between 70 mK and 140 mK as the size of the air bubble changes. That is, in a calibration temperature sweep of 50 K, the related error is below 0.15% which is negligible compared to other errors of the calibration process.

Still, it has to be emphasized again that the thermal transient measurement has to be carried out in cooling, at the same low measurement current where the calibration occurred.

It is advisable to compose the calibration curve of five or more TSP versus temperature pairs, in order to detect possible nonlinearity. The typical time needed to stabilize the temperature of both the thermostat and the PCB-mounted device is around 20 minutes for each set point in liquid bath. For devices attached to a TEC thermostat plate, 5 minutes is expected. Typical calibration software tools automatically check the stabilization of the thermostat temperature and of the TSP. These values are considered to be stable if their change remains within a predefined limit in a predefined time window, e.g., the change is less than 0.1 °C and 2 mV in 120 seconds.

5.7   Noise and Immunity in the Thermal Measurements

Thermal measurements operate on very low signals; accordingly they are most susceptible to noise and external interference. In a way they are prodigal, investing many watts or kilowatts for heating the thermal response can be just a few millivolts.

The electrical noise and the external interference in thermal measurements have several distinct sources.

One major source of the noise superposed on the actual measured thermally induced signal is the internal noise of the device (shot noise in case of diodes).

The amplitude of the shot noise can be derived from the equations of the thermodynamics. At a certain bandwidth of the measurement, it is

$$i = \sqrt{4kT\Delta f / R_D} \tag{5.6}$$

where k is the Boltzmann constant in joule per kelvin, T is the absolute temperature in kelvin, Δf is the bandwidth of the measurement and R_D is the dynamic electrical resistance of the device under test.

From (4.16), the differential resistance of a diode is inversely proportional to the I_M measurement current, $R_D = mV_T/I_M$, and such

$$i = \sqrt{4kT\Delta f \cdot I_M / mV_T} \tag{5.7}$$

the noise current grows with the square root of the I_M measurement current. This result is physically sound; a larger particle flow involves larger fluctuations.

However, the measurement records the resulting voltage fluctuation, which is

$$v = i \cdot R_D = \sqrt{4kT\Delta f \cdot mV_T / I_M}, \tag{5.8}$$

the internal noise in the forward voltage diminishes with the square root of the measurement current.

Selecting a higher measurement current is also advantageous for lower susceptibility to external interference. The thermal measurements are normally not carried out in a shielding cage; signals from powerful high-frequency sources such as broadcast stations, motors, and similar are demodulated and added to the temperature-related parameter change. The measurement current source of the thermal tester also produces some unavoidable noise current.

The noise components of the internal device noise, the external perturbations, and the tester yield an i_{noise} noise current compound. (The *power* of independent noise generators is to be added up which results in a square root growth of the resulting noise *voltage*.)

However, the testers measure the device voltage, along with the noise voltage. Supposing that the current generators in Fig. 5.10 yield both the necessary bias and the inevitable noise, we can state that the R_D dynamic resistance of the diode shunts the noise current. The resulting noise voltage can be given as

$$v_{noise} = i_{noise} \cdot R_D \tag{5.9}$$

Accordingly, the voltage of all perturbations shows square root decrease with higher measurement current.

Electrical noise calculation plays an important role in tester construction, but it also has importance for a broader audience when selecting the measurement current for thermal transient measurements. A higher measurement current results in lower

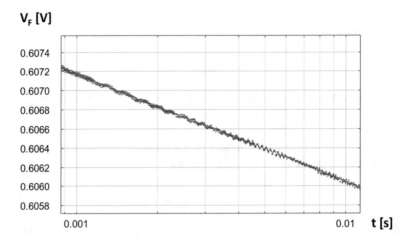

Fig. 5.26 High-frequency interference from an external source on a measured thermal signal

noise accordingly to (5.8). It is also advantageous for shorter electric transients as shown in Fig. 2.68.

A general practice is to select the lowest measurement current with already acceptable noise. For a broad category of silicon devices 1 mA and for LEDs, 10 mA is an acceptable value. Larger modules may need a measurement current of several amperes.

As it can be calculated from (4.16) at $I_F = 1$ mA and $m = 1$ ideality factor, the differential resistance of a pn junction is $R_D = 26\,\Omega$, at $I_F = 10$ mA it is $R_D = 2.6\,\Omega$, and so on. This low source resistance of the thermal measurements effectively shunts not only the noise content from the noise sources discussed so far but also the possible external interference superposed on the wires of the measurement arrangement by capacitive coupling. This makes the shielding of force or sense wiring in Fig. 5.6 actually unnecessary.

On the other hand, this low source resistance makes the loops formed by the wiring of the current source and the device under test (Fig. 5.6) susceptible to electromagnetic induction. External electromagnetic sources can cause observable interference (Fig. 5.26).

This can be prevented by clear separation of the force and sense wiring (Fig. 5.6), and twisting of the wires on the sense side.

Chapter 6
Thermal Transient Measurements on Various Electronic Components

Gábor Farkas, András Poppe, Zoltán Sárkány, and András Vass-Várnai

As exposed in Chap. 5, for thermal transient measurements, one or more heater elements and one or several temperature sensors are needed. In most cases, the heat source is a semiconductor, typically called the "chip" in the literature on system design and the "die" (plural dice, sometimes dies) in works on semiconductor technology and packaging. In discrete devices that have a single die, the powering and sensing can occur on the same device.

The heat source in powered semiconductor chips is in most cases the thin surface layer, the "junction." In many device categories (diodes, insulated gate transistors, thyristors), the heat source that is also serving as the sensor is in fact a shallow pn junction driven into forward operation. In this way, one can record the temperature change of the hottest region in an electronic circuitry.

In this chapter, the thermal transient measurement of the most important device categories will be discussed in detail, including:

- The electrical device characteristics of the device type and its temperature dependence
- The powering options of the device, the ways to achieve the needed high power level in the electrical device characteristics with proper current/voltage settings
- Safe operating area (SOA) of the device at high power

G. Farkas (✉) · Z. Sárkány
Siemens Digital Industry Software STS, Budapest, Hungary
e-mail: Gabor.Farkas@siemens.com

A. Poppe
Siemens Digital Industry Software STS, Budapest, Hungary

Budapest University of Technology and Economics, Budapest, Hungary

A. Vass-Várnai
Siemens Digital Industry Software, Plano, TX, USA

- The electrical stability of the device in the operation point of high power, suppression of potential oscillations, use of external circuitry for setting stable operation point
- The powering options of the device at low power level, the selection of the temperature sensitive parameter (TSP) of the device
- Electric signal distortion, superposed on the thermally induced parameter change measured through TSP
- Device noise superposed on the measured signal

The first, most detailed section in this chapter is Sect. 6.1. This section treats all topics listed above, focusing on the measurement of diodes. All techniques and ascertainments in it are valid for all other devices in which in one of the measurement modes a single pn junction can be used as heater and sensor. These devices include thyristors, MOSFET devices measured on their reverse diode, etc.

Electrical stability problems do not arise in the measurements where only two pins of the device are accessed. Safe operation typically means only a current limit, sometimes supplemented with a recommended derating of the maximum current at higher device package temperatures.

In the subsequent sections, only the topics of particular importance for a device category are discussed.

In Sects. 6.2, 6.3, and 6.5, the measurement techniques of different transistor types such as MOSFET, BJT, and IGBT devices are described. In order to avoid repetitions, the concept of safe operation area is discussed at MOSFETs, and not repeated for other types. Electrical stability questions, oscillation effects are expounded for BJTs, where the base current adds an additional factor to the stability. A separate Sect. 6.4 describes possible external regulator circuit schemes which can set an exact pair of voltage and current resulting in an operating point of predefined power.

Sections 6.7 and 6.8 give a short overview of the thermal transient measurements of passives, such as resistors and capacitors. Section 6.9 is dedicated to point out the differences regarding the measurement of wide bandgap devices.

Section 6.10 deals with the thermal measurements of LED devices in which the input electrical energy is converted into heat and light through two separate transport mechanisms. Finally, some techniques are presented in Sect. 6.11 which enable the thermal transient measurement of integrated circuits.

6.1 Measurement of Diodes

The diode characteristics and its temperature dependence was outlined in Chap. 4, Sect. 4.3. As many conclusions in this chapter will be built on these characteristics, it is worthwhile to summarize in Table 6.1 the basic equations and some of their consequences.

Table 6.1 The diode equation and some affiliated relationships

Equation	Reference in Chap. 4	Notes, typical values
Diode equation, *Shockley equation* for I_F (V_{Fpn})		
$I_F = I_0[\exp(V_{Fpn}/mV_T) - 1]$	(4.10)	V_{Fpn} typical value 0.5 to 0.8 V
Diode equation, *Shockley equation* for V_{Fpn}; *voltage drop* on the series resistance		
$V_F = V_{Fpn} + V_{FRs} = mV_T \ln \frac{I_F}{I_0} + I_F R_S$	(4.15)	V_{Fpn} typical value 0.5 to 0.8 V
Temperature dependence of the I_0 saturation current constant		
$I_0 = G_{sum} \cdot T^{\frac{3}{m}} \cdot \exp\left(-W_g/mkT\right)$	(4.13)	
Differential (electrical) resistance, $dV_F(I_F,T)/dI_F$		
$R_D = R_{Dpn} + R_S = m \cdot V_T/I_F + R_S$	(4.16)	V_T/I_F value at 300 K 28 Ω @ 1 mA, 28 mΩ @ 1 A, etc.
Temperature sensitivity, often used as TSP, $dV_F(I_F,T)/dT$		
$S_{VF} = \frac{dV_F}{dT} = \frac{V_F - 3V_T - W_g/q}{T}$	(4.17)	S_{VF} typical value −1 to −2.5 mV/K
Change in forward voltage at constant temperature		
$V_{F1} - V_{F2} = V_T \cdot \ln(I_1/I_2) + R_S \cdot (I_2 - I_1)$	(4.20)	Difference of $V_T \cdot \ln(I_1/I_2)$ at ratio 10:1 60 mV; 100:1120 mV, 1000: 1180 mV . . .

All equations referred to in this chapter can be looked up either at the place of their original definition or in the table.

6.1.1 Characteristics of the pn junction, Setting an Operation Point and Applying a Power Step

The temperature and current related characteristic equations of a diode were summarized in Eqs. (4.13) and (4.15). Curve tracer instruments plot these characteristics by setting an I_F forward current and displaying the corresponding V_F forward voltage values, while maintaining the temperature of the device. These appliances work in pulsed mode; they apply the current for a short time only in order to avoid the heating of the device and ensure that it stays at the programmed ambient temperature.

One can recognize that curve tracers actually realize a "short transient" operation, accordingly, for a precise recording of the V_F voltage value; the extrapolation scheme back to "zero time" proposed in Sect. 6.1.4 and illustrated in Fig. 6.14 is to be applied.

A simulation experiment can save the difficulties associated with the pulsed measurement of a temperature-dependent diode characteristics. In Example 4.1 of Sect. 4.3.2, the characteristics of a pn junction with a series resistance was generated in a broad current and temperature range.

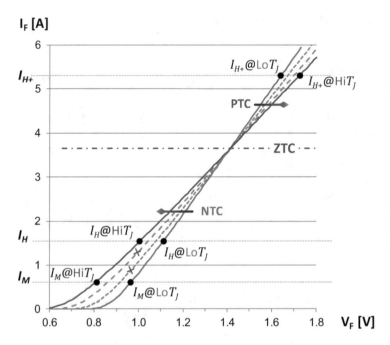

Fig. 6.1 V_F–I_F characteristics of a pn junction, "hot" and "cold" characteristics. Operation points of low I_M and high I_H forward current are shown in the negative temperature coefficient range. A higher I_{H+} heating current in the positive temperature coefficient range is also shown

Figure 6.1 replicates an excerpt of Fig. 4.5 from the example, annotated with characteristic operation points. The figure and Eqs. (4.13) and (4.15) represent *absolute* values, how the V_F forward voltage of the diode can be calculated at an I_F forward current and T_J temperature, and what is the $V_F = (I_F, T_J)$ function.

In many cases, the behavior of the device at small AC signals around an operating point is of interest. A related problem is stability analysis, which investigates how the device returns to its operating point after an external perturbation. In both cases, differential quantities around the operating point are to be calculated.

At small electrical changes, as defined in Eq. (4.16), the *differential resistance* is $R_D = \Delta V_{Fpn}/\Delta I_F = V_T/I_F$ for the internal pn junction.

The temperature coefficient of the device, denoted by S_{VF}, or TC is the total voltage change on the internal pn junction and possible series resistance, divided by the temperature change, $S_{VF} = \Delta V_F/\Delta T_J$. The temperature coefficient can be negative, or positive at an operating point, denoted as NTC and PTC, respectively.

For the internal pn junction, the $S_{VFpn} = \Delta V_{Fpn}/\Delta T_J$ coefficient defined in (4.17) is of negative thermal coefficient; it is -1 to -2.5 mV/K for the typical temperature ranges and current densities.

For comparison, we recall that resistive sensors have a "proportional" (exponential) positive thermal coefficient, in the range of 3000–5000 ppm/K.

In Fig. 6.1, the blue line in the chart corresponds to the diode characteristics at "cold" T_J junction temperature. As stated above, this can be measured only at constant device temperature and by applying an I_F forward current for a short time in order to prevent temperature elevation. Similarly, the red line corresponds to "hot" T_J pulsed characteristics. The plot confirms that at a low I_M *measurement current*, the diode has a negative temperature coefficient.

We already stated in Chap. 4 that the R_S series resistance could be of positive thermal coefficient at higher current. As a consequence, at an intermediate I_H forward current, a negative temperature coefficient can be observed, which turns positive at a higher I_{H+} forward current, with an operating point of zero temperature coefficient between the two regions.

A draft illustrating the temperature change at steady high and low current and the transients between the two was outlined in Sect. 5.4.1.

Figures 5.12 and 5.13 correspond to the lower part of Fig. 6.1; they present the *transient change* between the intermediate I_H and the low I_M current levels, when they are both still in the negative thermal coefficient range of the characteristics.

The figures indicate that the heating transient starts at the $I_H@LoT_J$ point in the characteristics and ends at $I_H@HiT_J$. Similarly, the cooling transient occurs from $I_M@HiT_J$ to $I_M@LoT_J$. The transients at fast switching between current levels occur as shown in Fig. 5.13.

The preferred thermal test method for diodes is to record the *cooling transient*, for several reasons.

First, the power on the diode changes substantially during the heating transient (Fig. 6.2). For example, applying a high I_H current in the negative thermal coefficient region of the characteristics, the power on the diode continuously diminishes and distorts the heating transient. This effect may be handled by the convolution methodology, but the process to provide a valid Z_{th} curve would involve a large mathematical overhead.

Another aspect supporting the advantage of the cooling measurements is that a valid calibration process can be carried out at low P_{0+} power only, at which the difference of the "set" temperature of the calibrator device and the DUT remains negligible (see Sect. 5.6.3).

When the heating current is selected in the I_{H+} range where the diode has already a positive thermal coefficient, the considerations exposed so far remain mostly valid.

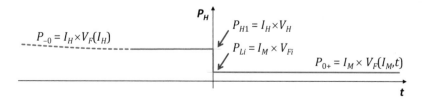

Fig. 6.2 Power change before and during the cooling measurement on a diode. The applied heating current corresponds to a I_H value in the negative thermal coefficient range of the diode characteristics

The power changes in hot state again, but unlike in Fig. 6.2 now the power grows until the diode reaches its hot steady state.

This positive thermal coefficient of the diodes at higher currents is advantageous in the construction of high current applications because this feature enables connecting several diodes in parallel. In a parallel scheme, the devices have approximately the same forward voltage. Because of the positive thermal coefficient of the characteristics, lower current flows through the hotter devices, and high current is diverted toward the cooler ones. However, the positive thermal coefficient of the devices can cause thermal stability problems as exposed in the next section.

In actual circuit realizations, external resistors of positive thermal coefficient can enhance the uniformity of current distribution among more diodes in parallel.

In the case when the characteristics is not taken by a curve tracer of short current pulses, a static V_F–I_F characteristics of the forward biased junction is recorded. The obtained curve depends on the actual T_J junction temperature of the device; consequently, it is influenced by the test environment, represented by an R_{thJA} junction to ambient thermal resistance in the simplest case:

$$T_J = T_A + (V_F \cdot I_F) \cdot R_{thJA}.$$

The thermal resistance of the mounting fixture of the test equipment depends also slightly on the T_A ambient and the T_J junction temperatures.

One such "static" curve can be plotted in Fig. 6.1, for example, starting at $I_M @ LoT_J$ and ending at $I_H @ HiT_J$ through the two points at intermediate temperature and current, denoted by an × mark.

6.1.2   Electrical and Thermal Stability of Diodes in Transient Processes

Figures 6.1 and 5.13 demonstrate the temperature dependence of the diode characteristics.

The figures show that one could select a fixed voltage and use the temperature-dependent current dictated by (4.10) for powering and sensing. As (4.10) and (4.15) are monotonous and dV/dI is always positive, the device is *electrically stable* when driven by any voltage or current. However, a small error in the applied voltage would cause an exponential change in the current according to (4.10). Conversely, applying a fixed current, a small error in its value causes only very flat logarithmic change in the voltage.

Moreover, as one can read in the lower part of Fig. 6.1 along any vertical line, at fixed voltage drive, a growing temperature causes growing current and power, which can be interpreted as a positive *thermal feedback* resulting in thermal runaway.

For all the reasons discussed so far, diodes are typically tested in *current jump* mode (Sect. 5.4.1). Using the continuous test principle, the device is heated by the

$I_H = I_{drive} + I_{sense}$ current until steady state is reached. At the beginning of the transient recording, the tester switches down to a small $I_M = I_{sense}$ current. As the charge in the junction has to be depleted; the travel between the steady voltage values corresponding to (4.15) is performed through an electrical transient, typically in the microsecond range.

The stability problems and the concepts of thermal overload and thermal runaway can be easily analyzed in the simple model of Fig. 2.6 in Sect. 2.3.2 where the thermal system is represented by a single R_{thJA} thermal resistance. The scheme implies for the momentary power and temperature values that

$$T_J(t) = P_{out}(t) \cdot R_{thJA} + T_A, \text{ and such } P_{out}(t) = (T_J(t) - T_A)/R_{thJA} \quad (6.1)$$

at T_A ambient temperature. The transient runs through these momentary values to a steady state where the generated and dissipated power are in balance, $P_{in} = P_{out}$, so the steady junction temperature is $T_{Jeq} = P_{in} \cdot R_{thJA} + T_A$ as stated in Eq. (2.10).

For diodes driven by a constant I_F forward current, the V_F forward voltage is temperature dependent, as defined by Eqs. (4.10), (4.11), (4.12), (4.13), (4.14), and (4.15). The power on the device is partly generated on the internal junction, $P_{inpn} = V_F \cdot I_F$, and partly on the R_S series resistance, $P_{inRs} = R_S \cdot I_F^2$. In many cases, the R_S series resistance reflects special recombination and ambipolar diffusion effects.

In discrete devices, the metallization pattern on the chip also contributes to the series resistance. For such constructions, the temperature of the series resistance is approximately the same as the junction temperature. However, in large modules, the wiring can be at a significantly lower temperature than the semiconductor device. A study on *hot* and *cold* resistor effects is given in [116].

In the usual case of a "hot" series resistor, the T_J junction temperature can be simply calculated from a $P_{in} = P_{inpn} + P_{inRs}$ sum and from the thermal resistance.

Reliability tests can conclude whether the device can withstand a certain T_J temperature for a prolonged time; this defines a T_{Jmax} maximum junction temperature. At silicon devices, T_{Jmax} is typically 150 °C – 175 °C; for wide band gap devices (Sect. 6.9) this limit temperature can be significantly higher. Although steady state can be reached at a junction temperature above T_{Jmax} for short periods, such *thermal overload* causes device failure after a certain time. The T_{Jmax} maximum temperature and the T_A ambient temperature determine the allowed maximum continuous power applied on the device,

$$T_{Jmax} = T_A + P_{inmax} \cdot R_{thJA} \text{ which results in } P_{inmax} = (T_{Jmax} - T_A)/R_{thJA}. \quad (6.2)$$

The maximum power is a major constituent in the safe operating area (SOA) diagrams published in data sheets. In case of diodes, the SOA diagram is mostly limited to specifying the maximum allowed I_F forward current of the device (see later in Fig. 6.17b).

At low ambient temperatures, the limiting factor of the allowed current is the electro-migration effect in the metallization on the die.

The P_{inmax} limit defined in (6.2) is to be abided at constant current powering; for pulsed excitations, the $P_{inmax} = (T_{Jmax} - T_A)/Z_{thJA}(D,t_p)$ formula, defined in Sect. 2.8.2, is valid, allowing a possibly much higher power at low D duty cycle.

The electric and thermal characteristics of diodes can be presented on actual measurements and on simulations. A simulation example on how the thermal dependence of the internal pn junction and the series resistance influence the diode characteristics was shown in Example 4.1.

A complete electrothermal simulation of the devices can be also realized if the scheme is complemented with an appropriate LTSpice model of the thermal environment. A study on such a simulation is presented in Sect. 4.3.3.

Electrothermal simulations are most important in analyzing thermal overload and thermal runaway effects, which are sometimes troublesome to follow experimentally due to their ephemeral nature.

Thermal runaway occurs, if the growth of P_{in} is above the growth of P_{out} at increasing temperature,

$$\Delta P_{in}/\Delta T_J > \Delta P_{out}/\Delta T_J. \tag{6.3}$$

The thermal coefficient of R_S is typically a positive $S_{RS} = \Delta R_S/\Delta T$ value. When the current is well above the ZTC zero thermal coefficient point, then $S_{RS} > |S_{VF}|$, which makes the total $S = \Delta V_F/\Delta T_J = S_{RS} + S_{VF}$ temperature sensitivity of the diode also positive.

In case of a fixed I_F driving current through the internal pn junction and the series resistance,

$$\Delta P_{in}/\Delta T_J = I_F \cdot \Delta V_F/\Delta T_J = I_F \cdot S \tag{6.4}$$

For the heat loss, by the definition of the R_{thJA} junction to ambient thermal resistance, we obtain

$$T_J = T_A + P_{out} \cdot R_{thJA}, \text{ thus } \Delta P_{out}/\Delta T = R_{thJA}. \tag{6.5}$$

From (6.3), (6.4), and (6.5), *the condition of imbalance causing thermal runaway is*

$$I_F \cdot S > R_{thJA}, \tag{6.6}$$

or for pulsed excitations

$$I_F \cdot S > Z_{thJA}(D, t_p). \tag{6.7}$$

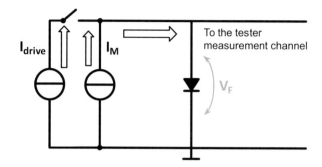

Fig. 6.3 Scheme for the thermal transient measurement of a diode

To the tester measurement channel

I_drive I_M V_F

6.1.3   Test Scheme, Measurement of a Discrete Device

As discussed earlier, testing a diode needs a powering circuitry and a data acquisition hardware. In most cases, the required sudden power change on the pn junction is realized in the aforementioned way, by switching down from a high $I_\mathrm{H} = I_\mathrm{drive} + I_\mathrm{M}$ heating current to a low I_M measurement current level in the simple arrangement of Fig. 6.3.

For illustrating the measurement sequence in a properly educational way, an appropriate specimen is to be selected. A device with a pn junction of fast operation guarantees that the early electric transient (demonstrated in Fig. 5.14) does not cover essential portions of the thermally induced change of the forward voltage. Moreover, a power device has to be chosen where the thermal behavior is relevant and the packaging enables meaningful application of the double interface methodology presented in Sect. 3.5.2. These criteria exclude using, e.g., rectifier diodes, which are optimized for high breakdown voltage and so are typically "slow" or signal diodes where the thermal aspects are irrelevant.

Example 6.1: Measurement of a Fast Medium Power Diode

A good selection which matches all above criteria is the inherent reverse diode (freewheeling diode) of a power MOS transistor device. The whole surface of the device chip is interwoven with a dense web of elementary "trenches," each of them with the vertical structure shown in Fig. 4.14. The reverse diode is formed between the n+ source diffusions and the p substrate (body) as indicated in the figure.

Using such a device in this example enables easy comparison with the thermal measurements of MOS devices presented in Sect. 6.2.

In the actual case, a medium power MOSFET (IRF540) was chosen, in a standard TO220 package. The device can withstand 28 amperes and 100 volts,

(continued)

Example 6.1 (continued)

accordingly significant power can be applied on its channel or on its reverse diode.

Figure 6.4 shows the MOSFET placed into the cavity of a thermoelectric cooler (TEC) appliance. This equipment provides fast and accurate temperature setting due to its well-balanced heating and cooling driven by delicate control. Both heating and cooling are ensured with a Peltier element, the former boosted by separate heating. The temperature difference needed for the Peltier element is provided by fan cooling on its backside; this also serves tempering the speed of the temperature change near to the set point.

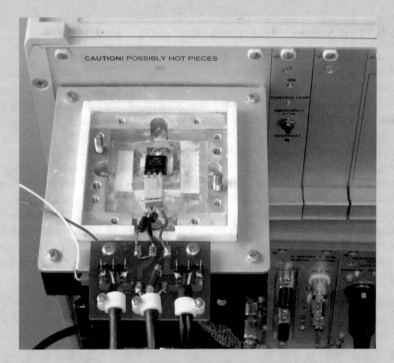

Fig. 6.4 IRF540 MOS transistor in the cavity of the TEC

The rough characteristics of an average reverse diode as presented in the data sheet of the manufacturer (Fig. 6.5) hints that at heating currents below the nominal 28 A rating, the device is in its negative thermal coefficient region.

Figure 6.6 shows the device mounted on cold plate, prepared for transient measurement.

(continued)

Example 6.1 (continued)

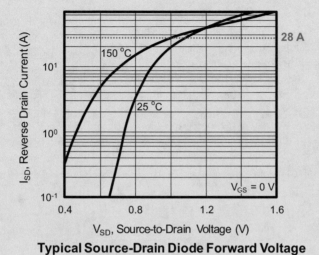

Typical Source-Drain Diode Forward Voltage

Fig. 6.5 Forward characteristics of the reverse diode in the module

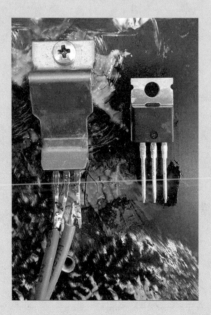

Fig. 6.6 The sample on cold plate, with leads attached

(continued)

Example 6.1 (continued)
Trial Measurements, Calibration, and Transient Test

With trial measurements, we found that a thermal steady state can be reached within 1 minute on the water-cooled cold plate. Applying a few amperes resulted in several centigrade temperature elevation which is more than enough for an adequate structure analysis.

We also tested several levels of measurement current. First, we present thermal transients measured on a dry cold plate surface. The recorded voltage transients of the diode at the heating current of $I_{\text{drive}} = 4$ A and measurement current of $I_M = 100$ μA, 1 mA, 10 mA are shown in Fig. 6.7. We can observe the voltage difference near the theoretical $\Delta V_F \cdot \ln(10) \approx 60$ mV value given in Table 6.1 between the lines belonging to the 1:10 current ratio in the plot. The noise of the transients was acceptable at all measurement currents.

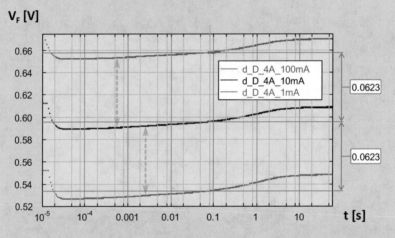

Fig. 6.7 Measured voltage transients of the diode at $I_{\text{drive}} = 4$ A, $I_M = 1$ mA, 10 mA, 100 mA

The calibration, that is, the forward voltage to temperature mapping, was done in a closed thermostat chamber [54]. Although typically this calibration process is done at one selected measurement current, now we carried out the mapping at four current values in a broader range, namely, at 100 μA, 1 mA, 10 mA, and 100 mA. The measurement standards prescribe a temperature range which is larger than the expected temperature change of the device during the transient and at least four or five temperature points in order to reveal eventual nonlinearity of the mapping.

(continued)

Example 6.1 (continued)

The voltage to temperature mapping is shown in Fig. 6.8. It has to be noted that due to its temperature control scheme a thermostat never reaches exactly a set point; instead it swings slightly around it.

For improved accuracy, the best practice is pairing the measured forward voltage with the actual measured stable temperature value of the thermostat, not with the set point. The temperature of the thermostat can be taken for stable if the change of the measured temperature and also the change of the measured forward voltage are below a predefined threshold for a predefined time.

Fitting a regression curve on the measured points, we find that the voltage to temperature mapping for this diode is fairly linear. The sensitivity is -1.94 to -2.5 mV/K in the 100 μA to 100 mA current range (Fig. 6.8). The slight decay of the S_{VF} sensitivity at higher measurement current corresponds to the physical expectations expressed in (4.17) and in Table 4.1. The distance between the curves belonging to different current slowly grows through the change of the $V_T = kT/q$ thermal voltage; as it is plausible from (4.20).

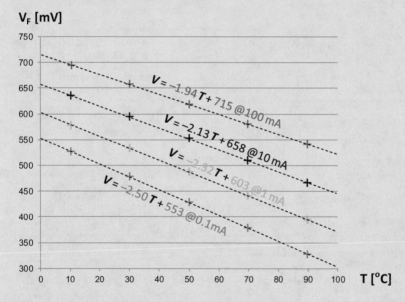

Fig. 6.8 Calibration curves, that is, voltage to temperature mapping at various sense current levels

In case of testing devices of extreme die size (processors, for example) or devices made of wide band gap semiconductors, the voltage to temperature mapping is nonlinear. Current evaluation software products can store measured voltage – temperature pairs of the calibration process. This facilitates

(continued)

Example 6.1 (continued)
calculating the absolute temperature change from the recorded transient voltage change using a polynomial or exponential interpolation [54].

In Fig. 6.9 the voltage transients of Fig. 6.7 are rescaled to a "quasi temperature" plot by the calibration curve. This plot corresponds to the T_J absolute temperature in the time range where the root cause of the voltage change is solely the change of the temperature. Other electric changes, typically at early times, are also shown as "temperature" in the chart. The section in which the voltage change is truly of temperature-induced nature can be ascertained as presented in Sect. 2.12.3.

The elevating section until 30 μs is not of thermal nature; it corresponds to an electric transient caused partly by the charge storage in the diode, partly by the internal relaxation process of the data acquisition channel of the measurement instrument. The correction of this early section is treated below in Sect. 6.1.4.

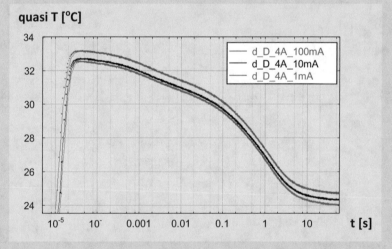

Fig. 6.9 Quasi temperature transients recorded on the reverse diode at 4 A heating current and several measurement current levels; converted from the voltage transients to T_J absolute temperature by the calibration curve

Higher I_M measurement current causes some temperature elevation at the low-powered state; accordingly, some upward shifting can be observed in the measured curves at elevating I_M, as expected. The captured transient curves are crisp with minor noise, which is acceptable at just 9 °C temperature elevation. Noise suppression can be improved by applying higher power step, but this was not necessary in the actual case.

(continued)

Example 6.1 (continued)

A slight improvement can be observed in the early transient behavior at higher sensor current; this improvement is much more expressed when measuring devices of large area.

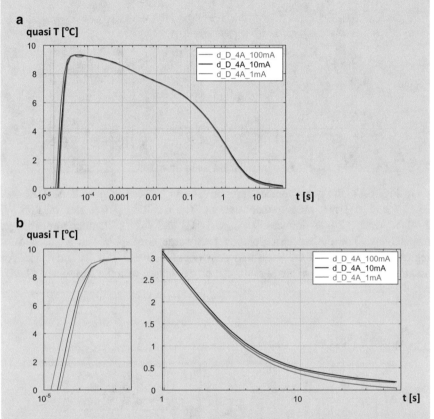

Fig. 6.10 ΔT_J quasi temperature change on the reverse diode, recorded at 4 A heating current and several measurement current levels; (**a**) the whole transient from steady state to steady state shown; (**b**) 10–50 μs and 1–60 s range enlarged. The curves fit perfectly in the 30 μs to 4 s time range

Fitting the converted voltage transients along their longest section, we gain the ΔT_J *temperature change* curves. In Fig. 6.10, it can be observed that the temperature change fits very accurately in the 30 μs to 4 s time range. A small difference can be identified in the electric transient zone, as expected when discharging the diode between dissimilar current levels. A few tenth of centigrade deviation can be observed in the several second time range, which definitely corresponds to a temperature change in the cooling mount, because the longest time constants of the package lay in the 0.1 s order of magnitude.

(continued)

Example 6.1 (continued)

Further analysis on the preferred heating and measurement current magnitudes will be given in Sect. 6.1.8.

Table 6.2 summarizes the measured hot and cold forward voltage and steady power level on the diode at several heating and measurement currents.

Table 6.2 Measured hot and cold forward voltage (V_{FH}, V_{FL}) and steady power level (P_H, P_L) on the diode at several heating and measurement currents

I_H [A]	I_M [A]	V_{FH} [mV]	P_H [mW]	V_{FL} [mV]	P_L [mW]	$\Delta P = P_H - P_L$ [mW]
2	0.001	777	1554	535	0.54	1553
3	0.001	801	2403	535	0.54	2402
4	0.001	819	3276	535	0.54	3275
4	0.01	819	3277	593	5.9	3271
4	0.1	819	3276	656	65.6	3210

In Fig. 6.11, the quasi temperature curves can be seen at $I_F = 2$ A, 3 A, 4 A heating and at 1 mA measurement current. The resulting power step of 1.55 W, 2.40 W, and 3.28 W induces 9 °C, 6.5 °C, and 4.1 °C temperature change, respectively. The dashed extrapolation backwards corresponds to a square root regression curve fit on the measured data points between 100 μs and 1 ms. An assumed starting temperature at $t = 0$ can be calculated with the formula of (2.26).

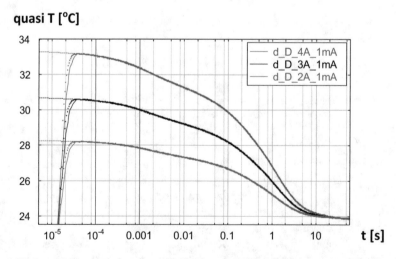

Fig. 6.11 Quasi temperature transients of the reverse diode at 1 mA measurement current and several heating current levels; the horizontal axis is the logarithm of the time, planned square root correction is indicated as dashed lines

6.1.4 Evaluation of the Measurement Results of a Discrete Device with Different Early Transient Correction Concepts

Figure 6.11 shows that many measured points of true temperature are known as early as 30 μs, and after this time the cooling curve is quite flat until 100 μs. It can be supposed that seeking an extremum in the 10–100 μs range and replacing the temperature change by a constant value does not significantly influence the assessed junction to case metrics of the device; or junction to ambient value of the assembly. Moreover, from the time range, it can be judged that the structural information lost by this simplification is related to the semiconductor chip only.

Transient Correction by Minimum Seek
Although the cooling curves have obviously a maximum in Fig. 6.11, the "quasi" Z_{th} curves gained by normalization are always of elevating nature in their thermally induced region. For this reason, this correction scheme is generally called minimum seek.

Cooling transients of the device mounted on dry and on wet cold plate at several heating and measurement currents are presented in Fig. 6.12. In the chart key, the prefixes d_ and w_ indicate dry or wet cold plate boundary, respectively. Solid curves belong to the dry condition and dashed ones belong to the wet condition. The heating current (4 A, 3 A, 2 A) and the measurement current (100 mA, 10 mA, 1 mA) are also encoded into the chart key.

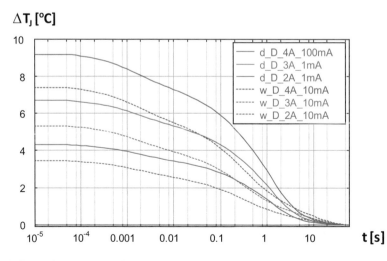

Fig. 6.12 Cooling transients of the device mounted on dry and on wet cold plate at several heating and measurement currents. Minimum seek correction between 5 and 100 μs is applied for correcting the early transient section. Solid lines belong to dry and dashed lines belong to wet boundary condition

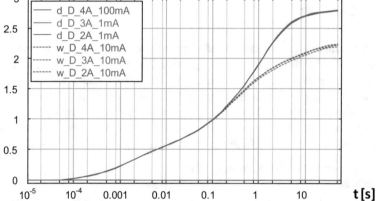

Fig. 6.13 Z_{th} curves at "dry" and "wet" boundary conditions, minimum seek correction between 5 μs and 100 μs applied

Normalizing the cooling curves by the applied power, the Z_{th} curves of Fig. 6.13 are gained. The chart proves that thermal impedance curves of the same arrangement fit perfectly at all heating and measurement currents. Moreover, the perfect fit also verifies that the calibration was executed correctly at each measurement current. Accordingly, with the differential measurement approach introduced in Sect. 3.1.5 and with accurate calibration process at all measurement currents, the widely used concept of "keeping the measurement current low" can be taken aside.

The chart confirms again that a signal is of true thermal nature in the time range where the Z_{th} curves at all heating and sensing currents coincide.

The Z_{th} curves of the plot can be converted to structure functions with the known procedure; these will be shown in Fig. 6.15 along with structure functions generated with an alternative early correction model. The R_{thJC} thermal resistance value is around 1.5 K/W when derived by the TDIM method, based on the point of divergence in the structure functions. A systematic approach for formulating a junction to case thermal resistance value from this divergence is presented in Chap. 7, Sect. 7.1.

Square Root Transient Correction
Those sections in the quasi temperature plot which obey the characteristic square root time dependence of homogeneous heat spreading can be best recognized in a plot in which the horizontal axis is scaled in the square root of the time. Figure 6.14 repeats the quasi temperature transients of Fig. 6.11, rescaled to square root of the time horizontal axis. The measured curves are straight between 100 μs and 1 ms; this signifies the optimal time range for a planned square root correction. The assumed starting temperature of the cooling at $t = 0$ is the intersection of the dashed lines indicating the square root regression and the ordinate.

quasi T [°C]

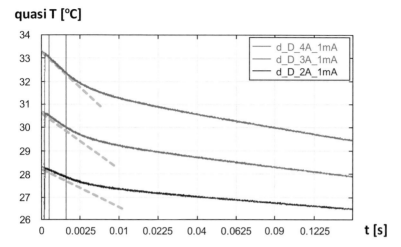

Fig. 6.14 Quasi temperature transients redrawn from Fig. 6.11, with the horizontal axis scaled in square root of time. The assumed starting temperature of the cooling at $t = 0$ is the intersection of the dashed lines indicating the square root regression, and the ordinate

The heat flux in the material stack of the packaged semiconductor remains homogeneous also in the die attach region and in the metal package base within a certain depth. Accordingly, further sections of the temperature change may also follow a square root of time shape. In Fig. 6.14, a corresponding straight section can be recognized between 10 ms and 160 ms indicating a different material layer with homogeneous heat spreading, with a different k_{therm} parameter in Eq. (2.26).

In actual measurements, it is often hard to separate the early transient section and the several sections obeying the square root rule. A feasible way to determine the best correction is to simulate the chip and die attach section in the assembly with approximate geometry and material parameters and comparing the measured and simulated results. A few simulations can already provide clear directions to the operator for applying the appropriate corrections on thermal transient measurements on all devices of similar composition.

Figure 6.15 compares the structure functions which were calculated from measured transients with minimum seek correction (d_D_4A_1mA and w_D_4A_10mA) and the ones with square root correction (d_D_4A_1mA_SQ and w_D_4A_10mA_SQ).

As already noted at Fig. 6.13, the different thermal metrics associated with distances in the structure function are not severely affected with the selection of one or other correction method, the R_{thJC} thermal resistance value is around 1.5 K/W, and the junction to ambient thermal resistances also match. The minimum seek correction inserts a break into the Z_{th} curve that generates a large false time constant. This is reflected in the initial distortion in the structure function, represented as two consecutive steps in the 0–0.5 K/W range. Still, the position of the end of this artificial flat plateau corresponds well to the die attach quality even using the minimum seek concept.

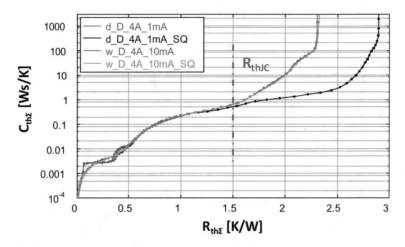

Fig. 6.15 Comparison of structure functions, curves corrected by minimum seek between 10 μs and 100 μs vs. by square root correction between 100 μs and 1 ms

When several measurement results are to be compared, it is reasonable to use minimum seek correction at a fixed time instant in order to restrict the comparison to actual measured data points. The square root correction concept helps restoring the part of the structure covered in the transients by the electric transient. On the other hand, it creates an extrapolated tail toward early time which enlarges small measurement inaccuracies and consequently inserts artificial differences.

The growth of C_{th} and the associated material volumes in Fig. 6.15 indicates that the spreading pattern in the package base corresponds to a truncated pyramid (see Example 7.1 in Sect. 7.1).

6.1.5 Measurement of Diodes in Power Modules

In modules the heating current(s) may flow through several devices, and more chips can be used for sensing, applying a measurement current on each of them. Details of the powering and temperature sensing principles are given in [95].

As an example, we present now the test of a SEMIKRON SKM150GAL12T4 power IGBT module (schematic in Fig. 6.16 and photograph in Fig. 6.18). The test was carried out on the high power diode parallel to the IGBT, which is realized as a separate chip.

The maximum ratings of the diode are 150 ampere and 600 volt. The simple safe operating area (SOA) diagram, that is, a current derating curve, is presented in the data sheet of the manufacturer (Fig. 6.17b). The figure indicates that the chips inside the module have a higher current limit than the internal wiring. The approximate characteristics of the average diode is also presented (Fig. 6.17a).

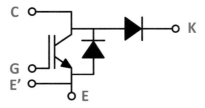

Fig. 6.16 Schematic of a high power module (SKM150GAL12T4). Letters E, E', G and C denote emitter and emitter sense, gate, and collector, respectively. An additional diode can be accessed between C and K

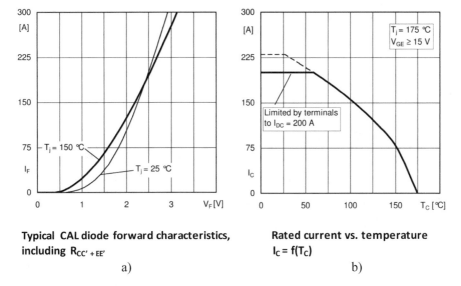

Typical CAL diode forward characteristics, including $R_{CC' + EE'}$

a)

Rated current vs. temperature $I_C = f(T_C)$

b)

Fig. 6.17 (**a**) Temperature-dependent forward characteristics, (**b**) current derating curve of the parallel diode in the module

The parallel diode is produced with the so-called Controlled Axial Lifetime (CAL) technology of SEMIKRON. A special $p+$ $n-$ $n+$ sequence in the doping profile of the chip ensures high breakdown voltage and soft recovery behavior during switch-off, resulting in lesser voltage overshoot at inductive loads.

Thermal transient tests were carried out on the module, first with careful four wire cabling (Fig. 6.19). The heating current was applied with thick cables terminated by compression lugs; the voltage of the diode was tapped at the E' pin on the emitter side. At the collector side, the lug of the sense wire was fixed directly on the force cable lug in order to avoid additional voltage on common cabling sections.

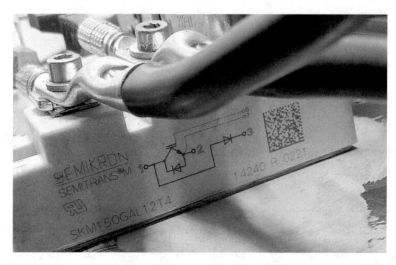

Fig. 6.18 High power module on a water-cooled cold plate, wetted by thermal grease. Force and sense wiring for the parallel diode attached

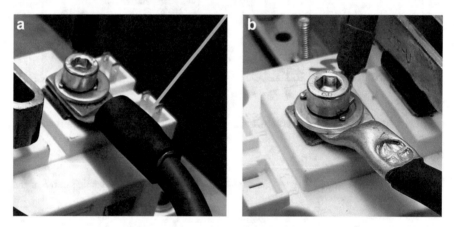

Fig. 6.19 An optimal four wire realization of the cabling for thermal transient measurement. Voltage drop measured on the compression lugs. The blue line in (**a**) indicates the proposed position of the emitter sense connection

6.1.6 Determining the Optimal Current Levels and the Timing for the Thermal Transient Test

In the daily practice of many laboratories, the applied heating and sensing currents are determined by tradition, experienced with low-power discrete devices, or by the poor accuracy of older thermal testers and also by some misconceptions.

Table 6.3 Calculated power on the module with steady I_H heating current, at cold plate temperature of 25 °C

I_H [A]	"Dry" cold plate ΔP [W]	"Wet" cold plate ΔP [W]
21	25.0	25.2
41	58.7	59.6
61	96.9	99.5
81	135.4	142.8

Fig. 6.20 Calibration of the diode at 1 A measurement current between 20 and 60 °C cold plate temperature, in 10 °C steps

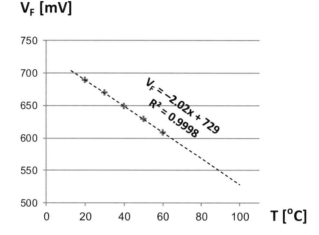

In the case of thermal transient testing, the power that develops on the device depends on one hand on the current applied, on the other hand on the actual thermal boundary. Better cooling lowers the chip temperature, resulting in a higher forward voltage at the same heating current and therefore in higher power on the device.

To test this effect, various I_H currents were applied on the module examined in Sect. 6.1.5 in two different cold plate arrangements, and the heating power (I_H forward current times V_F forward voltage) was calculated, as listed in Table 6.3.

The cooling transient was recorded at different measurement currents after switching off the heating current. We found that although the special doping profile of the CAL diode is advantageous in its intended operation, it influences the early electric transients in an undesirable way. Due to the long charge storage in the differently doped semiconductor regions, after switching between the current levels, the diode voltage jumps up and down, as charge carriers in different layers recombine with different timing.

The forward voltage to diode temperature mapping was constructed for each measurement current; it is shown for $I_F = 1$ A in Fig. 6.20.

The measured crisp, noise-free transients taken at 1 A measurement current are plotted in Fig. 6.21; the peculiar initial electric transients finish before 70 μs. The figure also shows the applicable square root correction between 500 and 5 ms. It can be observed that the applied high power at 80 A already stresses the cooling

quasi T [°C]

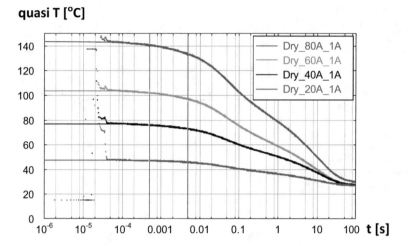

Fig. 6.21 Quasi temperature transients of the parallel diode on dry cold plate, heating current 20–80 A

quasi T [°C]

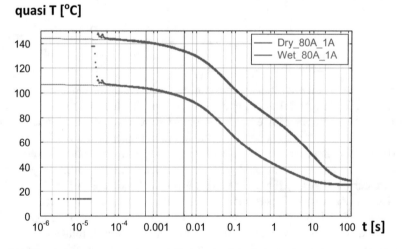

Fig. 6.22 Quasi temperature transients of the freewheeling diode on dry and on wet cold plate

capability of the cold plate; the transient finishes at a few degrees higher cold plate temperature.

Figure 6.22 presents the temperature transients at 80 A driving current at both boundaries. The temperature elevation was $\Delta T_{\text{Jdry}} = 142\ °C - 28\ °C = 114\ °C$ at 81 A/135 W powering on the dry surface. The improved cooling diminished this to $\Delta T_{\text{Jwet}} = 104\ °C - 25\ °C = 79\ °C$ at 81 A/143 W powering on the wet surface.

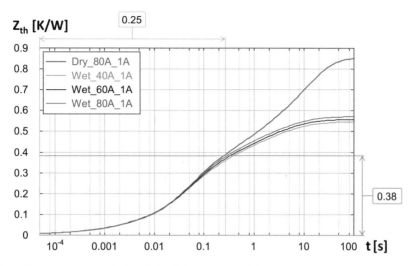

Fig. 6.23 Z_{th} curves at "dry" and "wet" boundary conditions, square root correction between 500 μs and 5 s applied

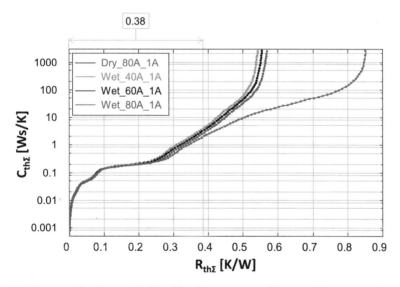

Fig. 6.24 Structure functions at "dry" and "wet" boundary conditions, at different powering

Figure 6.23 presents the Z_{th} curves, measured at correct four wire cabling, with square root correction between 500 μs and 5 s. The curves seem to coincide until 250 ms, in which time range the heat propagates within the module.

The structure functions in Fig. 6.24 offer a sharper view. A straight section between 0.25 and 0.38 K/W reveals concentric heat spreading in the base plate. In this section, already the impeded cooling toward the cold plate at higher applied power can be observed in the separated curves of the structure functions.

The steps of the derivation of standard JEDEC thermal metrics from the structure functions of Fig. 6.24 is given in Sect. 7.1, Example 7.2.

6.1.7   The Impact of Resistive Wiring Segments on Thermal Transients

Despite the high applied heating current, in the previous measurements on the IGBT module, the Z_{th} curves in Fig. 6.23 and the structure functions in Fig. 6.24 nearly coincide along the whole time range of the transient test. It is a general experience in a test carried out with sequentially increasing currents that the Z_{th} curves and structure functions show a characteristic shrinking, named "accordion" effect in [117]. The reason of the effect is that the temperature change in this curves is normalized by the power calculated as the product of the applied current, and the voltage measured on the external pins of the module.

At high heating current, the resistive segments in the current path cause a voltage drop in addition to the forward voltage on the internal semiconductor device. These resistive segments can be attributed to internal wiring in the module, embedded current measuring shunts, neglecting the use of the dedicated E' sense pin in the voltage measurement or simply incorrect cabling. The dissipated power on them is proportional to the square of the I_H current but normally does not participate in the heating of the semiconductor.

At low measurement current, these resistive segments "vanish"; there remains no perceivable voltage drop on them.

Because it is impossible to get power modules with "good" and "wrong" internal wiring, in the next example, we analyze the influence of resistive wiring segments in a semi-experimental way.

Example 6.2: Semi-Experimental Examination of the Impact of Resistive Wiring Segments on Normalized Transients
Suppose we repeat the previous set of measurements on the parallel diode of the IGBT module, but instead of measuring the voltage on the module at the optimal E' and C pins, an additional resistance of $R_S = 2$ mΩ is inserted in the current path between the access points. The $\Delta P_S = I_H^2 \cdot R_S$ term will add at the heating currents $I_H = 21$ A, 41 A, 61 A, and 81 A a ΔP_S power elevation of 2.2 W, 8.4 W, 18.6 W, and 32.8 W to the original ΔP power step values listed in Table 6.3.

The Z_{th} plot in Fig. 6.25 and the structure functions in Fig. 6.26 are calculated from the measured transients of Fig. 6.21 with the elevated power; the shrinking effect is conspicuous.

(continued)

Example 6.2 (continued)

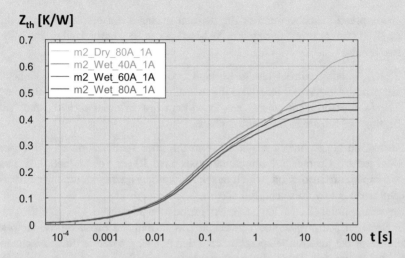

Fig. 6.25 The Z_{th} curves of Fig. 6.23 rescaled with the added ΔP_S power on the resistive section

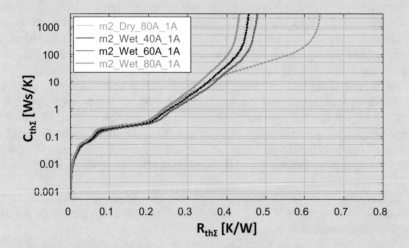

Fig. 6.26 The structure functions of Fig. 6.24 rescaled with the added ΔP_S power on the resistive section

6.1.8   Finding the Valid Heating and Sensing Current Magnitudes

The widespread misconception of the thermal management community regarding the required current levels during transient testing can be summarized in the following two points:

(1) The I_H heating current has to be high enough to induce at least 30–50 °C temperature elevation; otherwise the measurement will be of limited accuracy.
(2) The I_M measurement current has to be kept low in order to avoid the "self-heating" of the die in the cooling phase.

Postulation (1) is simply projected from the poor specification of old thermal transient testers. Current test equipment types have 10 μV voltage and 0.01 centigrade temperature resolution, which enables recording a temperature swipe of 5 or 10 centigrade at low noise and full accuracy.

Regarding point (2), we have to make further considerations.

In Sect. 3.1.5 a measurement principle was formulated based on the temporal difference of the junction temperature recorded at a single data acquisition channel of the test equipment. This technique is compliant with the present thermal transient testing standards [JEDEC] and is advantageous in many ways. First of all, all channel offset and gain errors of the equipment are cancelled-out during the voltage to temperature calibration process, so the only source of error can be the inaccuracy in the thermostat temperature. On the other hand, as the change of the R_{th} thermal resistance is mostly negligible in the span of the measurement, higher I_M just adds a constant (small) temperature shift to the actual transient, not influencing the calculations based on the *change* of the temperature.

The source of the misconception regarding the allowed measurement current is an incorrect interpretation of the standards [24, 29]. As it was seen in Chap. 3, Sect. 3.1.5, if the temperature difference at the junction is derived from the temperature of two separate points *in space*, e.g., from the voltage of the hot junction and some data from a different temperature sensor at a distant point, moreover, these data are measured on different tester channels: then offset and gain errors of the tester channels are added up.

In this case, a temperature surplus caused by higher I_M really adds to the measurement error; this is why some thermal measurement standards demand keeping the measurement current low.

The only prerequisite for a valid one-point measurement is the stability of the R_{th} thermal resistance during the thermal transient measurements. This is generally true; sudden change in R_{th} during a measurement occurs only in special cases, e.g., at phase change in the thermal interface materials. Such an effect may undermine the accuracy of the one-point measurement, but then the methods based on temperature measurements at distant points are equally invalid.

In the next example, a detailed analysis is presented on how the change of the magnitude of the measurement current influences the accuracy of powering during the measurement. It has to be noted that otherwise a higher measurement current is advantageous, because, as it was demonstrated in previous sections, it shortens early electric transients and diminishes device noise superposed on the measured voltage.

Example 6.3: Finding the Powering Error Magnitude During Transient Testing at Various Heating and Measurement Current Magnitudes

As an example, let us calculate the powering error in the transient test of the module presented in Sect. 6.1.5.

The equipment recorded $V_{F1} = 1.77$ V forward voltage on the "hot" device after a longer stabilization period; at $I_{drive} = 80$ A heating current, with $I_M = 1$ A added. This resulted in $P_1 = (I_{drive} + I_M) \cdot V_{F1} = 143.3$ W power on the device just before switching off the heating.

Just after switching to $I_M = 1$ A, the device forward voltage became $V_{F2} = 0.515$ V (140 °C point in Fig. 6.21, conversion by the mapping of Fig. 6.20). Then, in 100 seconds, this forward voltage grew by approximately 165 mV, to $V_{F\infty} = 0.680$ V (25 °C point in the charts).

The cooling started at $P_2 = I_{SENSE} \cdot V_{F2} = 0.5$ W power on the device. This means the power step was $P_1 - P_2 = 142.8$ W, as already shown in Table 6.3.

The only real source of inaccuracy that has an effect other than a constant shift in the junction temperature is the slow change of power during cooling, at the end of which the power grows to $P_\infty = I_M \cdot V_{F\infty} = 0.68$ W, and the change in powering during the full cooling period is $P_\infty - P_2 = 0.18$ W. Related to the $P_1 - P_2$ power step, this is 0.12% error in the measurement.

Representative voltage, temperature, and power values at characteristic time instants are listed in Table 6.4. The data belong to measurements at $I_{drive} = 80$ A heating current and $I_M = 4$ A, 1 A, and 0.1 A measurement current.

Table 6.4 Calculated voltage, temperature, and power of the SEMIKRON module at 80 A heating current and 4 A, 1 A, and 0.1 A measurement current. Row b, typeset in italics is explained in detail in the text

	I_{drive}	I_M	V_{F1}	V_{F2}	$V_{F\infty}$	P_1	P_2	P_∞
	A	A	V	V	W	W	W	W
a	80	4	1.778	0.551	0.716	149.35	2.204	2.864
b	*80*	*1*	*1.770*	*0.515*	*0.680*	*143.37*	*0.515*	*0.680*
c	80	0.1	1.763	0.455	0.620	141.22	0.046	0.062
	$T_1 =$ $P_1 \cdot R_{th}$	$T_2 =$ $P_2 \cdot R_{th}$	$P_1 - P_2$	$T_1 - T_2$	$P_2 - P_\infty$	Error: $(P_\infty - P_2)/(P_1 - P_2)$		
	°C	°C	W	°C	W	%		
a	82.14	1.21	147.15	80.93	−0.660	0.45%		
b	*78.85*	*0.28*	*142.86*	*78.57*	*−0.165*	*0.12%*		
c	77.98	0.03	141.17	77.64	−0.017	0.012%		

(continued)

Example 6.3 (continued)

Figure 6.27 below demonstrates the absolute error of the calibration in the 10–90 °C temperature range, at 4 A, 1 A, and 0.1 A sense currents. The maximum powering error over 80 °C temperature change is 0.2 °C.

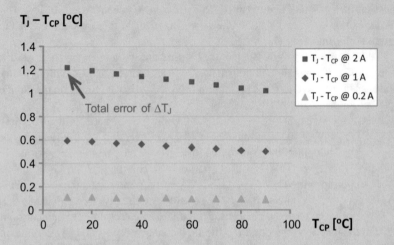

Fig. 6.27 Difference of the actual T_J junction temperature of the measured module and the T_{CP} cold plate temperature during calibration

This example explicitly demonstrates that the requirement formulated in several thermal measurement standards prescribing a low measurement current in order to avoid the self-heating of the device in thermal transient measurements can be overridden if the differential transient measurement approach is applied.

In the case when an exact T_J junction temperature has to be achieved at a high heating current, a corrective step in the cold plate temperature regulation can be performed to reach precisely the targeted temperature. This occurs at the combined thermal, radiometric, and photometric measurements of high-power LED devices, where the optical characterization takes place in the thermal steady state at accurate junction temperature before initiating the cooling (see Sect. 6.10). The R_{th} thermal resistance of the assembly can be determined for this correction by transient measurements before starting the actual test [118].

6.2 Measurement of MOS Devices

Power can be applied on devices having three pins in many different ways, and these devices also offer an abundance of temperature-sensitive parameters for transient temperature measurement.

These devices have typically a "control" type pin, which can be regulated at zero power (MOSFET, IGBT) or low power (BJT). Two other pins are constructed such that they allow to flow a certain current at certain voltage, as governed by the control pin. The device characteristics determine the power dissipation on the device and the selection of a temperature-dependent operating point for temperature sensing.

In this section, powering options and temperature-sensitive parameters will be defined for MOSFET devices. The treatment of the powering is completed with the definition of the safe operation area in their characteristics and with thermal stability criteria. The detailed investigation of the electrical stability is postponed to the exploration of bipolar junction transistors in Sect. 6.3 because the finite control current of these devices makes the analysis more complex. Still, all statements on the stability of bipolar devices apply equally to insulated gate transistors.

6.2.1 Powering and Temperature Sensing Options of MOS Devices

MOSFETs devices are tested in frequent cases like two pin devices of the previous section. Power MOSFETs have an inherent reverse diode (body diode) between their source and drain, as highlighted in Figs. 4.14 and 4.15. Thermal testing on this reverse diode (Fig. 6.28) is very simple but suffers from the same limitations regarding low power level and the length of the electric transient as other diodes discussed before.

The circuit scheme in the figure is based on the assumption that the typical enhancement mode (normally-off) silicon MOSFET device is in off state at zero gate-source voltage, lacking an inversion channel.

Depletion mode (normally-on) devices can also be measured on their reverse diode, but an external V_{GS} source has to be used to switch off the conduction path

Fig. 6.28 MOSFET powered and temperature measured on the reverse diode

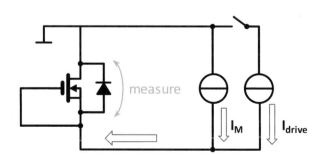

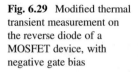

Fig. 6.29 Modified thermal transient measurement on the reverse diode of a MOSFET device, with negative gate bias

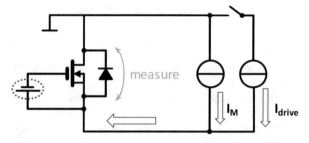

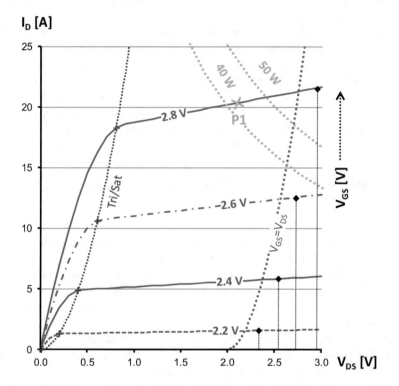

Fig. 6.30 Output characteristics of a MOSFET, triode, and saturation regions of operation are shown. The $V_{GS} = V_{DS}$ ("MOS diode") curve and constant power curves are constructed

parallel to the reverse diode (Fig. 6.29). The same scheme can be used for the measurement of silicon carbide MOSFETs, as discussed in Sect. 6.9.1.

In all other relevant powering modes of a MOSFET, the powering occurs on the conducting channel. For understanding these modes, the different regions of the device output characteristics have to be explored. For generating these charts, the LTSpice model constructed in Fig. 4.11 will be used again.

Figure 6.30 recapitulates the output characteristics of the device presented in Fig. 4.12, now in a narrower V_{GS} voltage range just above the $V_{th} = 2$ V threshold voltage.

The characteristic operation modes defined in Sect. 4.4 can be easily identified, the triode (ohmic, linear) region corresponds to the fast elevation of the curves until the V_{DS_sat} separation curve, denoted as Tri/Sat, and then the slope of the characteristic lines in the saturation (quadratic) region is related to the λ channel-length modulation parameter.

The hyperbolic dotted green curves represent constant $P_D = I_D \cdot V_{DS}$ values; two curves belonging to 40 W and 50 W power are plotted in the chart.

The figure also offers a twofold interpretation for the channel resistance. In a static approach, it is the V_{DS}/I_D quantity in the triode region of the curves. This R_{DSON} value decreases with growing V_{GS} and increases along a constant V_{GS} curve when the elevating V_{DS} moves toward the V_{DS_sat} separation curve. In a dynamic approach, an R_{DSON} quantity can be interpreted as the inverse of the slope of the constant V_{GS} curves near to the origin.

The g_m transconductance grows at higher current and higher gate voltage. This quantity corresponds to the distance of curves in the plot. The current grows from 1 to 5 A between the curves of 2.2 and 2.4 V, resulting in $g_m = 4\ A/0.2\ V = 20$ siemens, while the distance between the curves of 2.6 and 2.8 V results in $(20\ A - 12\ A) = 8\ A$ and corresponds to $8\ A/0.2\ V = 40$ siemens.

The growth of the g_m transconductance is of eminent importance in transient measurements. As expounded in Sect. 5.5, high amplification in the complex circuitry composed of the device under test, the tester, and the cabling makes the arrangement susceptible to oscillations. It is a frequent observation that oscillations can be observed on the early sections of a thermal transient, which decays after a short time. This phenomenon indicates that at high power the device had been oscillating, and the power step associated with the switching to low current is false.

Figure 6.31 summarizes the results of the simulation on the circuit model of Fig. 4.11, at 25 °C "cold" and 125 °C "hot" temperature. Blue curves represent the "cold" and red ones the "hot" characteristics of the device.

The chart can be used for examining the powering and sensing in various test schemes. It can be observed generally that in a transient test in which the power step is induced by switching from a higher current to a lower one (current jump), the curves imply positive thermal coefficient at high current, that is, the V_{DS} voltage increases at growing temperature. At low current, negative thermal coefficient can be observed; V_{DS} diminishes at growing temperature. Zero temperature coefficient belongs approximately to the $V_{GS} = 2.4$ V curve.

However, for similar reasons, growing temperature causes the current and the dissipation to shrink at fixed drain-source voltage and high current, resulting in negative thermal coefficient. At fixed V_{DS} and low current, the "hot" curve runs above the "cold" one; positive thermal coefficient can be observed.

Figure 6.32 presents the transfer characteristics of the MOSFET, the dependence of the I_D drain current from the V_{GS} gate-source voltage, at high V_{DS}, that is, in the saturation region. This chart can be constructed from the points of Fig. 6.31 or can be gained with a separate LTSpice simulation. Again, various domains of positive thermal coefficient, negative thermal coefficient, and a zero thermal coefficient

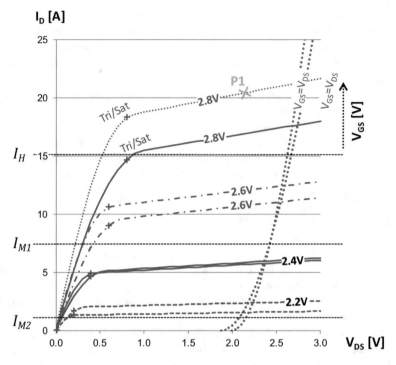

Fig. 6.31 Output characteristics of a MOSFET, blue curves represent the cold and red curves the hot characteristics. Positive thermal coefficient of the curves can be observed at high I_D current and negative thermal coefficient at low I_D current

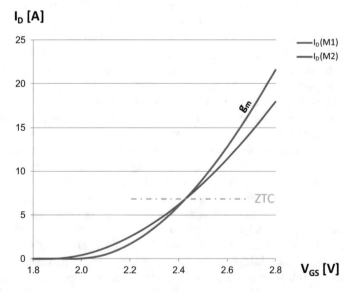

Fig. 6.32 Transfer characteristics of the MOSFET device, at 25 °C temperature (blue curve) and 125 °C temperature (red curve)

Fig. 6.33 MOSFET
powered and temperature
measured as "MOS diode"

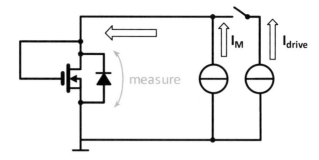

Fig. 6.34 MOSFET
powered at constant V_{GS}
voltage, temperature
measured on the conducting
channel

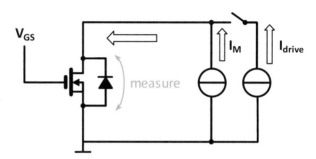

point can be observed. It can be stated again that the g_m transfer conductance (dI_D/dV_{GS} slope of the curve) grows at higher current.

A popular thermal transient measurement technique is converting the MOSFET into a so-called "MOS diode" by shorting its drain and gate (Fig. 6.33).

The $V_{DS}(I_D)$ function of the equivalent two-lead device can be constructed by connecting the $V_{DS} = V_{GS}$ points in the output characteristics (dotted blue curve in Fig. 6.30, dotted blue and red curves belonging to different temperatures in Fig. 6.31). As the typical threshold voltage of power MOSFETs is 3–5 V, this arrangement yields 6–10 times higher power than the reverse diode measurement of $V_F = 0.6$–0.8 V at the same current. This quadratic curve is much flatter than the exponential one of real diodes, so making an estimation on the $P_D = V_F \cdot I_F$ power in advance is harder.

Figure 6.31 implies that in the "MOS diode" measurements after a high I_H heating current, an appropriate low I_{M2} measurement current has to be in the cooling phase of a transient test, where the device has a high negative thermal coefficient, far from the zero thermal coefficient point at 2.4 V.

A more flexible power programming can be achieved by proper voltage control on the gate (Fig. 6.34). In the simplest case, a constant voltage is applied, and the resulting two-lead equivalent scheme behaves corresponding to one curve at constant gate-source voltage in the characteristics of Figs. 6.30 and 6.31.

It can be observed that in the triode domain the device with fixed V_{GS} can be driven by I_D current or V_{DS} voltage in the same way; small alterations in the excitation cause only minor deviation in the response. On the flat curves of the saturation domain, an applied V_{DS} voltage determines the drain current

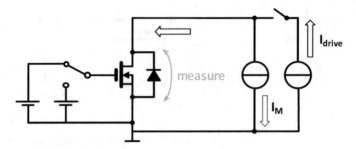

Fig. 6.35 MOSFET powered at constant V_{GS} voltage, temperature measured on the reverse diode

unambiguously, but small deviations in I_D and the device temperature would cause unpredictable change in the voltage.

This powering mode is of high importance mode in Active Power Cycling tests. Devices which normally work in switching applications are characterized by their "conduction loss"; they dissipate on their R_{DSON} channel resistance at high V_{GS} and low V_{DS}. The power on the device can be calculated by the $P_D = R_{DSON} \cdot I_D^2$ formula.

The temperature measurement can also occur on the conducting channel. Figure 6.31 indicates that at low current, in the triode domain, the $S = dV_{DS}/dT$ temperature sensitivity is proportional to the measurement current. At the low powered state of the thermal transient, a high sensitivity can be achieved only at a suitable high I_{M1} measurement current. The $\Delta P_D = R_{DSON} \cdot (I_H^2 - I_{M1}^2)$ power step can be suitably high at relatively low current ratio between I_H and I_{M1}, e.g., a current ratio of 1:5 results in a power ratio of 1:25.

In many cases, using the conducting channel as temperature sensor can be troublesome; at high current devices with low R_{DSON}, also I_{M1} has to be high for appropriate sensitivity. As shown in Chap 4 Sects. 4.6 and 4.7, the use of the channel resistance as TSP is limited for SiC MOSFET devices because of their non-monotonous temperature dependence and for GaN HEMT devices because the time variant current collapse effect.

Thermal testers not necessarily can yield a measurement current of many amperes for using the channel resistance as TSP. For this reason, a "switch-over" measurement mode is often preferred, in which the MOSFET is powered at constant V_{GS} voltage on the channel, and the temperature is measured on the reverse diode as shown in Fig. 6.35. This methodology is proposed for power MOSFET modules in the guidelines of [25].

A further elevation in power can be reached using a feedback scheme often called "magnified diode." This will be presented in detail in Sect. 6.3 on bipolar devices, but it is suited for insulated gate devices, too.

The best powering scheme can be achieved if one controls the gate by a more intelligent concept. With some exaggeration, one can add an *analog computer* to the test arrangement which "calculates" the preprogrammed power needed. As it will be shown in Sect. 6.3, the "computer" is an analog control loop resembling the scheme of Fig. 6.36 and comprises just a few components.

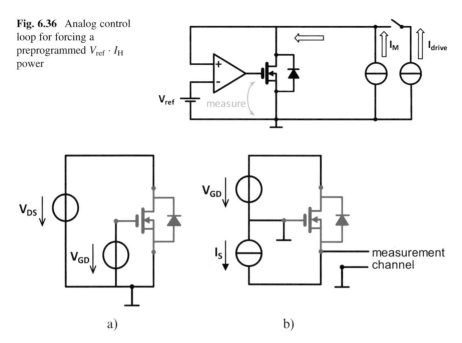

Fig. 6.36 Analog control loop for forcing a preprogrammed $V_{\text{ref}} \cdot I_{\text{H}}$ power

Fig. 6.37 MOSFET powered in (**a**) "common source" (**b**) "common gate" arrangement

All test schemes presented so far have one thing in common: the change of power is induced by a sudden change in the current applied at a certain point in the circuit. These methods can be described as *current jump* powering.

In a broader approach, three pin devices can have many appropriate powering modes. In each case, some pins are dedicated to device control and other ones to applying current and/or voltage for powering.

The control has to appear on two pins, which can be interpreted as "input port." The power has to be applied on other two pins, which can be called "output port." The devices have three pins; accordingly, one of the pins has to be shared between the ports.

Figure 6.37 shows two of the three possible configurations. Version (a) bears the obvious name "common source"; (b) is called "common gate" arrangement.

In the common gate version, there is some freedom in selecting the current through the device and the voltage applied on the device, $P = I_{\text{S}} \cdot (V_{\text{DG}} + V_{\text{GS}})$, where V_{DG} is dictated by the voltage source, but V_{GS} is act of God, or rather of the technologist at the implanter creating the doping profiles of the device. The power step can be realized either by current jump, as before, changing suddenly I_{S}, or by *voltage jump* where V_{DG} has to be changed.

The temperature sensor is present again; all curves in Fig. 6.30 are temperature dependent.

Electrical stability is not guaranteed in all arrangements, due to internal frequency-dependent feedback in the device, as detailed in the next sections.

More stability problems can be experienced in the "common source" arrangement. As the gate is insulated, it has to be driven by a voltage generator. Figure 6.30 implies that a current can be applied on the drain easily in the triode domain where the MOSFET behaves as a controlled resistor.

In case one needs higher voltage for higher power, then the flat section in saturation mode is reached, where the MOSFET behaves like a current generator. As there is a contradiction between the current of the external generator and the current dictated by the saturated device, the resulting voltage is unpredictable and temperature dependent.

For this reason, curve tracer equipment which produces the output characteristics similar to Fig. 6.30 typically uses voltage generators in both positions of Fig. 6.37a. The temperature dependence of the curves can be well seen also in such plots.

6.2.2 Safe Operation Limits Affecting the Thermal Measurement

Figure 6.30 also helps creating the chart of the safe operation area (SOA). This chart is mainly based on lifetime tests carried out at different device temperature and voltage and current levels. These tests are closely related to the thermal transient tests treated throughout this chapter.

In the most general form of the SOA, pulsed excitation is assumed, as defined in Chap. 2, Sect. 2.8.2. All calculations which yield the chart are related to the T_A ambient temperature and the junction to ambient $Z_{thJA}(D, t_p)$ pulse thermal resistance. These latter can be provided by transient tests on a complete assembly and by thermal transient simulations of device models of detailed geometry and material parameters. The SOA belonging to a static R_{thJA} thermal resistance belongs to the duty cycle $D = 1$.

The output characteristics in Fig. 6.30 already provides useful data for the SOA chart, namely, the initial section of the curves defines constraints between voltage and current through R_{DSON}, and a couple of the $P = V_{DS} \cdot I_D$ power limit curves can also be inherited. It was shown in Fig. 4.13 that all these curves appear as straight lines in a log-log output characteristics representation.

Adding further constraints as maximum device current, breakdown voltage, and thermal instability limit, the five limit-lines defining the SOA diagram (Fig. 6.38) are the following:

1. R_{DSON} limit (orange)
2. Current limit (black)
3. Thermal overload limit (green)
4. Thermal instability limit (red)
5. Breakdown voltage limit (blue)

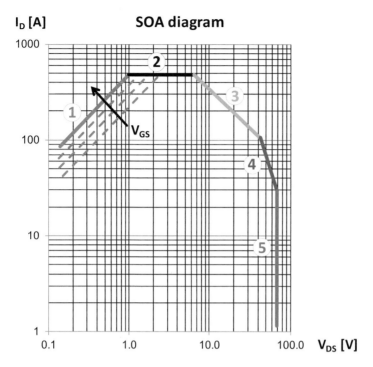

Fig. 6.38 SOA diagram of a power MOSFET

R_{DSON} **limit**   Strictly taken, this line is not related to the safe operation of the device; this is a direct plot of a portion of the device characteristics in the triode/ohmic mode of the MOSFET operation. The curves shown belong to the different V_{GS} gate-source values, as already presented in Fig. 6.30 and in the log-log chart of Fig. 4.13. The solid orange line belongs to the maximum allowable V_{GS} voltage:

$$I_{\text{DS}} = \frac{V_{\text{DS}}}{R_{\text{DSON}}(@V_{\text{GSmax}}, T_{\text{Jmax}})}$$

For lower V_{GS} voltages, this limit-line sinks lower since R_{DSON} increases. At lower T_{A} ambient temperature, T_{J} also shrinks in a test arrangement of a certain R_{thJA} junction to ambient thermal resistance, such this limit line moves up again.

Current limit   This line expresses the maximum current which the internal chip of a power device or a package or module can handle. In up-to-date devices, the chip limit is often higher than the package limit. This value is accordingly package dependent but does not significantly change when application conditions such as ambient temperature vary.

Thermal overload limit   This limit-line is calculated from the maximum power which the system is allowed to generate to reach a stable junction temperature T_{Jmax}

in thermal equilibrium (at a given T_A). It is obvious that the cooling concept of the system and therefore thermal variables will strongly impact this limit-line. These variables include the ambient temperature T_A and the pulse thermal impedance $Z_{th}(D,t_p)$ at duty cycle $D = t_p/T$ and pulse length t_p, as defined in Sect. 2.8.2.

The limit-line for a pulsed application (which is typical in active power cycling tests or DC/DC converters) can be calculated by assuming thermal equilibrium of the generated and dissipated power, $P_{diss} = P_{gen}$.

Applying higher P_{gen}, the system still remains in thermal equilibrium, but the junction temperature is higher than the absolute rating of the device resulting in poor reliability and shorter lifetime.

In case one can rely on the data sheet of a packaged semiconductor only, then T_J can be approximated using the junction to case pulse thermal resistance:

$$T_J = T_A + Z_{thJC}(D, t_p) \cdot P_{diss}$$

However, this approach neglects the energy storage in the environment around the device in the appliance or the test bench. A more accurate result can be gained using a measured pulse thermal resistance chart as defined in Sect. 2.8.2,

$$T_J = T_A + Z_{thJA}(D, t_p) \cdot P_{diss}.$$

The familiar hyperbolic $I_{DS}(V_{DS})$ limit curve is given by

$$I_{DS} = \frac{T_{Jmax} - T_A}{Z_{thJA}(D, t_p) \cdot V_{DS}}. \tag{6.8}$$

Thermal instability limit Following the maximum-power limit-line reveals a point at which the slope of the limit-line changes. This point indicates the start of the thermal instability (thermal runaway) limit-line.

To understand the origin of this limitation, it is necessary to consider the criterion for thermal instability. A MOSFET (or generally a system) is considered to be thermally unstable in case the power generation P_{gen} rises faster than the power dissipation P_{diss} over temperature,

$$\Delta P_{gen}/\Delta T > \Delta P_{diss}/\Delta T. \tag{6.9}$$

In this condition, the temperature of the system is not stable, and the system is not in thermal equilibrium, unlike in case of the thermal overload limit-line.

As $P_{gen} = V_{DS} \cdot I_D$ and $P_{diss} = (T_J - T_A)/Z_{thJC}(D,t_p)$, from data sheet, or rather $P_{diss} = (T_J - T_A)/Z_{thJA}(D,t_p)$ from thermal transient test, (6.9) can be rearranged as:

$$V_{DS} \cdot \frac{dI_D}{dT} > \frac{1}{Z_{thJA}(D, t_p)} \tag{6.10}$$

The thermal instability limit is obviously reached at higher V_{DS} voltage.

The $S = dI_D/dT$ thermal sensitivity changes with the applied drain-source voltage and can be read along a vertical V_{DS} line in Fig. 6.31. The corresponding red line of ④ can be constructed in an actual MOSFET characteristics.

Equation (6.10) indicates that thermal instability can arise only at positive S sensitivity values, that is, *below* the ZTC line in Fig. 6.32 with fixed V_{GS} and *above* the ZTC line at fixed current. An example of the construction of the thermal runaway limit in the transfer characteristics of an actual MOSFET type is given in [119].

Thermal instability can occur in high current MOSFET devices of low R_{DSON} value already at the typical voltages applied on them in the "MOS diode" style measurements. These devices are to be powered on the conducting channel or on the reverse diode in thermal transient tests. Nonlinearity and thermal stability issues related to the physical background of R_{DSON} are presented in depth in [178, 179].

Breakdown Voltage Limit

This vertical line defines the maximum allowable V_{DS} voltage on the device. This limit is related to the avalanche effect which starts when high voltage occurs between the pinch-off of the channel and the drain in Fig. 4.10b.

The multiplication of the current involved in the avalanche effect does not necessarily destroy the device, but it also has a positive thermal coefficient and results in constantly growing current at fixed high V_{DS} voltage, initiating a thermal runaway mechanism.

Example 6.4: Comparison of Thermal Transient Measurements on Reverse Diode and "MOS Diode"

In Example 6.1, the thermal transient testing of an IRF540 medium power MOSFET was presented. The measurement occurred on the reverse diode of the sample with the powering and sensing arrangement of Fig. 6.28.

After carrying out the test, the whole procedure including calibration, thermal transient measurements, and result evaluation was repeated without changing the mechanical arrangement, but setting all heating and measurement current values to negative. This turns the measurement scheme of Fig. 6.28 into the scheme of Fig. 6.33. If negative current values are not available in the actual tester equipment, just the leads toward the sample with connected pins have to be reversed, before applying I_{drive} and I_M.

The calibration procedure produced a voltage to temperature mapping of good linearity, similar to Fig. 6.8. The "absolute" mapping was used during the evaluation, but it can be noted, just for comparison, that the slope of

(continued)

Example 6.4 (continued)

the calibration chart of the "MOS diode" was $S_{\text{Vth}} = -3.7$ mV/K at 10 mA measurement current. as opposed to the $S_{\text{VF}} = -2.12$ mV/K slope at the same measurement current in the reverse diode measurement.

In fact, there is no reason to use the "sensitivity" or "K-factor" quantities in the measurement evaluation; up-to-date testers are able to provide an absolute temperature calibration. The only reason to use a sensitivity value instead is when many semiconductor devices of the same manufacturing batch are to be compared. Typically, the reproducibility of the samples is better than the repeatability of the calibration procedure; in such a way, forcing the same sensitivity value to all samples ensures a comparison of higher accuracy.

Several power levels were applied again, as previously in the reverse diode measurement. In Fig. 6.39, selected Z_{th} curves are shown. The powering of the presented curves was 4 A in the reverse diode measurements resulting in $V_F = 0.82$ V and $P_D = 3.28$ W, while in the "MOS diode" measurements, the powering was 1 A, resulting in $V_{\text{DS}} = 3.7$ V and $P_D = 3.7$ W.

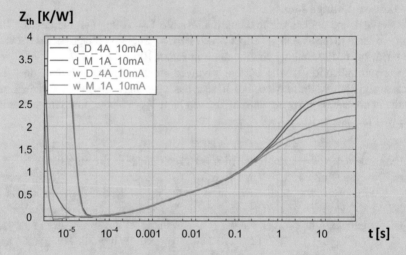

Fig. 6.39 Comparison of Z_{th} curves measured in different powering scheme, at different power levels and two boundary conditions

In the figure, all curves fit perfectly in the 30–200 ms time range; after this time, the variations in the boundary on the case surface cause a conspicuous divergence. The good match between 30 and 100 μs indicates that the power generation occurs in a shallow surface layer of the chip in both powering methods.

6.2.3   *Measurement of MOS Modules*

In the previous subsection in Example 6.4, it was emphasized that the measured normalized transients (Z_{th} curves) of a device are similar along the time range of interest, regardless of the selected powering and sensing method. Consequently, all other thermal descriptors derived from them such as structure functions are also very similar in the corresponding thermal resistance and thermal capacitance range.

In some cases, it is unavoidable to use a mix of powering and sensing methods. For example, in a MOSFET full bridge module (Fig. 6.40), all transistors are accessible in a full four-wire force and sense manner. If someone wants to measure all self- and transfer impedances, some devices have to be powered on their channel at fixed V_{GS} voltage or in MOS diode mode; others have to be biased on their reverse diode. The direction of heating and measurement currents is proposed in Sect. 5.3.4, Fig. 5.7.

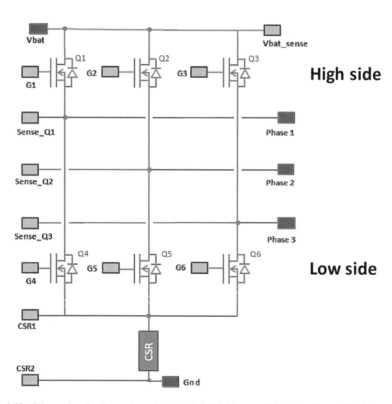

Fig. 6.40   Schematic of a three-phase MOSFET full bridge module. Transistors Q1, Q2, and Q3 constitute the hide side, Q4, Q5, and Q6 the low side of the bridge

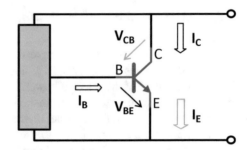

Fig. 6.41 Bipolar transistor in an operating point defined by voltages and currents. The "black box" contains external circuit elements, e.g., the circuitry in the thermal transient tester which force the currents and voltages

6.3 Measurement of Bipolar Transistors

Bipolar junction transistors (BJTs) are devices with three leads again, like their insulated gate relatives of the previous section, but their operating point is defined not only by the voltages across their leads but also by the currents flowing into their leads.

Figure 6.41 shows the voltages and currents which can be defined on the device. Of course, not all these quantities are independent, $V_{CE} = V_{CB} + V_{BE}$ and $I_E = I_C + I_B$. The "outer world" which defines the voltages and currents and represents the external circuitry around the device is denoted now by the "black" box on the left. In our case, this circuitry is the thermal tester.

When all values of the figure are positive related to the reference direction shown by the arrows, the transistor is in "normal active" state. This is the typical powering of the transistor in thermal testing. A power step can be applied on the device by a current jump or voltage jump as discussed in Sect. 5.5.

Several circuit schemes can provide the voltages and currents of Fig. 6.41. Some manufacturers of some testers, and unfortunately also some measurement standards, swear that one or other arrangement of sources and switches ensures the *only* correct powering and sensing. The truth is that the transistor *does not know* what is the content of the black box.

If the voltages and currents governed by one or other box are already set, then only parasitic effects influencing high frequency behavior can be different.

In normal active state, the collector is of "high impedance" (diode in reverse operation), while the emitter is of "low impedance" (diode in forward operation). In such a way, the easiest way to force power is putting a voltage source on the collector and a current source on the emitter. This can be done in three different ways (Figs. 6.42a, b and 6.43).

6.3.1 Electrical and Thermal Stability of the Measurement

Many sources in the literature and standards support the definition in Fig. 6.42a as the only one. This realization is simple but has some disadvantages.

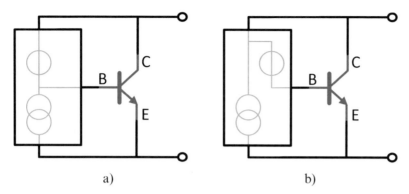

Fig. 6.42 Powering of a BJT device, (**a**) common base, (**b**) common collector

Fig. 6.43 Powering of a
BJT device, common
emitter. The transistor can
be in the same operating
point in all arrangements,
independently from the
connection of the sources in
the box on the left side

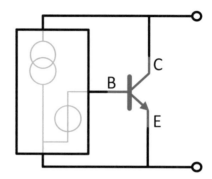

At first, the transistor operates in this setup as a common base mode amplifier. This circuit has very high cutoff frequency, and this way tends to oscillate at a frequency of many megahertz.

A typical collector voltage waveform is shown in Fig. 6.44. The actual device was powered in a thermal tester by $V_{CB} = 5$ V and $I_E = 2$ A. The collector voltage shows 1 V swing around the mean value, which is below the programmed 5 V.

The oscillation causes inaccuracy in the powering value and noise in the recorded transient.

Often the operator of the thermal tester is unaware of the oscillations of the voltage. The gain of the common base amplifier is larger at high voltage and current. Therefore, the oscillation vanishes when the device is switched down to the measurement current (or low voltage). The tester produces agreeable thermal transients, but the power used in the thermal resistance calculations will be incorrect.

Ferrite beads on the leads and capacitors between collector and emitter may suppress the effect. However, checking the waveform by an oscilloscope on the collector and on the emitter is an absolute must.

Another disadvantage of the arrangement in Fig. 6.42a is that it needs two supplies of high power as the current of the lower current source also flows through the upper voltage source, $I_E \approx I_C$.

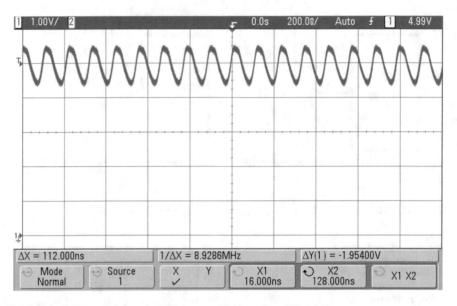

Fig. 6.44 Collector voltage of a BJT measured in the setup of Fig. 6.42a

A cost-effective solution can be gained by a simple rearrangement of Fig. 6.42a into Fig. 6.42b or Fig. 6.43.

Obviously, a high current power supply is needed to provide I_E or I_C (depending on which point one calls "ground" in the figure).

The V_{CB} source can be of much lower current, corresponding to the typical expected I_E and B values.[1]

The transistor operates now as a common emitter amplifier, having B times lower cutoff frequencies. In such a way, oscillations are less likely and easier to suppress.

6.4 Programmed Powering

In the tests shown in previous sections, the voltage and current sources setting the operating point of the device were programmed to a constant predefined value. The actual power level determined by the device under test (DUT) characteristics was slowly changing in the heating phase, and the power measured just before the start of the cooling was used in calculations.

Regulation to an even power level is an important test mode in active power cycling tests (Chap. 7, Sect. 7.4), but that regulation mode is also restricted to

---

[1] B is the transistor-specific ratio between I_C and I_B. The differential ratio for I_C and I_B is called beta, β.

Fig. 6.45 Powering a "magnified diode"

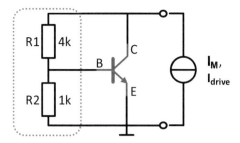

adjusting the sources to a new constant value before a power pulse, based on the experienced power change at preceding pulses.

The power can also be kept at a constant value during the heating transient by appropriate regulation of the sources. A fast circuitry should realize a control loop with time constants in the tens of microseconds range so that the regulation does not distort the time range of the transient holding relevant chip and die attachment quality information.

In this section, a power regulation scheme is introduced, in which one high current source delivers the actual power, and simple low power circuits provide additional voltage control. The DUT is an active component in the regulation. The circuit schemes are now constructed with bipolar DUT, because its finite base current imposes additional difficulties. Insulated gate devices can be used with no difficulty in the same circuits.

Figure 6.45 presents an easy way to increase the power on the device. The resistive divider ensures electrical and thermal stability because it realizes a negative feedback.

The arrangement is often called "magnified diode" because the V_{BE} emitter-to-base pn junction obeys the V–I characteristics of Eq. (4.10) at a given emitter current and device temperature.

V_{CE} is now V_{BE} multiplied by a factor of $(R1 + R2)/R2$ because from the circuit of Fig. 6.45,

$$V_{BE} = V_{CE} \frac{R2}{R1 + R2}. \qquad (6.11)$$

The solution is nearly perfect for devices with negligible input current (MOSFET, IGBT, Darlington transistor). With the actual resistor values, one has an approximately 5 times increase of the power on low input current devices.

The only drawback of the resistive divider is that it "steals" a fraction of the current which can be of interest at lower sense current levels.

At devices where the input current is high, the attenuation of the resistive divider is less accurate. This can be improved by lower resistor values, but then the distortion of the measurement current grows. Typically, this circuit is less usable for bipolar junction transistors.

Fig. 6.46 Improved driving for the bipolar transistor as "magnified diode"

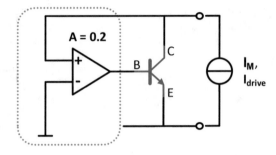

Fig. 6.47 Adding constant power to the transistor drive

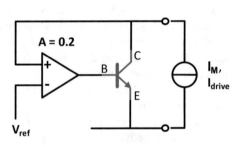

However, the idea can be improved at very low cost; simply, the amplifier (or rather attenuator) represented by two resistors has to be perfected.

In Fig. 6.46, the circuit is now enhanced by an operation amplifier. It has infinite input resistance, in such a way the sensor current is not influenced and its high output current can feed the base at any level needed.

For the power one gets, claiming that the operation amplifier produces A times the voltage difference on its inputs:

$$V_{BE} = A \cdot V_{CE}, \text{ and such } V_{CE} = \frac{V_{BE}(I_E, T)}{A}. \tag{6.12}$$

Similar equation was given for the circuit of Fig. 6.45, but now the A gain can be set in a precise way. Equation (6.12) clearly explains why the name "magnified diode" is invented. In Fig. 6.46, the actual gain is $V_{CE} = 5 \cdot V_{BE}$.

Adding a V_{ref} reference voltage to the scheme, a part of the temperature and current dependence can be eliminated, for Fig. 6.47:

$$V_{BE} = A \cdot (V_{CE} - V_{ref}), \quad \frac{V_{BE}}{A} = V_{CE} - V_{ref}, \tag{6.13}$$

$$\text{and such } V_{CE} = V_{ref} + \frac{V_{BE}(I_E, T)}{A} \tag{6.14}$$

As typical thermal testers have current and voltage output and often they are designed for driving diode-like structures, one can use the circuit of Fig. 6.47 in such

Fig. 6.48 Refined schematic with full suppression of the temperature-dependent component in powering

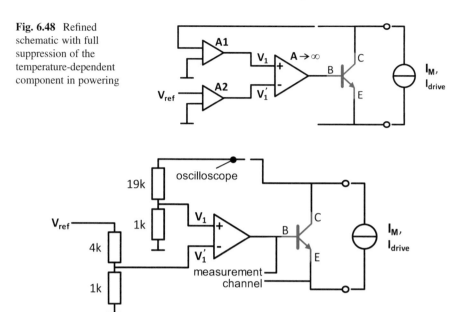

Fig. 6.49 Actual realization of programmed powering of a bipolar junction transistor

a way that the main tester output provides high current for I_{drive}, and an auxiliary low power voltage output yields V_{ref}.

Still in all previous attempts, the actual power level suffered from a fraction of uncertainty, because V_{BE} was far from being constant. Again a low cost addition can make the power fully programmable; trial steps can be saved.

For the circuit of Fig. 6.48, it can be stated that $V_1 = V_{\text{CE}} \cdot A_1$, and $V_1' = V_{\text{ref}} \cdot A_2$, but on the other hand $V_{\text{BE}} = (V_1 - V_1') \cdot A$, and A is very high, which forces $V_1 \approx V_1'$. So,

$$V_{\text{CE}} \cdot A_1 = V_{\text{ref}} \cdot A_2, \qquad V_{\text{CE}} = V_{\text{ref}} \cdot (A_2/A_1), \tag{6.15}$$

and the power step is

$$\Delta P = I_{\text{drive}} \cdot V_{\text{ref}} \cdot (A_2/A_1). \tag{6.16}$$

The actual realization is that simple as shown in Fig. 6.49.

With the actual resistor values $V_1 = V_{\text{CE}}/20$, $V_1' = V_{\text{ref}}/5$, $V_{\text{CE}} = V_{\text{ref}} \cdot (20/5)$, this sets the collector voltage to $V_{\text{CE}} = V_{\text{ref}} \cdot 4$.

However, V_{BE} remains dependent on current and temperature, and this is the quantity which one can record as thermal transient. Again, at some frequencies, the negative feedback of the circuit can turn to positive, but stability can be achieved by proper frequency compensation of the operation amplifier. An oscilloscope probe on the collector is needed for checking whether instabilities occur.

The simple circuits of Figs. 6.48 and 6.49 act as "analog computers" "calculating" quickly the V_{BE} value needed at certain current and temperature to achieve the exact V_{CE} dictated by V_{ref}.

When switching down from $I_{drive} + I_M$ to I_M, then the regulation loop of the circuit "relaxes" with slight wobbling, but the settling time is typically much shorter than the electric transient of the junction; the return of the collector voltage to the value programmed in Eq. (6.15) cannot be observed.

The power step can be achieved either in current jump mode, in switching between $I_{drive} + I_M$ and I_M, or in voltage jump mode with a fast alteration of the V_{ref} level.

With one of the previous circuit schemes, an operating point of fixed power can be easily programmed, even on the flat sections of the device characteristics where a given point is hardly hit by just controlling the current. For example, in the saturation domain of Figs. 6.30 and 6.31, a stable $P1$ point at $V_{DS} = 2$ V, $I_F = 20$ A cannot be reached with trial and error in gate voltage setting, and $P1$ gets immediately missing at temperature changes. Figure 6.31 indicates that V_{GS} is regulated to 2.8 V for reaching $P1$ at 25 °C and to higher gate-source voltage at higher temperatures.

6.5 Measurement of IGBT Devices

Insulated gate bipolar transistors (IGBTs) are minority-carrier devices with high input impedance. They combine the high switching speed of a MOSFET with the high conductivity characteristics of a BJT, resulting in low saturation voltages and high breakdown voltage. The structure of IGBTs and their operation principle is presented in Sect. 4.5.

6.5.1 Powering an IGBT

A typical output characteristics plot is shown in Fig. 6.50, apparently similar to the MOS characteristics of Fig. 6.30 but shifted to the right by a V_F voltage drop caused by the base-to-emitter pn junction of the parasitic transistor.

Unlike in case of MOSFETS, the region where the current stabilizes and does not grow further at increasing collector voltage is called *active region*, and in order to increase confusion, the region above V_F where the current increases quickly is called *saturation region*. Shortly, at MOSFETs, "saturation" is used for stating "the current does not significantly grow at higher voltage," and at IGBTs the word means "the voltage does not significantly grow at higher current."

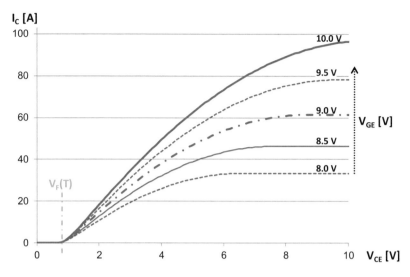

Fig. 6.50 IGBT output characteristics (I_C–V_CE)

Fig. 6.51 IGBT test setup using "saturation mode"

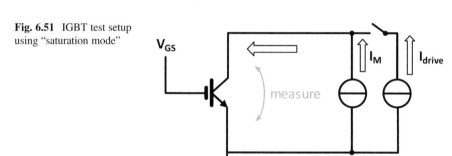

The output characteristics was constructed with LTSpice simulation, in the same circuit scheme as Fig. 4.11, replacing the VDMOS1 device with an IGBT model named IGBT2 and defined as

```
.MODEL IGBT2 NIGBT    TAU=63.552E-9 KP=14.397 AREA=16.000E-6 AGD=6.4000E-6
+ WB=117.00E-6 VT=5.3804 MUN=1.0000E6 MUP=150 BVF=9.9990
+ KF=.5005 CGS=38.737E-9 COXD=88.530E-9 VTD=-5
```

At high gate voltage, the channel resistance becomes very low, and the device characteristics will be dominated by the series parasitic pn junction (Sect. 4.5). This way the measurement becomes similar to the diode measurement in the previous section. This measurement mode (Fig. 6.51) is used nearly exclusively for testing high power IGBTs designed for switching applications.

The selection of the test parameters heavily depends on the datasheet values describing the IGBT component and the corresponding thermal test environment.

From the electrical perspective, the maximum absolute ratings and the device characteristics can help in designing the appropriate powering scenario. The V_{GE} voltage must be selected such that the device is driven to saturation, which can be typically achieved at $V_{GE} = 15$ V or higher. For selecting the I_{drive} current value, the maximum allowable DC collector current and the T_{Jmax} values give proper reference. The thermal test environment has to be sized to balance the input electrical power and keep the T_{Jmax} below the maximum limit. Thermal transient tests of IGBTs are typically performed in a forced convection environment, such as a water jacket, surface of a liquid-cooled cold plate, and in some cases fan cooling.

The I_M measurement current has to be selected such that the noise level of the measured thermal response stays reasonably low. In case the device consists of multiple parallel connected IGBT dice for a logical switch position, the sensor current needs to be increased proportionally. Increasing the sensor current may also reduce the length of the initial electrical transient as an additional benefit, as presented in Sect. 2.12.3 and especially in Example 2.13.

In the time range in which the device transient is of thermally induced nature, the selection of the measurement current does not affect the measurement results. Multiple $I_M - V_{GS}$ operating points will provide the same thermal data with consistent TSP calibration.

The connection toward the device pins however may affect the measured thermal resistance data for medium to high power IGBT devices. As described in Sect. 5.3, the accurate measurement of the IGBT chip's power dissipation is essential for a precise calculation of structure functions. In case of higher heating current values, the power dissipated on the electrical delivery network can be in the range of 30% of the total power budget ([120], Fig. 6.52, Table 6.5).

In order to address this problem, many manufacturers provide dedicated "emitter sense" and in some cases "collector" sense pins wire bonded directly to the IGBT-s corresponding terminals (Figs. 6.16 and 6.52).

Sensing the power step via these dedicated terminals will create a true force/sense wiring (Kelvin measurement setup), allowing the measurement of the real power dissipation of the chip without the rest of the power distribution network.

Table 6.5 Contribution to the total dissipated power of different circuit blocks in a 500 A IGBT device with 2 chips/switch position

Active layer Chip1	Active layer Chip2	Bond wires
296 W	291 W	43 W
Metallization Chip2	Metallization Chip1	Rest of the power delivery circuit
7 W	6 W	270 W

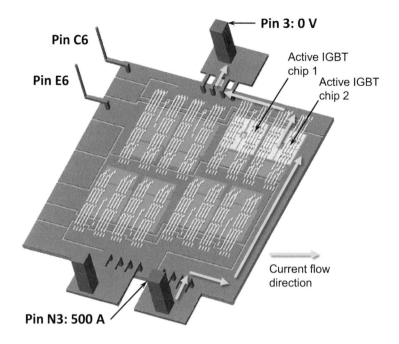

Fig. 6.52 Power module with dedicated emitter and collector sense pins

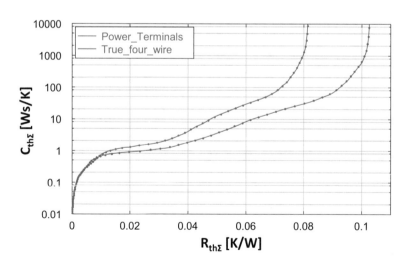

Fig. 6.53 Comparison of structure functions, measurement with appropriate force/sense wiring and powering/sensing on the main power terminals only

Using the real power values, the measured structure functions will also be re-scaled. Figure 6.53 shows that in a true four-wire setup, the measured dissipation is lower than the one including the joule heating of the copper traces in the module. This way the measured thermal resistance values will also be proportionally higher.

Using incorrect wiring in the thermal transient measurement can lead to improper estimation of the thermal performance.

The structure functions were created from transients with square root correction of the initial section between 0.6 and 5 ms. This approach is fully valid for comparing different devices. In case the internal structure of the chip and the die attach is to be analyzed based on structure functions, the square root correction is to be used with caution. In IGBT devices, the power generation occurs in depth, filling a significant volume of the chip. A true restoration of the early transient can be done via comparison of measured and simulated results as highlighted in [121].

The shrink of the normalized thermal resistance expressed in Fig. 6.53 corresponds to a an actual measurement at one specific powering. A more generalized discussion on the related "accordion" effect is presented in Sect. 6.1.7 and Example 6.2.

A standard building block in electronics is a pair of transistors between to two power rails. This scheme is used to connect actively one of the rails to an output at low resistance and low voltage drop. Several variants of this scheme are called push-pull output stage in amplifiers, totem pole output in transistor logic, and half bridge in power electronics. Two or three pairs of the transistors constitute a full bridge (Fig. 6.40). In the next example, the thermal characteristics of an IGBT full bridge will be investigated.

Example 6.5: Determining the R_{thJC} Junction to Case Thermal Resistance Metrics of an IGBT Module Using Test and Simulation

A commercially available half bridge IGBT inverter module (Infineon F3L300R12MT4) was selected for measurement (Fig. 6.54). Both high side and low side IGBT functionalities are realized physically as two chips connected in parallel. The IGBTs are complemented with equal number of reverse diodes.

In order to demonstrate the validity of the R_{thJC} values for a complex package, both simulation and measurements with the JEDEC JESD 51-14 method [40] were used and compared to each other [122]. The layer structure used for the simulation is shown in Fig. 6.55.

To make sure that the test and the simulation are comparable, the simulation model of the component was calibrated to real physical test data [123]. The process started with a thermal transient test to obtain baseline information on the behavior of the selected IGBT chip, in our cases two physical dies connected in parallel. After creating the first version of the thermal model of the component, applying the same power step on the same dice in the simulation environment, the resulting transient curves or the corresponding structure functions can be compared.

(continued)

Example 6.5 (continued)

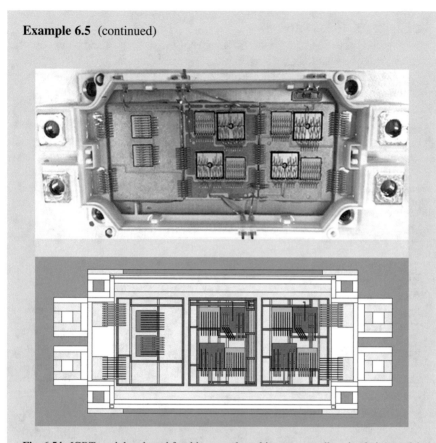

Fig. 6.54 IGBT module selected for this example and its corresponding simulation model

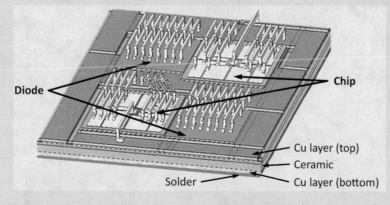

Fig. 6.55 IGBT module's layer structure used for simulation

(continued)

Example 6.5 (continued)

Typically for the first attempt, the shapes of the test-based and simulation-based structure functions show similarities; however, the magnitude of certain partial thermal resistance sections does not match perfectly. As some of the material parameters were best guess values in the beginning, this is a normal situation. To overcome this, a set of experiments were designed, mainly focusing on the variation of the thermal conductivity coefficients of materials in the model. Normally, these have the highest influence on the results.

Another important parameter was the active area size, the coverage of the heated area on the die. Table 6.6 summarizes the set of variables applied in this example:

Table 6.6 Set of calibration parameters

	Initial value	Parameter range	Calibrated value
Active area [mm^2]	81	64~81	79
Die adhesive [W/mK]	33	30~35	33
Ceramics [W/mK]	25	25~35	34
Solder [W/mK]	40	35~45	45

The simulation software was set to create scenarios within the predefined parameter ranges and run a set of simulation experiments to check the effect of these parameters on the final results. Based on the analysis of these simulation experiments, we could select the best set of parameters which make the tested and simulated curves overlap.

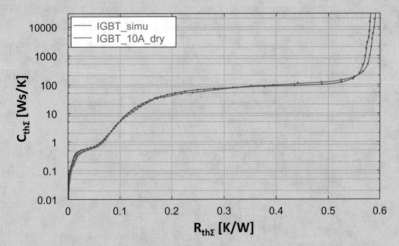

Fig. 6.56 Comparison of structure functions, from actual test and the calibrated model

Figure 6.56 demonstrates the final, good fit between the measurement and the simulation response in structure function space. By increasing the thermal

(continued)

Example 6.5 (continued)

conductivity coefficients of the ceramic and the solder layers, the points of the simulated structure function shifted toward the origin over the R_{th} scale, and finally the two structural models showed good alignment.

A calibrated simulation model can help obtain characteristic package metrics, or it can help verify them if they were previously defined using tests, such as by the JEDEC JESD 51-14 method (although it is defined for discrete, single heat-source packages).

In case of this sample, the R_{thJC} measurement of the IGBT was performed using a combination of tests and simulation. First, the so-called thermal regions were defined over the IGBT die surface and over the package surface, and then the simulation determined the average temperatures in both regions.

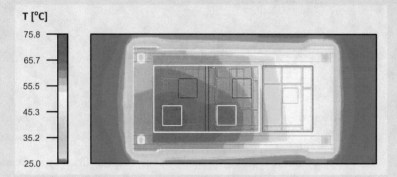

Fig. 6.57 Baseplate temperature plot of the package, with component outlines turned on. Thick white lines indicate the tested regions

Figure 6.57 illustrates the distribution of the temperature over the package surface. The final R_{thJC} will definitely depend on the selection of the regions or perhaps individual isotherms (which is even harder to control). In case of this package, the following data were obtained (Table 6.7):

Table 6.7 Maximum and average region temperatures

Temperatures	Maximum	Mean
Base plate [°C]	70.34	62.90
IGBT active area [°C]	75.83	74.29
R_{thJC} [K/W]	0.079	0.163

The power step used for the calculations was 69.5 W. Not surprisingly, the difference between values using the maximum temperature values of the chip and the base plate and using the mean values is large, about of 1:2 ratio in this case. Using the maximum temperature values will result in a clear underestimation of the R_{thJC} value, while the mean values depend on the actual size of the base plate region selected for the averaging.

(continued)

Example 6.5 (continued)
Figure 6.58 shows the divergence of the structure functions when the dual interface method is applied.

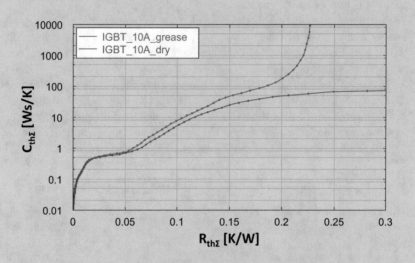

Fig. 6.58 Dual interface results taken on the package from the IGBT-s point of view at wetted by grease and dry boundaries

The separation of the two curves is monotonous, starting from approximately 0.05 K/W to a very explicit separation around 0.17 K/W. This is a commonly seen phenomenon in case of large area, multi-heat-source packages [69]. The thermal boundary condition change between the surface of the power module and the heat-sink changes the spreading angle of the heat-spreading cone. The higher is the thermal resistance, the wider the angle becomes; in other words, the heat spreads more horizontally in the module's baseplate.

Overall, condensing a complex, three-dimensional heat spreading problem into a single R_{thJC} value is not possible; therefore, there are multiple definitions to choose from. Looking at Fig. 6.57, it is obvious that there is a wide range of isotherms crossing the case; the selection of a proper value (min, max, or average) has to be defined by individual companies based on their own standards.

6.6 Measurement of Thyristors (SCR) Devices

Thyristors (silicon controlled rectifier devices) can be measured as simple diodes after having been ignited once on their gate electrode.

In reference [117], thermal transient tests of an electronic circuit breaker module are presented. The module contains two thyristors and some control circuitry on a ceramic direct-bonded copper (DBC) substrate. Due to the module geometry determined by the form factor of standard circuit breakers, the measurements showed strong accordion effect.

6.7 Resistors

Resistive heaters and sensors are widespread in electronics and also in general heating. They are also broadly used in thermal transient testing.

As an example, the thermal test chips and thermal test vehicles presented in Sect. 5.3.1 are built nearly exclusively on resistive heaters and often possess resistive sensors.

When separate resistors are used for heating and temperature sensing, optimal resistance values can be selected to match the specification of the voltage or current source used for heating and to conform the sensor bias and data acquisition characteristics of the test equipment.

Resistive sensors are often thin metal layers. As expounded in Sect. 4.2, these have a "proportional" (exponential) thermal behavior, with a positive thermal coefficient in the range of 3000–5000 ppm/K.

For popular 100 Ω platinum sensors (PT100), this value is 3850 ppm/K or similar. This means that these have a temperature sensitivity of ~3.85 mV/K around room temperature when 1 V is maintained by 10 mA on a PT100; for different current, it is proportionally higher or lower. Higher resistance values are not recommended for sensors in thermal transient testing, because of the noise considerations depicted in Sect. 5.7, in Eq. (5.8).

When the device under test in the transient measurement is a resistive element and it is used for heating and for temperature sensing at the same time, an optimal resistance value cannot be ensured. The related problems have been partly described in Sect. 6.2.1, in connection with powering and sensing on the conducting channel of MOS devices operated in the triode (linear) domain.

One important restriction in these cases is the ratio of the I_H heating and the I_M measurement current. For example, in case of a device of 10 Ω resistance, applying 2 A heating current, the resulting power will be $P_D = R \cdot I_H^2 = 40$ W.

In order to match the input range of the data acquisition channel of a typical test equipment, a few mV/K thermal sensitivity is to be achieved.

Diode type sensors of Sect. 6.1 are advantageous in this aspect, a ratio of 100:1 between the I_H heating and the I_M measurement current results in only 120 mV decrease in the forward voltage. As shown above, reaching proper thermal sensitivity needs a voltage of 0.5–1 V on a resistive device. For a device of 10 Ω, this would require 0.1 A measurement current.

However, when a device channel of a MOSFET or HEMT device is of 10 mΩ R_{DSON} resistance, then 50 A is needed to reach 25 W heating power. At 5 A measurement current, the power dissipation shrinks to 0.25 W, and the voltage on the channel is 50 mV.

The power step is $\Delta P_D = 24.75$ W, causing, for example, 9.9 K temperature elevation on a device with 0.4 K/W thermal resistance.

Supposing R_{DSON} has a positive thermal coefficient of 4000 ppm/K, the thermal sensitivity at the 5 A/50 mV operating point of the resistive device will be 0.2 mV/K, suitable only for advanced thermal testers. At the actual powering, the induced voltage change during the thermal transient will be about 9.9 K · 0.2 mV/K, that is, a bit less than 2 mV. This underlines the need for even higher measurement currents, hardly available at present-day testers.

6.8 Capacitors

Beside active semiconductor components, the performance and lifetime of passive components such as resistors and capacitors is affected by the temperature as well.

As the structure of the capacitors is rather different from the semiconductor components, some aspects of the thermal transient measurement discussed in Chap. 5 needs to be reconsidered.

Unlike in case of semiconductor components where the heat is generated in a well localized thin layer inside the package, in capacitors the heat generation is distributed in almost the whole device. A schematic representation of a ceramic and an electrolytic capacitor is shown in Fig. 6.59. The majority of the heat is generated in the dielectric material and the electrodes in the device core, but in case of high current loads, even the joule heating on the leads is to be considered. As a result, at least the active core of the device needs to be considered as a single volume heat source; the detailed internal structure cannot be distinguished on thermal transient measurement results. Accordingly, the required time resolution can be significantly reduced depending on the size and type of the targeted capacitor.

Temperature-Dependent Electric Parameter
In order to measure the thermal transient, first, an adequate temperature-dependent electric parameter needs to be found. In Chap. 5, the most commonly used

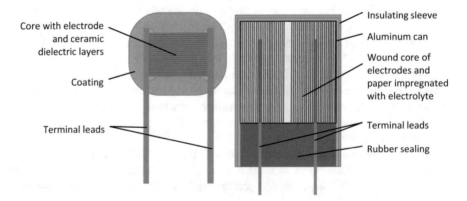

Fig. 6.59 Schematic structure of a ceramic (left) and an electrolytic capacitor (right)

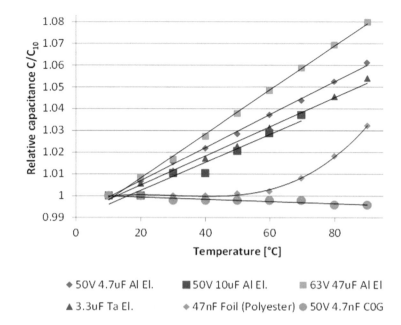

Fig. 6.60 Temperature dependence of various capacitors measured with a simple multimeter

temperature measurement modes are discussed for semiconductors and resistive structures. Common property of those temperature-dependent parameters is that they can be measured as voltage change after applying DC current and voltage on the device.

Various capacitor types use very different materials and have slightly different structures but can be described with identical parameters. Besides the main quantity, that is, their capacitance, parasitic effects cause energy loss and so they contribute to the heating of the component. Such parameters can be expressed as the equivalent serial resistance (ESR), the dissipation factor (DF, tan δ) which is defined as the ratio of the ESR and capacitive reactance, leakage current, etc. As an example, Fig. 6.60 shows the temperature dependence of the capacitance of several different electrolytic, foil, and ceramic capacitors in the 10–90 °C temperature range.

All these parameters can show significant temperature dependence, but none of them can be measured with simple DC methods. Because of the reactive nature of a capacitor, a periodic excitation needs to be applied and measured with high bandwidth and resolution.

This chapter focuses on the capacitance as temperature-sensitive parameter and presents one possible measurement method.

Temperature Measurement and Powering

Measurement of the capacitance with adequately high bandwidth and accuracy is a challenging task. However, for the temperature measurement, it is sufficient to generate a capacitance and hence temperature-dependent signal; the absolute value of the capacitance is not important.

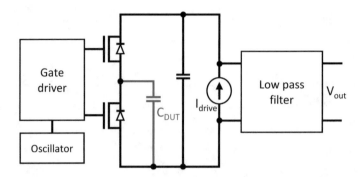

Fig. 6.61 Block diagram of switched capacitor measurement setup

One possible measurement method is to construct a switched capacitor circuit as it is shown in Fig. 6.61. With the repeated charging and discharging of the capacitor C_{DUT}, the transported average current is proportional to the applied voltage, the switching frequency, and the C_{DUT} capacitance as follows:

$$I_{H\_pulsed} = f \cdot C_{DUT} \cdot V_{DUT} \tag{6.17}$$

With forced current powering, the average voltage drop V_{OUT} on this switched capacitor network will be proportional to the drive current (I_{H_pulsed}), the switching frequency (f), and the DUT capacitance (C_{DUT}).

In an adjustable switching frequency range of 10 kHz–1 MHz range, this measurement setup can be used to measure capacitors in the range of about 0.1 nF–1 μF.

The combination of load current and switching frequency can set a desired output voltage level and dissipation for an optimal measurement.

Example 6.6: Measurement of a Ceramic Capacitor with Different Lead Lengths

The thermal transient response of a 100 nF ceramic capacitor (with X7R dielectric material) was measured, soldered to a PCB with two different lead lengths, representing different heat paths. In order to simplify the settings, a heating transient was measured, and hence only one operating point had to be set. The balance between the applied current and used switching frequency needs to be found to provide adequate dissipation and temperature sensitivity. Low switching frequency will result in high output voltage and high temperature sensitivity but limits the achievable measurement bandwidth. In case of the current example, the filter cut off frequency was 1 kHz; hence, the switching frequency had to be kept at least above 100 kHz for acceptable V_{out} signal quality. At a selected $V_{out} = 5$ V voltage level and 100 kHz switching frequency, the resulting current is 50 mA. Based on trial measurements, finally 70 mA driving current and 4 V voltage drop was selected, and the switching frequency was adjusted accordingly ($f = 140$ kHz).

(continued)

Example 6.6 (continued)

The temperature sensitivity was calibrated in a thermal chamber. At each temperature stage, a short heating transient was measured, and the signal was extrapolated to $t = 0$ s with a fitted exponential curve. The calibration curve could be well described with a second-order polynomial.

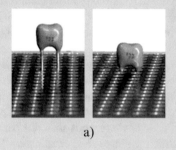

a)

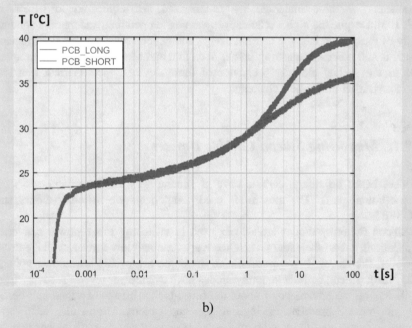

b)

Fig. 6.62 (a) Through-hole ceramic capacitors mounted on PCB with long and short leads, (b) measured thermal transient curves of the two mountings

The resulting heating transient curves are shown in Fig. 6.62. It is clearly visible that the two curves fit well as the heat spreads in the package and the leads, until about 2 seconds where in case of the shorter lead the heat reaches the PCB and the two heat paths start to differ, demonstrating the change in the structure.

6.9   Devices Made of Wide Bandgap Semiconductor Materials

Silicon technology dominated the semiconductor market in most application areas in the last more than 60 years, primarily because the abundant occurrence of silicon in nature and its relatively simple manufacturing technology. A constant development pushed the limits of the Si-based technology further and further; however, there were always applications where special material properties were required and hence other semiconductor materials such as gallium arsenide had to be used for LEDs or high-frequency amplifiers.

In the last two decades, the demand for even smaller, lighter, and more efficient power converters accelerated the research and development of wide bandgap compound semiconductor technologies. The most promising wide bandgap semiconductor materials are silicon carbide (SiC) and gallium nitride (GaN). Thanks to their band structure properties, their advantageous material parameters allow devices to operate at higher temperatures, voltages, and frequencies [124]. The main physical parameters of these materials are summarized and compared to silicon in [17, 103]. Despite the higher efficiency, operating temperature, and thermal conductivity of these materials, the thermal design and hence accurate thermal characterization is still crucial in ensuring proper operation and adequate lifetime. Moreover, due to the special electrical behavior of these devices, their thermal transient characterization brings new challenges.

6.9.1   Measuring Silicon Carbide Devices

Devices based on silicon carbide have in general very similar structure to their silicon counterparts. The most often used components are Schottky diodes and MOSFETs.

Silicon Schottky diodes are widely used in switching mode power converter applications due to their low reverse recovery time and low forward voltage drop; however, the narrow bandgap of silicon limits the achievable breakdown voltage [125, 126]. Silicon carbide Schottky diodes have even lower reverse recovery time while having the advantage of about ten times higher breakdown field and hence they are more appropriate for high voltage applications. Meanwhile, the reverse current is lower, and the maximum operating temperature is much higher than in case of silicon components.

SiC Schottky diodes can be measured using the simple current jump method just as in case of classic Si pn diodes as it is described in Sect. 6.1. No special measurement conditions are required.

SiC MOSFET devices however require special attention. The general problem is the immature manufacturing technology resulting in charge trapping phenomena at the semiconductor-gate oxide surface. The charge accumulation in the traps causes

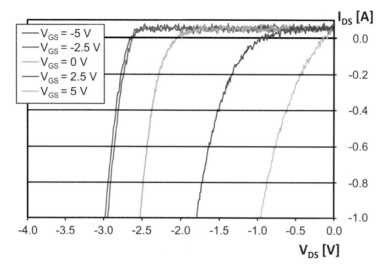

Fig. 6.63 SiC MOSFET characteristics with negative current load at various gate-source voltage levels. (After Ref. [128])

the threshold voltage to change in a time scale overlapping with thermal transient measurements and hence distorting the measurement results [127].

As a consequence, the classic measurement modes utilizing the gate threshold voltage as temperature-sensitive electric parameter cannot be used.

Body Diode Measurement
As discussed in Sect. 6.2, the most basic solution for the measurement of Si MOSFETs is using their parasitic body diode for both temperature sensing and heating. This method can be applied in case of SiC MOSFET devices as well; however, it was shown by several studies [128] that a proper negative V_{GS} voltage is often mandatory to properly close the gate and cut the parasitic current flow through the channel (see Fig. 6.29). In Fig. 6.63, the characteristics of a SiC MOSFET device is shown at negative currents with various V_{GS} voltages. It can be observed that there is a significant difference between the characteristic curves at $V_{GS} = 0$ V and $V_{GS} = -2.5$ V. After a certain gate-source voltage, the effect of the MOSFET channel diminishes, and the diode characteristic stabilizes.

> **Example 6.7: Body Diode Measurement with Different Negative V_{GS} Biasing**
> The selected SIC MOSFET (C2M0025120) in TO247 package was fixed on a large fan-cooled heatsink with a thermal gap pad between the cooling surface of the device and the heat sink. The fan was turned on to provide a constant air flow.

(continued)

Example 6.7 (continued)

In order to demonstrate the effect of the gate-source voltage on the thermal transient measurement results, first the optimal negative V_{GS} voltage range had to be determined for the tests. In practice the recommended method is to apply the measurement current on the drain and source pins of the device and starting from $V_{GS} = 0$ V gradually decrease the gate-source voltage and measure the V_{DS} voltage. In Fig. 6.64, the measured drain-source voltage is shown as the function of the set gate-source voltage, at $I_D = -100$ mA. It can be clearly seen that initially the V_{DS} voltage drops quickly, and then below ~4.5 V, it stabilizes as the MOSFET channel fully closes. For proper thermal transient measurement, a gate-source voltage of -5 V or lower needs to be selected. It also has to be considered that the maximum negative gate voltage rating must not be exceeded.

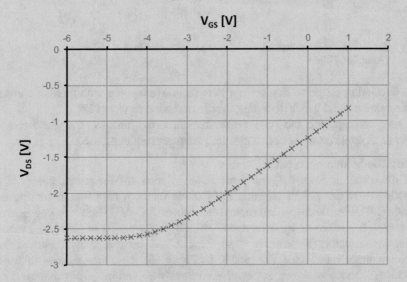

Fig. 6.64 V_{DS} vs V_{GS} characteristics of the SiC MOSFET at -100 mA load current

With trial measurements, the heating current was selected to be -10 A resulting in $V_{DS} = -3.2$ V voltage drop and 32 W power dissipation. The high V_F forward voltage which characterizes SiC diodes can be observed here as well.

(continued)

Example 6.7 (continued)

Fig. 6.65   Measured uncalibrated quasi-temperature transients at various V_{GS} voltage levels

The thermal transient measurement was repeated with identical heating and sensing current settings but with different gate voltage levels between 0 and −6 V. The uncalibrated "quasi-temperature" transient curves with assumed −2 mV/K temperature sensitivity are shown in Fig. 6.65.

Curves corresponding to gate-source voltages of 0 V (black) and −2.5 V (green) are elevating in the early section up to several hundred microseconds, which indicates obviously a non-thermal nature. Curves corresponding to gate-source voltages of −5 V (blue) and −6 V (red) start at higher "quasi" temperature and overlap perfectly. The electrical transient ends before 60 μs and a clean thermal signal can be seen afterwards. As it was shown in Fig. 6.64, the electrical behavior of the device does not change below −5 V, and hence the thermal transients are identical as well.

After the measurements, temperature sensitivity calibration was run at −5 V gate voltage. The recorded calibration curve was of good linearity, and the resulting temperature dependence was 2.09 mV/K.

Using the calibration curves, the transient response graph can be rescaled as it is shown as the BodyDiode_Vgs = -5 V curve in Fig. 6.66.

(continued)

Example 6.7 (continued)

quasi T [°C]

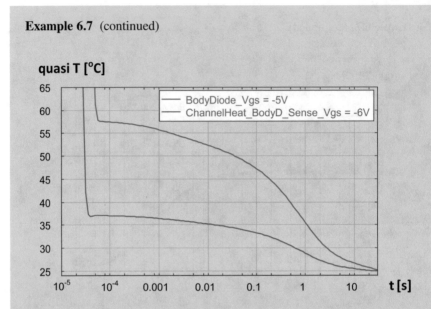

Fig. 6.66 Calibrated thermal transient curves measured in body diode heating and on-state heating configuration

On-State Heating, Body Diode Measurement

As it was demonstrated, the SiC MOSFET body diode has a much higher forward voltage than in case of silicon devices. As a result, the maximum current on the body diode is often lower than that the MOSFET channel can handle due to the dissipation limit of the device. Sometimes, especially in case of power cycling tests, it is required to heat the device close to its current limit, and hence heating on the body diode is not acceptable. In such cases, the MOSFET channel needs to be used for the heating. However, because of the low R_{DSON} channel resistance, it cannot be efficiently used for temperature sensing. The solution is to use the scheme of Fig. 6.35, in which the MOSFET channel is used for heating only; the thermal transient response is measured on the reverse (body) diode.

Example 6.8: On-State Heating and Transient Measurement on the Body Diode

The previously measured SiC MOSFET device was used for this test as well. As the same body diode is used for the temperature measurement, the previously captured calibration data are still valid at the gate-source voltage of -5 V and measurement current of -100 mA. For the heating state, the MOSFET device was turned on with $V_{GS\_heat} = 15$ V and a heating current of

(continued)

Example 6.8 (continued)

20 A was applied. It has to be observed that the direction of the heating current source is now opposite to the measurement current.

The thermal transient curves measured on the heated MOS channel and on the body diode are compared in Fig. 6.66.

The different power step makes it hard to compare the two curves, but after initial transient compensation (square root fit between 60 and 200 μs) and evaluation, the generated structure functions can be seen in Fig. 6.67. According to the device datasheet, the junction to case thermal resistance of this component shall be 0.27 K/W marked with the red vertical marker. The two curves fit perfectly in the early section, and only moderate difference can be seen above 0.5 K/W. The section between 0.27 K/W and about 0.8 K/W corresponds to the thermal gap pad material between the device package and the heat sink. The small shift to higher thermal resistance at the far end of the structure function was caused by the variation of this layer between the two measurements.

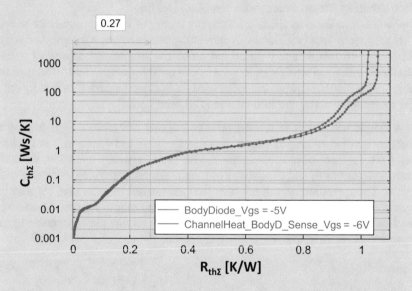

Fig. 6.67 Structure functions of a C2M0025120 MOSFET measured with body diode heating and with channel heating

6.9.2   Measuring Gallium Nitride Devices

Unlike in case of SiC material, the most widely used devices based on GaN technology utilize special device structures, high electron mobility transistors

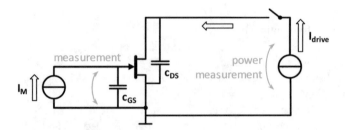

Fig. 6.68 Schematic of HEMT measurement setup with heating on on-state channel resistance and temperature sensing on gate Schottky diode

(HEMTs). Their structure and the physical background of their operation is presented in Chap. 4, Sect. 4.7.

Thermal transient testing of HEMT components is a challenging task. On one hand, the HEMT structure has no parasitic body diode to be utilized; on the other hand, the channel resistance is usually too low for temperature sensing. The depletion mode operation further complicates the measurement task.

Gate Schottky Measurement with On-State Heating

A possible solution is to utilize the Schottky barrier of the gate contact for temperature sensing. A small measurement current connected between the gate and source pins can open this Schottky diode and provide a temperature-dependent voltage parameter with good sensitivity, low noise, and good linearity in normal temperature ranges. This diode, however, cannot be used for heating the device, as usually only a few milliampere maximum load current is allowed on it.

A simple solution is to use the on-state channel resistance for the heating by applying a switched heating current between the drain and source as it is shown in Fig. 6.68. As these devices are optimized for high-frequency operation, the added parasitic inductances and capacitances of the measurement wiring can cause oscillations. These oscillations can not only fake the measurement results, but they can even destroy the component. It is hence always important to monitor the measurement using an oscilloscope during the first experiments and use appropriate bypass capacitors (C_{GS}, C_{DS}) and ferrite beads if needed.

Example 6.9: Channel Heating with Gate Sense Current Applied
A 25 W broadband RF amplifier HEMT (CREE CGH40025F) was selected as a typical example. The tested transistor is in a screw-down flange package. For proper cooling and connections, the package was fixed to an aluminum block with screws, and a PCB was created to implement the measurement connections and the additional components. The drain and gate connections were soldered to the PCB surface. This transistor is designed to be used in a wide frequency range from 6 GHz down to DC. The maximum allowed gate current

(continued)

Example 6.9 (continued)

is 7 mA. In order to have sufficiently low noise, the measurement current was selected to be relatively high, namely, 5 mA. The temperature sensitivity calibration showed good linearity and gave a -5.07 mV/K temperature sensitivity (Fig. 6.69).

Based on trial measurements, $C_{GS} = 100$ pF and $C_{DS} = 4.7$ nF bypass capacitors were used, which provided low noise during the measurement and eliminated the oscillations in all tested operating points. The maximum specified drain current of this device is 3 A. The built-in current source of the measurement system could only provide up to 2 A heating current; hence, transient measurements were run at 2 and 1.8 A as it is shown in Fig. 6.69. The applied power steps of 1.64 and 1.3 W resulted in 11.25 and 8.9 °C temperature change, respectively.

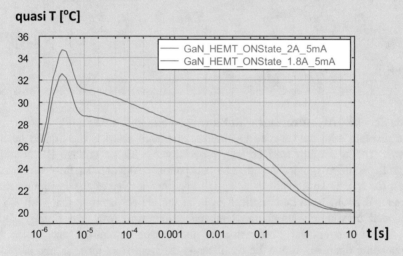

Fig. 6.69 Raw thermal transient measurement results of the GaN HEMT device at 2 and 1.8 A

The transient measurements showed low noise and short electric transients, less than 20 μs. The measurement was repeated at two different heating current levels in order to check if the transient response scales with the heating power. As it can be seen in the Z_{th} curves (Fig. 6.70), the two measurement curves fit perfectly after being normalized by the power step and hence can be considered a valid thermal transient.

(continued)

Example 6.9 (continued)

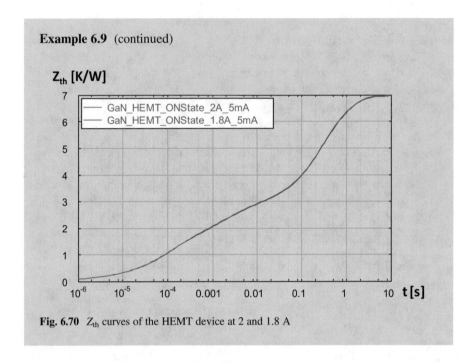

Fig. 6.70 Z_{th} curves of the HEMT device at 2 and 1.8 A

Gate Schottky Measurement with Elevated Voltage Heating

The problem with the on-state heating measurement setup is the limited heating power. The maximum load current, especially for higher voltage RF amplifier HEMTs, is often rather limited. The dissipation achievable in fully on state is often too low to sufficiently heat up the component even at the maximum rated current. It could have been noticed even in the above example that with 2/3 of the rated maximum current, the temperature elevation was just slightly above 10 °C. Moreover, the distribution of the power generation and hence the temperature distribution may be significantly different from those in real amplifier applications.

In order to increase the dissipation and approach the real application conditions, the drain-source voltage needs to be increased during the heating. This of course can be achieved by decreasing the gate voltage, but it still needs to be ensured that during the cooling transient the sensing current can flow through the gate to enable the measurement of the temperature. On the other hand, as the gate voltage starts to approach the threshold voltage, the operating point becomes more and more instable. The schematic circuit shown in Fig. 6.71 provides a solution for both issues.

This circuit is the modified version of the one used for on-state measurement and shown in Fig. 6.68. The V_G voltage source is controlled; it is capable to switch between two preprogramed voltage levels synchronously with the heating current.

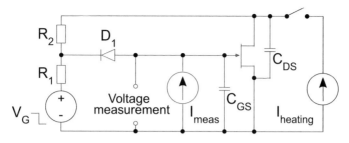

Fig. 6.71 Schematic of HEMT measurement setup, heating with elevated drain-source voltage and temperature sensing on gate Schottky diode

During the cooling, the V_G voltage source is programmed to a positive voltage, e.g., 5 V. In this state, the D_1 diode is reverse biased, and hence the voltage source has no effect on the gate-source voltage. The I_{meas} current can flow through the gate Schottky diode, and the temperature-dependent voltage signal can be measured.

During the heating, the V_G voltage is programmed to a negative voltage. Through the R_1 resistor and D_1 diode, the I_{meas} current is shunted, and the gate of the HEMT is pulled to a negative voltage relative to the source. The R_2 resistor is responsible to provide a negative feedback from the drain of the HEMT and hence stabilizing the operating point.

Example 6.10: Channel Heating with Elevated Drain-Source Voltage
The same GaN HEMT that was used in Example 6.9 was tested with elevated drain-source voltage according to schematic in Fig. 6.71. The previously used measurement setup was extended with the additional components of $R_1 = 100\ \Omega$, $R_2 = 1\ k\Omega$, D_1: 1N4148 diode, $C_{GS} = 100\ pF$, and $C_{DS} = 4.7\ nF$.

During the sensing, the V_G voltage source was set to 10 V. For the heating, the gate voltage was adjusted to reach approximately $V_{DS} = 20\ V$ voltage at $I_H = 500\ mA$ heating current, which means 10 W dissipation. The resulting V_G voltage was approximately $-5.9\ V$. The turn off transient signals of V_G (green), V_{GS} (red), and V_{DS} (blue) are shown in Fig. 6.72. It can be seen that the switching between the heating and sensing states happens in less than 3 μs, with no oscillations visible either during the heating, that is, before switching, or in the cooling state, that is, after switching.

The measured total temperature change was about 100 °C. The Z_{th} curve of the captured transients with two different boundary conditions are shown in Fig. 6.73. The two curves fit well up to 1.4 seconds, where the junction to case thermal resistance can be read as $R_{th,JC} = 7.7\ K/W$.

(continued)

Example 6.10 (continued)

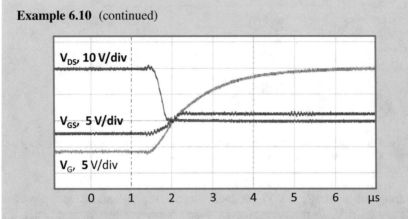

Fig. 6.72 Turn off transient signals of V_G (green), V_{GS} (red), and V_{DS} (blue) measured by oscilloscope

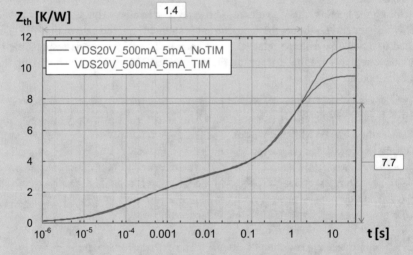

Fig. 6.73 Z_{th} curve of the GaN HEMT heated at $V_{DS} = 20$ V, $I_D = 500$ mA, with (blue) and without TIM material (red curve)

Measurement of Enhancement Mode GaN HEMTs

GaN devices have high potential in power electronics applications, but recent technologies regularly have produced normally-on devices. For safer operation and simpler gate drive circuits, there is extensive development of modified structures to achieve optimal performance with enhancement mode (normally-off) operation [129].

Recent advancements as mentioned in Chap. 4, Sect. 4.7 include an additional gate dielectric layer (MIS HEMT) which reduces gate leakage current and decreases switching and conduction losses [129, 133].

Fig. 6.74 Schematic of HEMT measurement setup with heating on on-state channel resistance and temperature sensing using gate leakage current

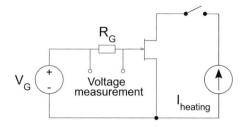

As the result of the increased threshold voltage and an order of magnitude reduction of gate leakage current, the temperature measurement using the threshold voltage of the gate Schottky diode at constant current biasing becomes impractical or sometimes impossible. Even if the diode-like I–V characteristic of the gate still exist, the source resistance often reaches tens of $k\Omega$ range. This high source impedance is comparable to the input resistance of thermal transient tester input impedances, which are optimized for low noise and low source impedance.

A possible measurement approach is to alter the configuration and use the gate leakage current as temperature-dependent electric parameter with constant gate to source voltage biasing. A schematic circuit diagram is shown in Fig. 6.74. A properly sized R_G resistor is connected in series with the gate in order to convert the leakage current into a voltage signal. The R_G resistor shall be sized to reach 50–100 mV voltage change at the desired measurement temperature range. The noise characteristics can be further improved by using lower resistor values and introducing additional signal amplifier.

In case of MIS or MOS device structures, the leakage current can be low enough to consider using measurement techniques used for classic Si MOSFET devices (Sect. 6.2).

Further suggestions on measuring the thermal transient behavior of devices made of wide bandgap materials are given in [16].

6.10   Measurement of LED Devices

Measurement of the thermal impedance/thermal resistance of LED (light emitting diode) packages[2] relies on the principles of the thermal testing methods already shown for semiconductor diodes in Chap. 5 (Sects. 5.3.2 and 5.4 in particular) and in Sect. 6.1 of this chapter.

---

[2] In this context and also in the context of optical measurements of solid-state light sources, the correct terminology is to use the term *LED package* to refer to a packaged LED chip as a functional light source. Frequently, when it does not result in ambiguity, we shall simply use the term *LED* as a packaged, functional device. An LED package is often mounted on a star-shaped metal core printed circuit board for an easier handling, e.g., during different laboratory measurements. Sometimes such an LED assembly will be also referred to simply as an LED.

We assume that there is only a single LED chip in the package. There are, however, LED packages that contain multiple chips, such as integrated RGB LEDs where separate red, green, and blue LED chips are encapsulated in the same package, with individual access for each chip to be driven. In such cases, we consider every single chip as an individual device to be tested separately, though the mutual thermal interactions among the LED chips are also of interest. Regarding the self-heating and transfer heating in multi-die packages, refer to Sect. 2.7 of Chap. 2, and see Sect. 3.2 of Chap. 3 for the thermal metrics of such packages.

Another typical multi-die LED package configuration is when multiple similar LEDs chips are electrically connected in series, with a single cathode and anode electrode for the entire string, with the same forward current powering all the chips of the string. We consider such an LED package as if it had a single "super LED chip," where the so-called *ensemble characteristics* such as the overall forward voltage or the total emitted flux of light and the generated heat cannot be broken down to the contributions of the individual chips. Thus, such LED packages are characterized by a larger forward voltage, e.g., if five phosphor-converted white LEDs are connected in series, then their overall forward voltage would be around 15 V. A typical configuration is when this voltage reaches, e.g., the mains voltage (110/230VAC at 60/50 Hz). These are specifically called as *high voltage LEDs*. Regarding thermal and optical testing, however, the same principles are applicable for such high voltage LED strings as for single die LED packages, but all characteristics, including the thermal resistance or impedance, are related to the virtual "super LED chip." We restrict our discussion to LED packages driven by a constant DC forward current, though high-voltage LEDs can also be directly driven by the AC voltage of the mains supply. The thermal impedance, as a single thermal property would also be characteristic to such *directly AC-driven LEDs*; in such cases, the frequency domain representation of the thermal impedance (see Sects. 2.8.1 and 2.9 of Chap. 2) is best used. Testing of AC-driven LEDs is beyond the scope of this subsection; for a detailed discussion on this topic, refer to the relevant chapter of a book on LED thermal management [8] or see some conference papers such as [73]; regarding optical testing of AC-driven LEDs, refer, e.g., to a related technical report of CIE [47].

Last, but not least, we do not distinguish between the different heating mechanisms in an LED package such as heating resulting from losses due to optical absorption within the package or resulting from conversion losses in phosphor layers of phosphor-converted white LEDs, etc. Neither do we distinguish between the actual blue chip junction temperature and the temperature of the phosphor layer covering the blue chip. Thus, for phosphor-converted white LEDs, we assume a single junction temperature. Regarding detailed studies on the possible effects of phosphor layers in different LED package architectures, refer to the various technical papers published, such as [134] or [135], etc.

6.10.1   Introduction to LED Thermal Testing

Until high-efficiency LEDs appeared, most of the supplied electrical power to a conventional indicator LED was converted to heat; only a few percent of the electrical power was converted to light. Thus, measuring the thermal resistance of an LED package as if it was just a package of a normal silicon diode was not a problem. However, since high energy conversion rates were achieved by modern LEDs, the amount of the energy emitted in form of light has become an issue from the perspective of thermal testing. Figure 6.75 is the counterpart of Fig. 5.1 of Chap. 5, adapted to high power LED packages, indicating the concurrent power conversion from electricity to heat and light. For conventional diodes calculating the heating power was simple: it was equal to the supplied electrical power: $P_H = P_{el} = V_F \cdot I_F$. In case of an LED, obeying the principle of energy conservation, one has to subtract the emitted optical power from the supplied electric power when calculating the heat remaining in the package:

$$P_H = V_F \cdot I_F - \Phi_e, \tag{6.18}$$

where Φ_e denotes the *total emitted radiant flux* of the LED, also known as the *emitted optical power*, P_{opt}, measured in watts.

As hinted by Fig. 6.75, the amount of the emitted light changes with the LEDs' T_J junction temperature: the hotter the junction is, the less light is emitted. The relationship between the forward current, the junction temperature, the heating power, and the emitted optical power of an LED is rather complex; all these quantities mutually depend on one another, as illustrated in Fig. 6.76.

This mutual dependence can be well expressed by the *energy conversion efficiency* of an LED defined as

$$\eta_e = \frac{P_{opt}}{P_{el}} = \frac{\Phi_e}{V_F \cdot I_F} \tag{6.19}$$

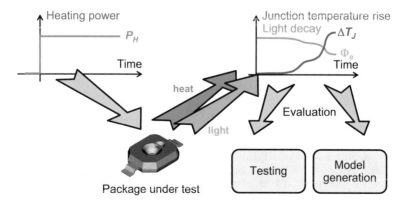

Fig. 6.75   Thermal transient testing of LEDs (with heat generation and light emission)

Fig. 6.76 The mutual dependence of the emitted light, the thermal and electrical properties of an LED

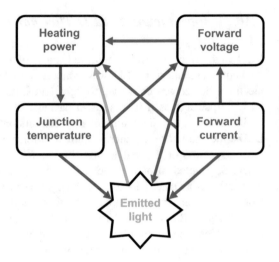

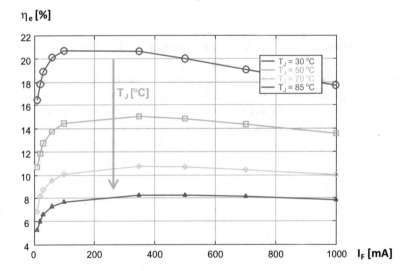

Fig. 6.77 Measured temperature and forward current dependence of the radiant efficiency of an amber LED (Cree XPG)

that is, a dimensionless quantity, usually given in percentage. η_e is often called *radiant efficiency* or *wall plug efficiency* or simply *efficiency*.[3] As illustrated in

---

[3] In the notation, we use the conventions of lighting by using index *e* that refers to a property of light from an energetic point of view. If light is characterized from the point of view of visual perception by the human eye, an index *V* indicates this. Thus, the visual counterpart of the total emitted radiant flux, Φ_e, is the total emitted luminous flux, Φ_V, measured in lumens (lm), the counterpart of the energy conversion efficiency; η_e is the *luminous efficacy*, η_V, measured in lumen/watt (lm/W).

Fig. 6.77, the η_e energy conversion efficiency of an LED is a function of the junction temperature, T_J, and the forward current, I_F, supplied to the device.

The $\eta_e(T_J)$ dependence is a nonlinearity that needs to be accounted for, when the measured $\Delta T_J(t)$ junction temperature transient of an LED is converted into $Z_{th}(t)$ thermal impedances. To get the actual, real thermal impedance that represents the thermal properties of an LED's junction to ambient heat conduction path, one has to normalize the junction temperature transient with the change of the real heating power of the LED:

$$Z_{\text{th\_real}}(t) = \frac{\Delta T_J(t)}{V_F \cdot I_F - \Phi_e} = \frac{\Delta T_J(t)}{P_{el} - P_{opt}} = \frac{\Delta T_J(t)}{P_{el} \cdot (1 - \eta_e)} \tag{6.20}$$

The above so-called *real thermal impedance*, $Z_{\text{th\_real}}(t)$ of an LED (defined in the JEDEC JESD51-51 standard [43]) is by and large independent of its junction temperature, T_J and forward current,[4] I_F. Later, we shall refer to the actual (I_F, T_J) pairs as different *operating points of an LED*.

As seen in Eq. (6.20), for obtaining the real thermal impedance of an LED package, besides the supplied electric power we need to know the optical power emitted by the device, or at least, we need to know its radiant efficiency, η_e. As a simple rule of thumb, if the radiant efficiency is known, we can measure an LED as if it was a normal diode without emitting any radiation:

$$Z_{\text{th\_el}}(t) = \frac{\Delta T_J(t)}{V_F \cdot I_F} = \frac{\Delta T_J(t)}{P_{el}} \tag{6.21}$$

and from this so-called *electrical only thermal impedance* of an LED (also defined in the JEDEC JESD51-51 standard [43]). From this, and the energy conversion efficiency, we can calculate the real thermal impedance:

$$Z_{\text{th\_real}}(t) = \frac{Z_{\text{th\_el}}(t)}{1 - \eta_e} \tag{6.22}$$

If the physical properties of an LED need to be modelled and simulated, the thermal model of its package derived from the $Z_{\text{th\_real}}(t)$ real thermal impedance are needed. Using $Z_{\text{th\_el}}$, however, is justified in situations when measuring the emitted optical power of LEDs is not possible, thus, the energy efficiency under real

---

[4] In luminaires, LEDs are mostly driven by a constant DC forward current; thus, here we focus on thermal testing of constant DC-driven LEDs. When an LED is dimmed with a pulse-width modulated square signal, its thermal impedance is best represented by the set of pulse thermal resistance diagrams introduced in Sect. 2.8.2 of Chap. 2. Some aspects of thermal testing of directly AC mains driven LEDs will be discussed later in this section.

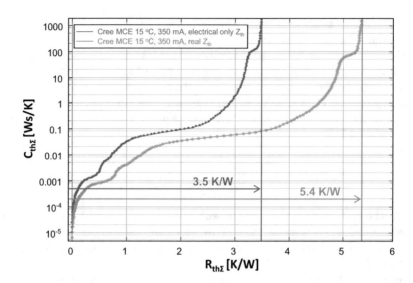

Fig. 6.78 Structure functions of a 10 W phosphor-converted white power LED package measured at 15 °C cold plate temperature and 350 mA forward current: blue curve scaled in R_{th_el}; green curve scaled in R_{th_real}

operating conditions is not known. A typical example is the inline testing where the measurement of Z_{th_el} is the only realistic target.

As discussed in different sections of Chaps. 2 and 3, the thermal resistances are the steady-state values of the thermal impedance functions; thus, the LED thermal testing standard [43] also defines the R_{th_real} and R_{th_el} "real" and "electrical only" thermal resistances of the LED packages. Both the measured thermal impedance and thermal resistances refer to a specific test environment.

As obvious from Eq. (6.22), the real thermal impedance/resistance values of LED packages are always bigger than the electrical only ones, e.g. in case of an LED with 50% energy conversion efficiency, $R_{th\_real} = 2 \cdot R_{th\_el}$. Therefore, the fair thermal metrics to be reported, e.g., in an LED data sheet are the real thermal impedance/resistance values. The relation of the "real" and "electrical only" thermal impedances is also well illustrated in Fig. 6.78. From the overall "real" and "electrical only" thermal resistances of the measured LED package (3.5 and 5.4 K/W, respectively), the energy conversion efficiency calculated back by Eq. (6.22) is 35% for the given operating point of the device.

The effect of correcting the measured thermal impedances by the emitted optical power of LEDs is also demonstrated in Fig. 6.79. The structure functions presented in the figure were created for an amber LED exhibiting a temperature and current dependence of its energy conversion efficiency presented in Fig. 6.77 earlier. In Fig. 6.79a, the structure functions scaled in R_{th_el} are provided (no correction by P_{opt}); therefore the strong temperature dependence of the η_e efficiency affects all the

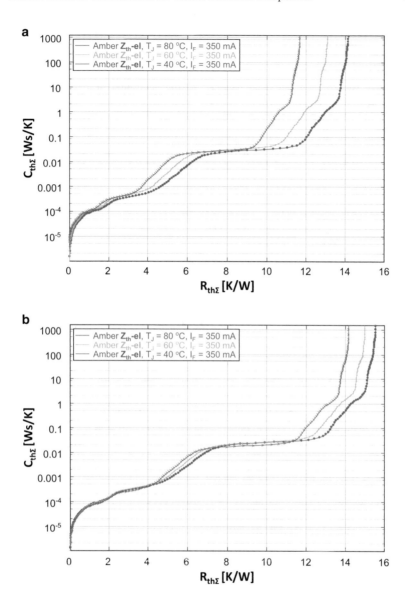

Fig. 6.79 Structure functions of an amber LED (Cree XPG), measured at (80 °C, 350 mA), (60 °C, 350 mA), and (40 °C, 350 mA) operating points: (**a**) without correction by the emitted optical power (scaled in R_{th_el}); (**b**) corrected by the emitted optical power (scaled in R_{th_real})

structure function regions. If we apply the correction of the heating power by the emitted optical power (thus, the strong temperature dependence of the η_e efficiency is considered), only two regions of the LED assembly show slight temperature dependence; see Fig. 6.79b. Note, that the initial regions of all structure functions nicely overlap, proving that within the LED package, no strong thermal nonlinearities are present.

6.10.2   Recommended Simple Thermal Metrics for Single Die
LED Packages

For power LEDs, both the LED thermal testing standards [43, 44] and the most recent recommendations on optical measurements of high power LEDs [45] recommend temperature-controlled cold plates as (thermal) test environment that allow setting the LED junction temperatures to a specific value. Therefore, for power LEDs, the meaningful *thermal metrics are the junction to case thermal resistances/impedances*. For the inclusion of an R_{thJC}-type single thermal resistance value in the datasheet of an LED package, combining the test procedures of [40, 43, 44] standards is recommended, i.e., it is advised to apply the *transient dual thermal interface* method *for the real thermal impedances* obtained from the thermal transient measurements of LEDs.

6.10.3   Using the Electrical Test Method for LEDs

Equation (6.20) can be considered, as a rough set of instructions for the thermal impedance measurement of LEDs, as follows:

1. A stepwise change between two stable states in the electrical powering of the LED package under test needs to be applied for the thermal transient measurements.
2. The transient response of the LED's junction temperature has to be recorded.
3. The optical power of the light emitted by the LED under test has to be measured, and the applied change in the electrical power has to be corrected with that when calculating the real thermal impedance (or real thermal resistance) of the device. The real thermal impedance/resistance has to be calculated based on (6.20).

The standard realization of the first two items is based on the diode measurement techniques described, e.g., in the classical standards, such as JEDEC JESD51-1 [30] and MIL-STD-750 Method 3101.3 [23].

As recommended by these standards, junction temperature of LEDs can also be measured with the electrical test method using the four wire test setup. The test setup already shown for the general diode measurements in Sect. 5.4.1 of Chap. 5 completed with LED-specific additions is illustrated in Fig. 6.80.

In the schematic shown in Fig. 6.62, we also indicated the most important elements of the test environment:

(i) The cold plate to provide the nearly isothermal surface for R_{thJC}-type measurements and for controlling the LEDs' junction temperature
(ii) The integrating sphere as an optical test environment to measure the total emitted radiant flux of the DUT LEDs, which will be discussed later

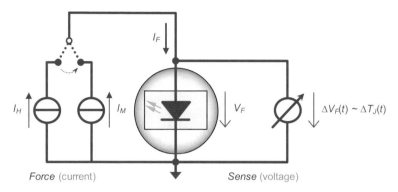

Fig. 6.80 LED measurement scheme as defined by the JESD51-51 standard [43]

The basic scheme of the test waveforms of the electrical signals (forward current, forward voltage) is shown in Fig. 6.81. This scheme corresponds to the continuous measurement of the cooling transients of diodes, presented earlier in Sect. 5.4.1 of Chap. 5.

Using the notations of Fig. 6.81, the difference between the V_{Fi}, initial LED forward voltage, under low current (I_M) conditions (after switching the device from the I_H heating current to the I_M measurement current) and the V_{Ff}, final voltage, measured at I_M current is, as a first approximation, directly proportional to the junction temperature change caused by the high current (I_H) heating. This approximately linear dependence of the forward voltage on the junction temperature can be described as follows:

$$V_F(I_M,t) = V_{Fi}(I_M) + S_{VF} \cdot [T_J(t) - T_J(t=0)] \qquad (6.23)$$

where $S_{VF} = S_{VF}(I_M)$ is the *temperature sensitivity of the forward voltage* measured at I_M current; its reciprocal is the *K-factor*: $K = 1/S_{VF}$. (See also Sect. 6.10.5 as well as Sect. 5.6 of Chap. 5). It has to be noted, however, that if the ΔT_J junction temperature rise caused by the I_H heating current is large, the above linear dependence of the V_F forward voltage at constant I_M measurement current no longer can be assumed to be linear [81, 82] (see also Example 2.12 in Sect. 2.12.2 of Chap. 2).

In principle, one has to wait infinite time for the junction temperature/forward voltage to get stabilized. In practice, however, with a few trial measurements, one can identify the t_M length of the measurement time window where at the end any further change of the forward voltage shrinks below the voltage change detection limit of the test equipment used. Thus, from Eq. (6.23), it follows that the final value of the forward voltage is

$$V_{Ff} = V_F(I_M, t=t_M) = V_{Fi}(I_M) + S_{VF} \cdot [T_J(t=t_M) - T_J(t=0)]. \qquad (6.24)$$

Recalling that $\Delta T_J(t) = \Delta V_F(I_M, t)/S_{VF}(I_M)$ and $K = 1/S_{VF}$ and recalling also the alternate definition of the standard thermal metrics based on the temporal difference

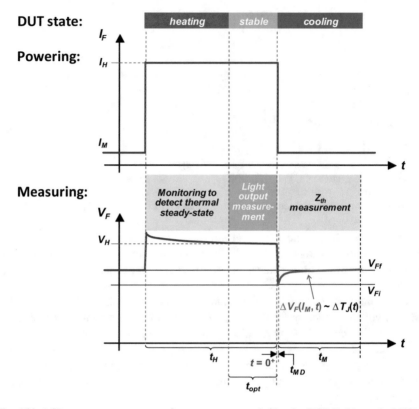

Fig. 6.81 LED measurement test waveforms as recommended by the JESD51-51 standard [43]

of the junction temperature (see Sect. 3.1.5 of Chap. 3) and assuming that the stepwise change of the actual heating power of the LED under test is known, the real thermal resistance the LED package can be expressed as

$$R_{th\_real} = K(I_M) \cdot \frac{V_{Fi} - V_{Ff}}{\Delta P_H} = K(I_M) \cdot \frac{\Delta V_F(I_M)}{\Delta P_H} \qquad (6.25)$$

and the real thermal impedance of the LED package tested is

$$Z_{th\_real}(t) = K(I_M) \cdot \frac{\Delta V_F(I_M, t)}{\Delta P_H} \qquad (6.26)$$

where $K(I_M) = 1/S_{VF}(I_M)$, using the notations of JEDEC's thermal testing standards.

Note that due to the initial parasitic electrical transients inherently present in the recorded values of the $\Delta V_F(I_M, t)$ function, the value of V_{Fi}, the forward voltage at the I_M measurement current at the $t = 0^+$ time instance, when "switching off" the power of the LED, is not known. Therefore, as mentioned in the general description of diode measurements (Sect. 5.4.1 of Chap. 5), the values of the $\Delta V_F(I_M, t)$ function during the t_{MD} period must be discarded and the value of the $V_{Fi} = \Delta V_F(I_M, t = 0^+)$ voltage needs to be back-extrapolated from the $\Delta V_F(I_M, t > t_{MD})$ values based on the $\Delta T_J(t) \sim \sqrt{t}$ approximation given by Eq. (2.26) provided in Sect. 2.4.2 of Chap. 2. This correction process, called *initial transient correction*, is described in detail, e.g., in Sect. 4.1.3 of the JEDEC JESD51-14 standard [40] and is also discussed in Sect. 6.1.4.

The largest thermal time constants of LED packages attached to cold plates are in the range of ~30 to ~120 s, that is, the cooling transient of the LED's junction temperature spans over this range. The proper length of the t_M measurement time window indicated in Fig. 6.81 can be determined by a few trial measurements. As mentioned already, this is the time period during which the V_{Ff} final steady-state value of the forward voltage is reached.

The t_H heating time is advised to be at least 1.5 times larger than the measurement window, $t_H > 1.5 \cdot t_M$, in order to achieve high repeatability in measurements following each other, thus, to avoid superposition of prior heating and cooling cycles to appear in the test results.

The best practice is to combine the above thermal transient measurement of LED packages with the optical measurements of the LED package under test. According to guidelines of CIE[5] on optical measurements of LEDs [45, 46], the light output properties of (high power) LEDs must be measured in a stable state, i.e., both their forward current and junction temperature should be steady; the stability of the junction temperature is indicated by the stability of the forward voltage. These requirements for light output measurement (including the measurement of the total emitted radiant flux) can be satisfied during the measurement of the LEDs' real thermal resistance/impedance: the stable state at the end of heating the LED under test by the I_H heating current opens a suitable window to perform the optical measurements. The t_{opt} time is part of the t_H heating period; therefore, the $t_H > 1.5 \cdot t_M$ requirement of the LED thermal testing standard [43] can be easily met.

For the calculation of the real heating power of LEDs, let us introduce the notations of Fig. 6.82. The heating power when I_H heating current is applied is

$$P_{H1} = V_H \cdot I_H - \Phi_e(I_H, T_{J1})$$ (6.27)

where V_H denotes the LED's steady forward voltage at the constant I_H heating current (see also Fig. 6.81), and T_{J1} denotes the steady-state junction temperature that develops under these circumstances. After switching to the low I_M measurement current, the real heating power is

$$P_{H2} = V_{fi} \cdot I_M + \Delta V_F(t) \cdot I_M - \Phi_e(I_M, T_J(t)).$$ (6.28)

---

[5]CIE: International Commission on Illumination (Commission Internationale de l'Eclairage)

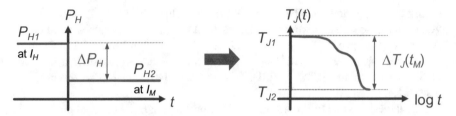

Fig. 6.82 The change of the real heating power of LEDs and the corresponding junction temperature response (Fig. 3.11 adapted to LED measurements, using notations of Fig. 6.81)

Thus, the change of the real heating power of the LEDs, ΔP_{H}, can be calculated as follows:

$$
\begin{aligned}
\Delta P_{\mathrm{H}} = P_{\mathrm{H1}} - P_{\mathrm{H2}} = &[V_{\mathrm{H}} \cdot I_{\mathrm{H}} - \Phi_{\mathrm{e}}(I_{\mathrm{H}}, T_{\mathrm{J1}})] \\
&- [V_{\mathrm{Fi}} \cdot I_{\mathrm{M}} + \Delta V_{\mathrm{F}}(t) \cdot I_{\mathrm{M}} - \Phi_{\mathrm{e}}(I_{\mathrm{M}}, T_{\mathrm{J}}(t))].
\end{aligned}
\tag{6.29}
$$

In practice, neglecting the term $\Delta V_{\mathrm{F}}(t) \cdot I_{\mathrm{M}} - \Phi_{\mathrm{e}}(I_{\mathrm{M}}, T_{\mathrm{J}}(t))$ results in less than 1% error; therefore, it is sufficient to calculate with

$$
\Delta P_{\mathrm{H}} = V_{\mathrm{H}} \cdot I_{\mathrm{H}} - V_{\mathrm{Fi}} \cdot I_{\mathrm{M}} - \Phi_{\mathrm{e}}(I_{\mathrm{H}}, T_{\mathrm{J1}}).
\tag{6.30}
$$

Using Eq. (6.30), the final formulae for the LEDs' real thermal resistance and real thermal impedance are

$$
R_{\mathrm{th\_real}} = \frac{K(I_{\mathrm{M}}) \cdot (V_{\mathrm{Fi}} - V_{\mathrm{Ff}})}{V_{\mathrm{H}} \cdot I_{\mathrm{H}} - V_{\mathrm{Fi}} \cdot I_{\mathrm{M}} - \Phi_{\mathrm{e}}(I_{\mathrm{H}}, T_{\mathrm{J1}})}
\tag{6.31}
$$

and

$$
Z_{\mathrm{th\_real}}(t) = \frac{K(I_{\mathrm{M}}) \cdot [V_{\mathrm{Fi}} - V_{\mathrm{F}}(I_{\mathrm{M}}, T_{\mathrm{J}}(t))]}{V_{\mathrm{H}} \cdot I_{\mathrm{H}} - V_{\mathrm{Fi}} \cdot I_{\mathrm{M}} - \Phi_{\mathrm{e}}(I_{\mathrm{H}}, T_{\mathrm{J1}})},
\tag{6.32}
$$

using again the $K(I_{\mathrm{M}}) = 1/S_{\mathrm{VF}}(I_{\mathrm{M}})$ notation of the thermal testing standards. As it will be highlighted later, both $S_{\mathrm{VF}}(I_{\mathrm{M}})$ temperature sensitivity of the forward voltage and its reciprocal, the K-factor, are also temperature dependent. Therefore, if a wide junction temperature range is to be covered by the measurements, instead of single value of S_{VF}, or the K-factor is to be replaced by the $S_{\mathrm{VF}}(I_{\mathrm{M}}, T_{\mathrm{J}})$ *calibration curve* or its higher order polynomial approximation.

If the DUT LED exhibits considerable nonlinear junction temperature dependence of the forward voltage (as TSP), then the above equations should be rewritten as follows:

$$R_{\text{th\_real}} = \frac{K(I_M, V_{Fi}) \cdot V_{Fi} - K(I_M, V_{Ff}) \cdot V_{Ff}}{V_H \cdot I_H - V_{Fi} \cdot I_M - \Phi_e(I_H, T_{J1})} \qquad (6.33)$$

and

$$Z_{\text{th\_real}}(t) = \frac{K(I_M, V_{Fi}) \cdot V_{Fi} - K(I_M, V_F(t)) \cdot V_F(t)}{V_H \cdot I_H - V_{Fi} \cdot I_M - \Phi_e(I_H, T_{J1})}. \qquad (6.34)$$

Regarding the choice between Eqs. (6.31) and (6.33) for the real thermal resistance or Eqs. (6.32) and (6.34) for the real thermal impedance, further suggestions are given in Sect. 6.10.5 regarding the K-factor calibration of LEDs.

6.10.4   Combined Thermal and Photometric/Radiometric Measurements

Since the radiant flux (emitted optical power) and energy conversion efficiency of any LED depends on the junction temperature (see Fig. 6.77), special attention should be paid to assure consistency between thermal measurement results and light output measurement results. This means that the radiant flux (emitted optical power) should be measured at the very same junction temperature which develops when the I_H heating current is applied at thermal tests.

Thus, as seen in Sect. 6.10.3 before, the recommended implementation of the electrical test method for the thermal (transient) measurement of LEDs can accommodate the measurement of the emitted total radiant flux as well: when the test LEDs are switched on (i.e., the applied steady heating current, I_H, is equal to their nominal forward current) and their T_J junction temperature is stabilized, the most important requirement of the recommendations on optical testing of LEDs [45, 46] is met. Therefore, it is straightforward to combine the thermal and optical testing of power LED packages. Such a combined test setup (shown in Figs. 6.83 and 6.84) assures faster and more reliable testing.

In these combined measurements, first the device is mounted on a temperature-stabilized plate, as anyway required in optical measurements of high power LEDs [45], and then this plate is fixed to an integrating sphere equipped with detector, filters, spectroradiometers, etc.

Both the JEDEC JESD51-52 standard [44] and the CIE 225:2017 technical report [45] recommend the so-called "2π" geometrical arrangement of integrating spheres. This means that we take the advantage of the fact that a high-power LED package, due to its exposed cooling surface at the bottom, emits light only in half of the total space, i.e., in 2π sr solid angle. Therefore, it can be attached to the side of an integrating sphere without obstructing any portion of its emitted light, and there is sufficient space outside the integrating sphere for the temperature-controlled cold plate that holds the package.

JEDEC JESD51-52: CIE 127:2007 compliant optical measurement system
CIE 225:2017: junction temperature control / 2π geometry

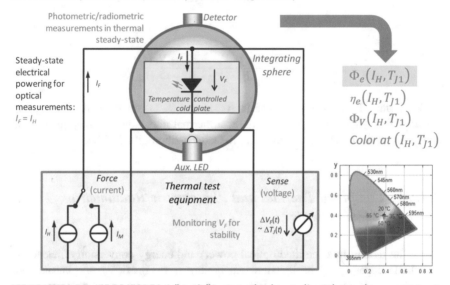

JEDEC JSD51-51: JEDEC JESD51-1 "static" test method compliant thermal measurement

Fig. 6.83 Photometric/radiometric measurement of an LED device in a combined arrangement

JEDEC JESD51-52: CIE 127:2007 compliant optical measurement system
CIE 225:2017: junction temperature control / 2π geometry

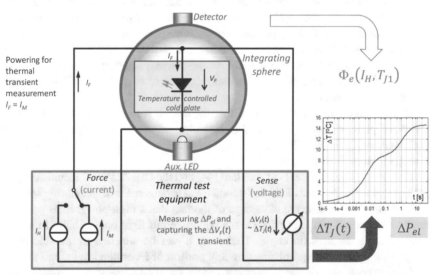

JEDEC JSD51-51: JEDEC JESD51-1 "static" test method compliant thermal measurement

Fig. 6.84 Thermal transient measurement of an LED device in a combined arrangement

In the first phase of a combined measurement (Fig. 6.83), the device is heated by appropriate heating current: $I_F = I_H$. When steady state is reached (no change in the LED's forward voltage), the emitted optical power and other light output properties of interest, like the F_V luminous flux, the CIE xy color coordinates, etc., are provided by the optical system. The light output properties of LEDs make sense only when they are measured at a known T_J junction temperature. The CIE 225:2017 document [45] provides a direct method for setting the junction temperature; the JEDEC LED thermal testing standards provide an indirect method to identify the junction temperature from the results of the combined thermal and radiometric measurements. The recommendations on the optical calibration, self-absorption correction, etc. provided by the CIE documents [45] and [46] need to be obeyed; the JEDEC JESD51-52 standard [44] provides detailed guidance on this.

The forward voltage of the device at the heating current is measured by the thermal test equipment. The measured value is stored and used for power calculation. The radiometric measurement provides the $\Phi_e = P_{opt}$, the *emitted optical power* as the other ingredient for the calculation of the ΔP_H calculation.

In the second phase of the combined thermal and radiometric/photometric measurement of power LEDs (Fig. 6.84), the thermal transient test equipment switches the forward current supplied to the device under test to the measurement current, $I_F = I_M$, and continuously measures and records the $\Delta V_F(t)$ transient resulting from the cooling off of the LED. Using the $V_F(T_J)$ calibration information and the measured forward voltage values, the first step of the thermal transient measurement data post-processing procedure results in the $\Delta T_J(t)$ transient and the final value of the ΔP_H power step.

Figure 6.85 shows a practical realization of a combined thermal and optical LED measurements station. The device has to be mounted on the temperature-stabilized cold plate only once. The system automatically measures all temperature and forward current-dependent LED parameters, going through a user-defined set of temperatures and currents. After an initial step needed for the optical measurements, the automated measurement is carried out in three embedded loops. The complete measurement sequence is as follows

1. Measurement of the dark offset of the photodetector(s) with DUT and auxiliary reference LEDs off.
2. Self-absorption correction with DUT LED off and auxiliary reference LED on.
3. New temperature is programmed (this can be changed in the slowest way); the system waits for temperature stabilization.
4. New current is programmed; the system waits for voltage and temperature stabilization.
5/a. Radiometric and photometric properties are measured by the optical detector(s)
5/b. Switch from programmed forward current down to the measurement current: thermal properties are measured by the thermal transient tester
Next current is programmed.
Next temperature is programmed.

298 G. Farkas et al.

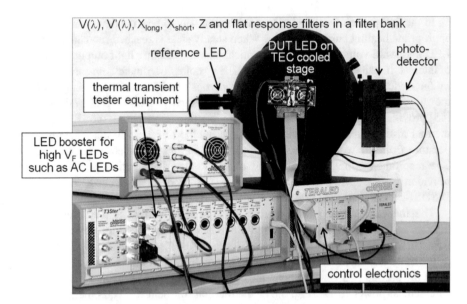

Fig. 6.85 Realization of a combined thermal and radiometric/photometric LED measurement station

In step 1, the DUT LED is switched completely off: the dark level offset measurement of the photodetector (and/or spectroradiometer) takes place. In step 2, the measurements needed for the self-absorption correction of the DUT LED are performed using an auxiliary reference LED with known and stable parameters.

In a post-processing step after making the measurements for every (I_F, T) operating point, the real heating power is calculated from the electrical power-measured radiant flux as presented in Eq. (6.30).

With these data, the R_{th_real} real thermal resistance can be gained according to equation from which also the real junction temperature is calculated back. Plots of junction temperature and forward current-dependent LED parameters then can be created, such as already shown in Fig. 6.77.

If such a combined LED testing station is not available, then it must be assured that the same thermal environment is used both for thermal and optical (radiometric) measurements of the LEDs, including the same current sources and voltage meters [140].

6.10.5 Calibration of the Temperature-Sensitive Parameter of LEDs

The precise calibration process is of high importance because this step influences the overall accuracy of the measurement. Besides the general considerations made for K-factor calibration in Sect. 5.6 of Chap. 5 already, the following further considerations need to be made for LEDs:

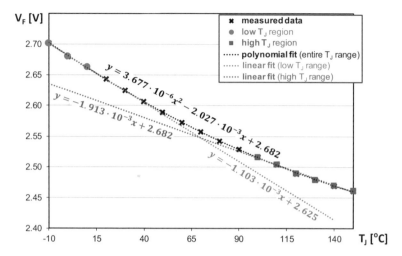

Fig. 6.86 Inter- and extrapolation of the temperature dependence of the forward voltage fitted to measured data in the 15–115 °C range, after [82]. (**x** markers, measured data; • markers, quadratic extrapolation toward low temperatures; ■ markers, quadratic extrapolation toward high temperatures). Linear interpolation on low and high temperature ranges shown

For K-factor calibration, it is recommended to cover the entire junction temperature range of the thermal transients. The linearity/nonlinearity of the measured $V_F - T_J$ relationship should be checked. Figure 6.86 presents an example for nonlinear $V_F - T_J$ relationship and the possible errors if linearity is assumed by means of a constant K-factor determined simply as the average slope of a straight line fitted to the measured (T_J, V_F) data points.

If justified, a robust method of establishing a constant K-factor is to apply linear regression to the measured data points and calculate and use the slope of the regression line as a constant K-factor. The low value R^2 from the linear regression calculation suggests that nonlinearity should be assumed for the $V_F - T_J$ relationship.

The general requirement for the K-factor calibration of diodes is that the temperature-controlled environment provides isothermal conditions around the package. The LEDs to be measured should be mounted to a cold plate with accurate temperature control, and it should be assured that the air surrounding the LED device is at the same temperature as the temperature of the cold plate. In order to avoid contamination of the lens of the LED device, liquid (oil) bath as temperature-controlled test environment must be avoided.

Using the temperature-controlled stage (cold plate) of the integrating sphere in a combined thermal and optical LED test setup (see Figs. 6.83 and 6.84) also for the K-factor calibrations a common laboratory practice. This is a deviation from the above recommendation, therefore during this practice special care should be taken. Therefore, the calibration in an isothermal environment advised previously should also be carried out for at least once for an LED package-type tested. According to the practice, if the difference of the fitted curves does not exceed 1 °C at any value of the

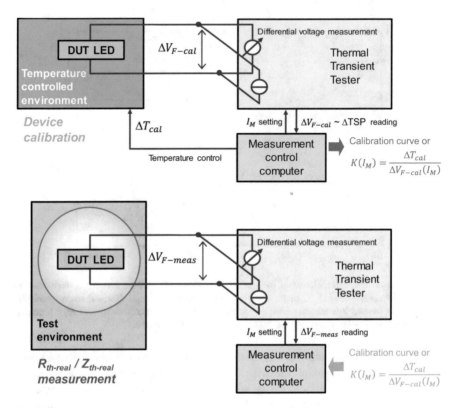

Fig. 6.87 Using the same current sources and voltage meters during temperature calibration and the combined thermal and radiometric/photometric testing of LEDs

forward voltage, performing K-factor calibration in the integrating sphere is permitted. Otherwise, only using isothermal calibration environment leads to correct T_J readings [81, 82].

Note that the temperature range that can be covered during LED measurements is limited by practical considerations. The lower temperature limit is determined by forming of dew at the DUT and measurement apparatus that has to be avoided, e.g., in order to protect the reflective internal coating of the integrating sphere. The highest temperature is limited by safety considerations. Also, if liquid-cooled thermostats are used for setting the temperature, the type of the coolant fluid (e.g., water or oil) sets temperature limits (freezing/boiling).

To assure the highest possible overall accuracy of the measurements, it is advised to use the same measurement current source and the same voltage meter during K-factor calibration and actual measurement; also use these devices during the optical testing [140], as illustrated in Fig. 6.87.

By having the same current source for providing the I_M measurement, it is assured that the S_{VF} temperature sensitivity of the LED is the same during thermal test as it was during temperature calibration. Using the same voltmeter assures that the offset error and the scale error of the voltage meter cancels out from the value of the identified junction temperature change, since

$$\Delta T_{\text{measured}} = K(I_\text{M}) \cdot \Delta V_{\text{F}-\text{meas}}(I_\text{M}) = \Delta T_{\text{cal}} \cdot \frac{\Delta V_{\text{F}-\text{meas}}(I_\text{M})}{\Delta V_{\text{F}-\text{cal}}(I_\text{M})} \tag{6.35}$$

where

- $\Delta V_{\text{F}-\text{meas}}(I_\text{M}) = V_{\text{Ff}} - V_{\text{Fi}}$ is the measured forward voltage change as a response to the change in DUT LED's real heating power (see Fig. 6.81).
- ΔT_{cal} is the temperature range of the calibration that was actually reached by the temperature-controlled calibration environment.
- $\Delta V_{\text{F}-\text{cal}}(I_\text{M})$ is the total change of the forward voltage achieved during calibration.

Note that if a multi-chip LED package is characterized with multiple LED chips connected in series, then the temperature calibration process results in the ensemble temperature sensitivity (or ensemble K-factor) as follows:

For an LED string, the junction temperature change induced variation of the forward voltages of the individual LEDs add-up; therefore, temperature sensitivity of the overall (ensemble) forward voltage of the LED line is the sum of the temperature sensitivities of the individual forward voltages:

$$S_{\text{VF}-\text{ensemble}} = \sum_{i=1}^{n} S_{\text{VF}i} \tag{6.36}$$

where $S_{\text{VF}i}$ denotes the temperature sensitivity of the forward voltage of the i-th LED in the string and n is the number of the LEDs connected in series. Thus, the ensemble K-factor would be $K_{\text{ensemble}} = 1/S_{\text{VF}-\text{ensemble}}$. The dependence on the I_M measurement current and the temperature dependence discussed above also holds for $S_{\text{VF}-\text{ensemble}}$ or K_{ensemble}.

6.10.6   The Junction Temperature of LEDs

If the $R_{\text{th}\_\text{real}}$ real thermal resistance and the $P_\text{H} = V_\text{H} \cdot I_\text{H} - \Phi_\text{e}(I_\text{H})$ actual heating power of an LED are known at a given reference temperature of the cold plate (I_H, heating current; V_H, the forward voltage when heating current is applied; see Fig. 6.81), then based on Eq. (3.5) presented in Sect. 3.1.5 of Chap. 3, the real junction temperature can be calculated as:

$$T_\text{J} = P_\text{H} \cdot R_{\text{th}\_\text{real}} + T_{\text{cp}}, \tag{6.37}$$

where R_{th_real} is the value measured as defined by Eq. (6.31) or Eq. (6.33) if nonlinear K-factor is to be used and T_{cp} is the cold plate temperature that was actually set (and achieved) for the combined thermal and optical measurement.

Due to the possible temperature dependence of the *total-junction-to-environment* X thermal resistance, the light output characteristics of LEDs should always be reported as function of the real junction temperature calculated for example as per Eq. (6.34) or according to CIE's method described in the CIE 225:2017 document [45].

The reason for the recommendation of reporting junction temperature according to Eq. (6.34) is that this calculation relies on precisely measured parameters – in which data correction accounting for parasitic electrical transients is inherently involved. Equation (6.34) can always be evaluated as part of the test data post-processing; thus, for the entire set of data obtained during a measurement, the corresponding junction temperature value can always be associated with.

If the DUT LED is measured in multiple operating points with the fully auto-mated multi-domain LED characterization process described at the end of Sect. 6.10.4, then it is advantageous to set the cold plate temperature for each measure-ment, such that the actually targeted T_J junction temperature is achieved as closely as possible. This obviously needs a pre-characterization step. The forward current, the forward voltage, and the emitted total radiant flux can be measured instantaneously. Therefore, if Eq. (6.34) is used to find the actual T_J junction temperature, a good estimation of the R_{th_real} real thermal resistance is needed. A good practical solution is to measure the LED package under test in four corner points of the domain of interest of the operating points; thus, obtain the $R_{th\_real}(I_{F\_min}, T_{J\_min})$, $R_{th\_real}(I_{F\_min}, T_{J\_max})$, $R_{th\_real}(I_{F\_max}, T_{J\_min})$, and $R_{th\_real}(I_{F\_max}, T_{J\_max})$ values and apply, e.g., a bilinear formula to obtain a good approximated value of R_{th_real} for the other forward current and junction temperature values within the range of interest. With such an approximated value to be used in Eq. (6.34) during a measurement, the difference between the targeted and actually achieved junction temperature is usually less than 1 °C.

The other method for setting LEDs' junction temperature for optical measure-ments is based on the prior measurement of the $V_F - T_J$ relationship when the nominal value of the forward current is applied. (This nominal value of the forward current is the heating current during thermal testing of an LED.) In other words, this process is the K-factor calibration of the LED for the heating current. This process starts with the DUT LED switched off; thus, Eq. (6.34) reduces to $T_J = T_{cp}$; the LED is in thermal equilibrium with its environment, so its junction temperature is equal to the controlled temperature of the cold plate. If a short pulse of the nominal forward current is applied, when the electrical transients due the switching vanished, the measurable forward voltage would be determined by the $T_J = T_{cp}$ temperature. However, even if the forward current pulse is short, self-heating begins; thus, the forward voltage starts shrinking with increasing junction temperature. Therefore, the actual $V_F(t)$ transient has to be captured, and after the initial transient correction, the measured voltage transient needs to be back-extrapolated based on the $\Delta T_J(t) \sim \sqrt{t}$

approximation to obtain the forward voltage value at the instance of switching that would correspond to the set T_{cp} cold plate temperature.

The $V_F - T_J$ calibration data at the nominal forward current gathered this way serves as a lookup table to find the right cold plate temperature, T_{cp} for the actual optical measurements at the required junction temperature.

For further details of this method, refer to the CIE 225:2017 document [45].

6.10.7 Some Recommendations on LED Test Data Reporting

Controlling the cold plate temperature, the LED junction temperature can be set to the desired values as discussed in Sect. 6.10.6. With the automated, combined thermal and radiometric/photometric measurements described in Sect. 6.10.4, the electrical characteristics and light output properties can be identified as functions of the junction temperature and forward current.

It is foreseen that LEDs' future electronic data sheets will include typical data of $\Phi_e(I_F, T_J)$, radiant flux values, $\eta_e(I_F, T_J)$, energy conversion efficiency values, and $\Phi_V(I_F, T_J)$, luminous flux values, $\eta_V(I_F, T_J)$, luminous efficacy values, as numerical tables in machine readable format, e.g., in XML files as already suggested by the Delphi4LED project [141].

According to the JEDEC JESD51-52 standard regarding thermal test results of LEDs, the measured thermal metrics should be reported together with details of the actual test environment, including the reference temperature. The R_{th_real}, real thermal resistance (Z_{th_real}, thermal impedance), must always be reported.

Figure 6.90 presents a Spice-like chip level multi-domain LED model according to [142]. The parameters of the entire LED pn-junction model are identified from the isothermal electrical characteristics; the parameters for the light emission (in green in the figure) are identified from the isothermal radiant flux characteristics. The compact thermal model of the LED package is to be connected to the model node denoted by J.

The R_{th_el}, thermal resistance value that is the "electrical-only" thermal resistance, can be optionally reported with a clear indication that the emitted optical power was *not considered*. Reporting the R_{th_el}, thermal resistance is meaningful only when the heating current and the corresponding forward voltage value to which the R_{th_el} value is related are also reported. The same applies if thermal characterization parameters (without considering the emitted optical power) are reported.

An emerging application area where such electronic LED data sheets are likely to play a key role is the *virtual prototyping in LED luminaire design* [143, 144]. The key elements of the proposed LED luminaire design flows are the Spice-like, chip level multi-domain LED models [142] and the dynamic compact thermal models of LED packages [145, 155].

Figure 6.91 presents the recommended minimal set of operating points for the isothermal characterization of mid-power and high power LEDs aimed as input for multi-domain modelling as suggested in [142, 146]. The chart shows the general

Table 6.8 Recommended I_F, T_J pairs

T_J [°C]	30	50	70	85	110
I_F [mA]	20	20	20	20	–
	30	30	30	30	–
	60	60	60	60	–
	100	100	100	100	–
	350	*350*	*350*	*350*	*350*
	500	500	500	500	500
	–	700	700	700	700
	–	–	1000	1000	1000

Binning values are typeset in italics

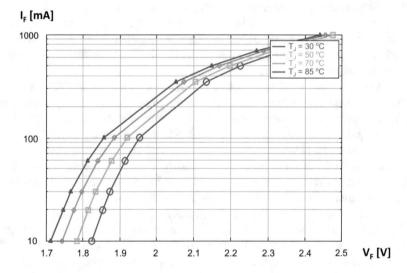

Fig. 6.88 Measured junction temperature dependence of electrical characteristics of an amber LED (Cree XPG)

recommendations on the forward current values to be chosen to assure proper coverage of the low current and high current regions of the relative efficacy curve. The recommended I_F, T_J pairs are given in Table 6.8.

The input to the generation of the Spice-like, chip level multi-domain LED models such as shown in Fig. 6.90 are the isothermal electrical and radiant flux characteristics for which examples were shown in Figs. 6.88 and 6.89. A machine readable reporting of such LED characteristics is essential from the point of view of the automated generation of parameter sets for the multi-domain Spice-like model of LEDs (Fig. 6.90). According to the findings of the Delphi4LED project [146], there is a suggested minimum set of operating points that should be covered by the isothermal IVL characteristics aimed as input for multi-domain LED modelling [142, 146]; see Table 6.8 as well as Fig. 6.91.

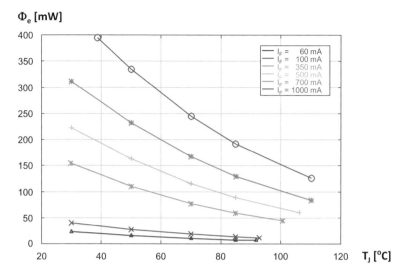

Fig. 6.89 Measured junction temperature and forward current dependence of the total emitted radiant flux (optical power) of an amber LED (Cree XPG)

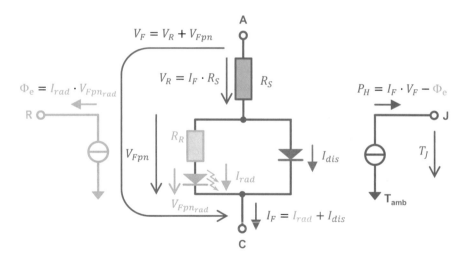

Fig. 6.90 Spice-like chip level multi-domain LED model

6.10.8   In-Line Testing of LED Devices with Short Pulses

In a prior section of this chapter, LED measurements aimed for laboratory testing were described. The optical measurements of LEDs aimed for in-line testing of LEDs are detailed in the CIE's technical report 226:2017 [48]. According to these procedures, LEDs are driven with a forward current pulse that is "on" for about

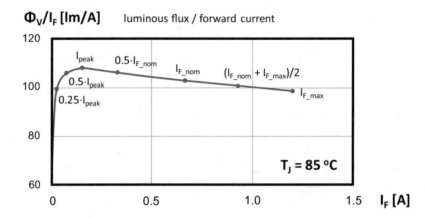

Fig. 6.91 The recommended distribution of forward currents along the relative efficacy curve at a given junction temperature: for the poroper choice of the minimal set of operating points for the isothermal characterization of mid-power and high power LEDs

100 ms, and all electrical and optical properties of interest are measured during this period. Usually, the spectral power distribution of the emitted light is measured with a spectroradiometer, using an integration time matched to the length of the current pulse.

This cca. 100 ms time-frame that is widely used by the LED manufacturers for the in-line optical testing of LEDs is in the same order of magnitude as the length of the thermal transient that already contains sufficient information about the die attach thermal resistance of LED packages, as illustrated in Chap. 3, Sect. 3.5.4. Therefore, it is feasible to also implement the die attach qualification through thermal transient measurements as an additional measurement during in-line testing of LEDs.

The question also arises if short pulse testing can also be used for the combined thermal and radiometric/photometric measurements yielding the isothermal IVL characteristics of LEDs that can be used as input for generating multi-domain simulation models of LEDs, possibly increasing the throughput of the present laboratory test methods (see Sect. 6.10.4). This topic is currently (as of February 2022) subject of research [147, 148]. Also, a recent technical committee of CIE is currently active in working out a new standard for optical testing of LEDs with short pulses, both for in-line and laboratory measurements [149], with the aim of combining the recommendations of CIE's prior technical reports 225:2017 [45] and 226: 2017 [48].

6.11 Measurement of Integrated Circuits

So far in this book, we have discussed thermal transient measurement of discrete devices with access to some pins where power can be applied and also to control pins through which a proper operating point can be set. In case of integrated circuits, the situation is more complex, and we usually have to make thorough considerations to decide how to characterize their thermal behavior.

Moreover, in many cases, more than one chip resides in a package, multi-core SoC (System on Chip) solutions, and multi-die packaging technologies are becoming common. Thermal management of the new technologies, the use of chiplets, 2.5D, and 3D packaging, requires advanced thermal characterization and design approaches.

As there is no single measurement scheme which can be applied to all configurations, in subsequent sections, some typical cases will be elaborated.

6.11.1 Measurement of Large Processor Chips

As discussed in earlier chapters, thermal transient testing requires electrical connections to the measured component or a sub-circuit, in order to enable applying heating power and to record the resulting thermally induced signal. In case of a component hosting millions of transistors, this may not be an obvious task. Fortunately, there are now common approaches which may allow us to perform this job in most cases.

Applicable Powering and Sensing Strategies in Case of Processor Type Integrated Circuits
The simplest and most commonly used approach is testing a device that is manufactured with CMOS (complementary metal-oxide-semiconductor) technology. CMOS logics are based on the combination of p-type and n-type MOSFET-s to fulfil certain logical functions and operations. A frequent structure of a simple CMOS inverter can be seen in Fig. 6.92a. It can be observed that the p-type substrate ("bulk," "body"), in which the n-type transistors reside, and the "well," the large n-type diffusion which hosts the complementary transistors, form a large area pn junction evenly distributed over the surface of the chip.

In normal operation, this pn junction is reverse biased; however, for thermal transient testing purposes, it can be opened by connecting the "ground" pins of the transistor to the positive terminal of the thermal transient tester and connecting the V_{DD} voltages to the ground. The benefits of this approach is its simplicity, as these pins are generally available both on typical test boards and on actual application-specific boards. Applying power on this reverse structure, the surface of the chip is heated up in a uniform way, and in a nondestructive manner, the devices remains functional. The method is not suitable for measuring junction temperatures, while the component is operating.

In the cases when connection to the substrate diode is not feasible or it is important to measure junction temperature values while the IC is functioning, the electrostatic discharge (ESD) protection diodes at the I/O pins of a digital circuit can serve as a perfect temperature sensing element (Fig. 6.93). Each digital I/O is equipped with such diode clamps to prevent an accidental electrostatic discharge, which could easily punch through the gate oxide of the transistors.

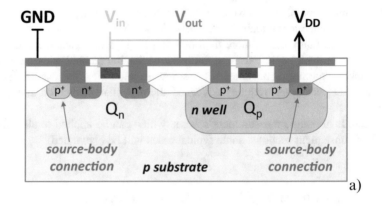

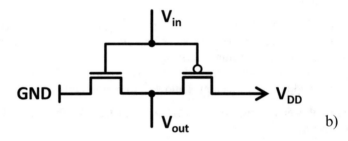

Fig. 6.92 Structure of a CMOS inverter (**a**) and its electrical schematic (**b**)

Fig. 6.93 ESD protection
circuit at an I/O pad

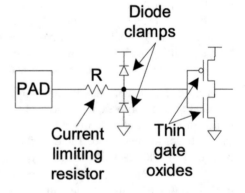

The measurement current source of a thermal transient tester can be connected to an ESD protection diode of an unused I/O pad to test the junction temperature, while the IC performs its normal operation. As the I/O pads are relatively small, the sensor current they can continuously take is also small; it is typically in the sub-mA range. For this reason, it cannot be used for creating the power step for the thermal transient measurement, which would defeat the purpose anyway. To heat up the IC following its operating profile, software support is required. For example, in a microprocessor or microcontroller system, power can be generated issuing continuous interrupt to a CPU or reading and writing a memory chip over and over. The timing of the software operation has to be carefully aligned with the heating and cooling phases of the thermal transient test. To calculate the power dissipation, the current consumption of the IC has to be monitored during this process separately. The benefit of this method is that it can help in obtaining junction temperature values directly related to real device operation. However, beside its complexity, the most significant drawback of the method is that the location of the ESD diodes is not optimal. As parts of the I/O-s, these are typically in the so-called I/O ring, the outer perimeter of the chip, which can be significantly cooler than the core.

Example 6.11: Thermal Characterization of a Multi-Core SoC
The following measurement example demonstrates the characterization of a - multi-core SoC (system on a chip), where each core was individually tested [150].

The test environment could be either a test board (similar to the description in the JEDEC JESD 51-9 standard) sized properly to accommodate the IC or it could be part of a system. A sketch of the SoC on application board is shown in Fig. 6.94.

All tests were carried out on reverse diodes, biasing them between the individual V_{DD} supply of the core and the common ground. One of the tested cores had an embedded capacitor used for the suppression of the power supply's noise. As the capacitor could increase the length of the initial electrical transient, the sensor current had to be carefully selected, as higher sensor currents can help reduce the switching time.

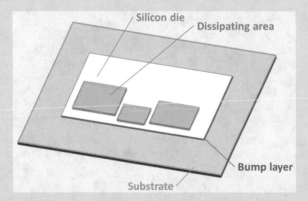

Fig. 6.94 Multi-core SoC-integrated circuit on an application board

Trial transients were recorded at several sensor current values between 5 and 200 mA prior to the calibration of the temperature sensitivity of the

(continued)

Example 6.11 (continued)
substrate diodes. Increasing I_M from 5 to 100 mA resulted in a 10:1 reduction
in the length of the initial electrical transient, from 170 μs to approximately
18 μs.

The calibration of the reverse diodes was carried out in two steps. First, just
the S_{VF} sensitivity factors belonging to several measurement currents were
established based on a few temperature points, as listed in Table 6.9 for Core1.

Table 6.9 S_{VF} scaling factors for converting the measured transient voltage change on
Core1 of the multi-core SoC to "quasi" temperature

I_M [mA]	S_{VF} [mV/K]	I_M [mA]	S_{VF} [mV/K]
5	2.62	50	2.13
10	2.32	100	2.06
25	2.21	200	2

Figure 6.95 presents the "quasi temperature" curves calculated from the
thermal transients of Core1 of the SoC at 2 A heating current, using the fixed
S_{VF} sensitivity factors of the table.

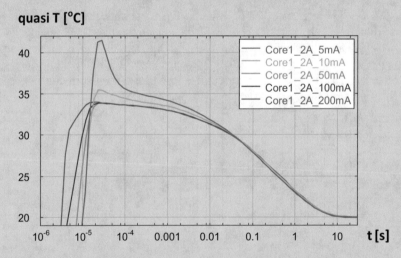

Fig. 6.95 "Quasi temperature" curves calculated from thermal transients of the integrated
circuit at 2 A heating current and different measurement current values, with fixed scaling
factors taken from Table 6.9

The device then was calibrated between 25 and 85 °C in fine 5 °C steps.
The resulting calibration curves for 100 and 200 mA are shown in Fig. 6.96.

(continued)

Example 6.11 (continued)

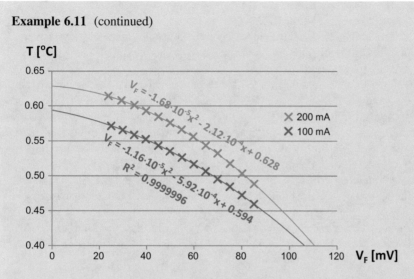

Fig. 6.96 Nonlinear calibration curve of `Core1` at 100 and 200 mA measurement current

The nonlinear characteristics observed in the temperature-voltage relationship can be well approximated with a second-order polynomial. The data were highly repeatable in this range.

For powering the core, 2 A heating current was selected. This resulted in an approximately 50 °C temperature elevation from the ambient, which was proven to be perfect for inducing a noise-free signal (Fig. 6.97).

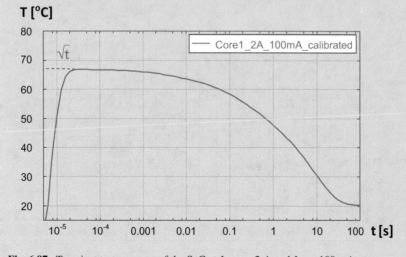

Fig. 6.97 Transient test response of the SoC at $I_{drive} = 2$ A and $I_M = 100$ mA

(continued)

Example 6.11 (continued)

The initial electric transient was corrected using a square root approximation between 30 and 260 microseconds, with the fitted curve perfectly matching the tested data points.

Two identical samples were tested, one mounted on a high conductance and one on a low conductance PCB. Calculated structure functions for Core1 are presented in Fig. 6.98.

The resulting separation point provides the R_{thJB} junction-to-board thermal resistance of the selected component, as defined in Example 3.1 in Chap. 3, Sect. 3.1.2. It has to be noted that the measured $R_{thJB} = 2.25$ K/W value is to be interpreted from the actual measured core toward the PCB. If the sample has multiple cores, each having its own V_{DD} voltage and a common ground, it is often possible to repeat the measurement for every single core in separate test rounds. The measured data would depend on the location and size of the corresponding heat sources. Figure 6.99 presents the dual interface structure functions for all the three cores.

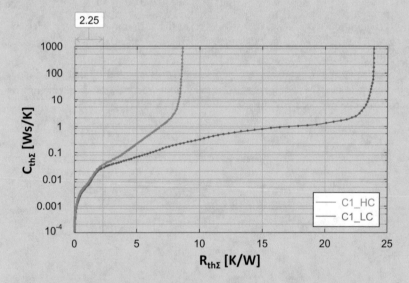

Fig. 6.98 Structure functions corresponding to boundary conditions realized by printed boards of high and low conductivity

(continued)

Example 6.11 (continued)

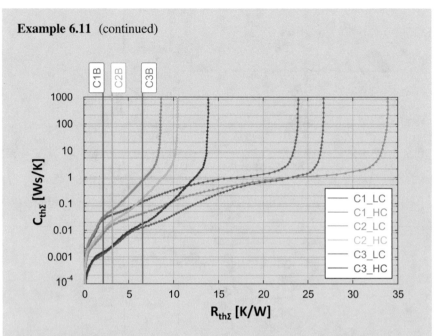

Fig. 6.99 Structure functions outlining the R_{thJB} values of all three tested cores

Considering the heat source area of the individual cores, the measured thermal resistances were inverse proportional to the heat source sizes; larger heat sources showed lower "apparent" R_{thJB} values, as proved in Table 6.10.

Table 6.10 Core sizes in the multi-core SoC vs. measured R_{thJB}

		Summary of each core's characteristics	
	Key in Fig. 6.99	Area [mm^2]	R_{thJB} [K/W]
Core1	C1B	21	2.25
Core2	C2B	15.5	3.0
Core3	C3B	9.5	6.5

6.11.2   Measurement of Operational Amplifiers and Voltage Stabilizers

Operational amplifier and voltage stabilizer devices have similar internal circuit construction. Basically, they are composed of a number of transistors which amplify a reference voltage by a fixed gain and an output stage which transmits the amplified voltage at high load capacity toward the external capacitive and resistive loads. The reference voltage is the input signal for operational amplifiers and a fixed voltage at voltage stabilizers.

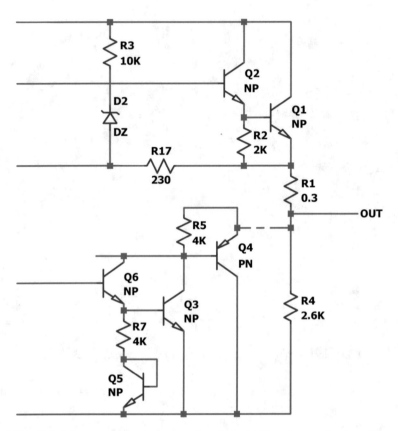

Fig. 6.100 Circuit diagram of an output stage of a voltage stabilizer. In operational amplifiers, the dashed connection between Q4 and OUT is also realized

Because of their purpose, these devices have an internal feedback, and they "resist" the current or voltage change applied on them, regulating in the opposite direction with a time constant in the microseconds or milliseconds range. For successful transient testing, they have to be driven out of the normal operation to a state where their function is paralyzed.

The output stage in devices based on MOS technology does not differ from the scheme of Fig. 5.7 or Fig. 6.40; two large transistors pull the output to the supply rail or to the ground rail (push-pull output). Accordingly, the reverse diode of the output MOSFETS between the drain electrodes and their respective substrate is accessible and can be used for thermal transient measurement.

The push-pull output of bipolar circuits is typically similar to the scheme of Fig. 6.100. Amplifiers have to assure both "source" capability, that is, current feed from the supply toward the output and "sink" capability, current from output toward the ground, equally. In voltage stabilizers, only one of the functions is fully realized, according to the polarity of the stabilizers.

Experience shows that the output transistors (Q1 or Q4 in the figure) can be well accessed from the external pins of the integrated circuit. Transistor Q1 can be inserted into the scheme of Fig. 6.41, using the supply rail as collector pin and the output as emitter pin. It is easy to feed the tiny base current of Q1 through the chain of the remaining circuit elements, and this is given from the supply through Q2 without further action. The remaining circuit components represent the "black box" in the scheme of Fig. 6.41. The amplifier or voltage stabilizer is degraded to a two-pin device, and the selected transistor corresponds exactly to the surface generating most heat in normal operation. Further control pins of such devices can be connected either to the supply or the output, preferably that arrangement is to be chosen which yields the lowest voltage at the measurement current.

6.11.3   Measurement of Multichip Modules

As it was stated previously, the inherent reverse diodes between the supply pins of CMOS or bipolar integrated chips can be used for both powering and temperature sense purposes. However, in the modules with multiple chips, these reverse diodes are connected in parallel; only an average characterization of all chips would be possible when powering all of them simultaneously.

In this case, selected transistors in output stages still can be used as suitable heaters and sensors. In the following example, a thermal transient measurement of a module with multiple chips is presented.

Example 6.12: Measurement of a RAM Module

We have chosen for this measurement a commercially available DDR2 memory module, with eight separate packaged RAM chips, plugged in an IBM PC-compatible motherboard. The measurement was done in a one cubic feet JEDEC standard still-air chamber (Figs. 3.2 and 5.4).

We wanted to trace the temperature change of each chip separately. Most signals are bus-like (address, etc.), but all outputs of all chips go separately to the edge connector of the module.

Figure 6.101 shows the available signals on a RAM chip. All inputs have protection diodes suitable for sensing. The outputs have a circuitry similar to Fig. 6.92. For applying power on the whole surface on the chip, the eight outputs belonging to one chip were tied together in this measurement.

(continued)

Example 6.12 (continued)

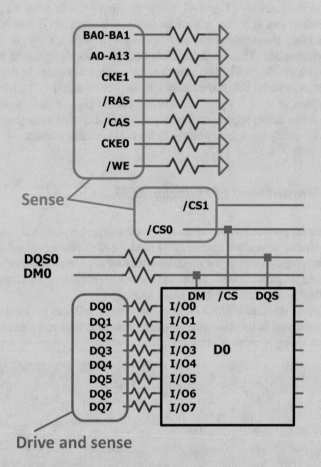

Fig. 6.101 I/O signals of a RAM chip

Applying power in a "dull" way, just sending a negative current into the output of the scheme in Fig. 6.92 in order to forward bias the diode between the n+ drain diffusion of the lower transistor and the substrate, we got the transient signal of Fig. 6.102.

(continued)

Example 6.12 (continued)

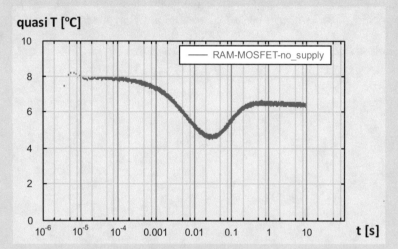

Fig. 6.102 Measured transient on the reverse diode of the lower MOSFET in the output stage of the RAM chip. No external supply connected

We found that the numerous npnp structures between the core of the RAM and the output amplifiers were not properly isolated, as the appropriate depletion layers were not reverse biased. Internal cross-action between inner structures resulted in electric transients in the 10–100 ms range.

Adding an external voltage source of 3.3 V and using it as VDD of the RAM module, we gained the proper thermal signal of Fig. 6.103.

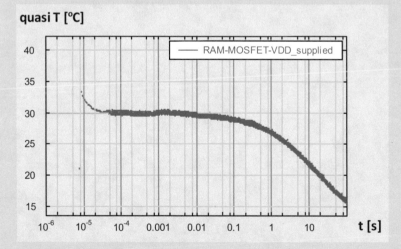

Fig. 6.103 Measured transient on the reverse diode of the lower MOSFET in the output stage of the RAM chip. VDD supply connected

(continued)

Example 6.12 (continued)

It is important to note that for acceptable noise level in the measurement, this auxiliary VDD has to be grounded correctly to the tester and must have very low noise. Further results of the measurement are presented in Sect. 2.7, Example 2.8.

Chapter 7
The Use of Thermal Transient Testing

Márta Rencz, Gábor Farkas, Zoltán Sárkány, and András Vass-Várnai

The main reason of the popularity of the thermal transient testing methodology is the broad range of its applicability. It is embracing the testing of packages and modules, determination of failure locations, structure identification, thermal qualification, and characterization of packages. With the help of it, one can increase the precision of the simulation models, and it enables more accurate methodologies for measuring thermal material parameters than the steady state methods.

In this chapter, several use cases are presented, with an emphasis on the transient dual interface method (TDIM). These use cases demonstrate how different views of the thermal transients, such as Z_{th} curves, derivatives, and structure functions, offer a simple and well repeatable methodology to distinguish the device under test from the measurement environment and to determine such important thermal metrics for packages as the junction to case and junction to ambient thermal resistance.

M. Rencz (✉)
Siemens Digital Industry Software STS, Budapest, Hungary

Budapest University of Technology and Economics, Budapest, Hungary
e-mail: rencz.marta@vik.bme.hu

G. Farkas · Z. Sárkány
Siemens Digital Industry Software STS, Budapest, Hungary

A. Vass-Várnai
Siemens Digital Industry Software, Plano, TX, USA

7.1 Thermal Qualification and Structure Identification, Use of the TDIM Method

As discussed in details in Chap. 2, the structure function methodology helps identify the constructional details of a packaged semiconductor. In Chap. 3, it was shown that the TDIM (transient dual interface method) methodology is a standard technique for the thermal qualification of packages. Below, we demonstrate in several examples how the method can be used in practice.

Example 7.1: Structure Functions of a MOS Transistor at Different Boundary Conditions and the Derivation of the Junction to Case Thermal Resistance
In Fig. 7.1, the measured thermal impedance functions and in Fig. 7.2 the calculated structure functions of a MOSFET device on cold plate are shown. This assembly has been used formerly in Chap. 2 Sect. 2.4.2 as an example for a distributed thermal system.

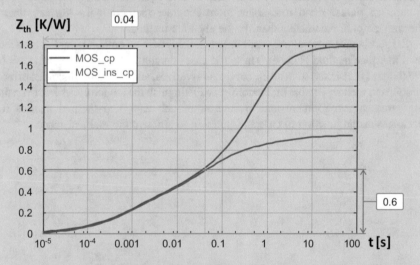

Fig. 7.1 Measured thermal impedance function of a MOSFET package at two different boundary conditions

As expounded in Sect. 2.4.2, the curve MOS_cp was derived from a thermal transient test with the device mounted on a water-cooled cold plate wetted by a good-quality thermal paste. Then, the transient measurement was repeated inserting a ceramics sheet of 2.5 mm thickness between the package and the cold plate, yielding the Z_{th} and structure function of MOS_ins_cp.

(continued)

Example 7.1 (continued)

In the actual case, it was easy to measure the geometry of the standard TO-220 package which hosts the semiconductor chip (Fig. 7.3).

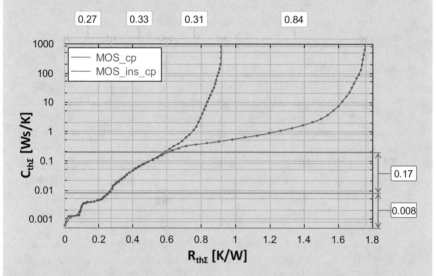

Fig. 7.2 Structure functions of a real distributed parameter system (MOSFET on cold plate, different TIM qualities) with characteristic R_{th} and C_{th} values

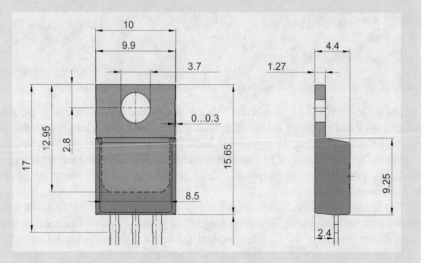

Fig. 7.3 Dimensions of the TO-220 package hosting the MOSFET device

(continued)

Example 7.1 (continued)

The size of the chip was determined by sectioning the package after the transient test. Based on the chip dimensions, the first section of the structure functions until 0.27 K/W and 8 mJ/K was identified as the chip itself (steep starting section in the structure function) and the die attach (flat plateau afterwards). The next section with 0.33 K/W partial thermal resistance and 0.17 J/K thermal capacitance can be attributed to the heat spreading in the copper tab of the package.

It has to be noted that from the dimensions in Fig. 7.3 the copper volume in a TO-220 package is approximately 140 mm^3, but the measured 0.17 J/K thermal capacitance corresponds to approximately 50 mm^3 of copper only. The one-third ratio is reasonable based on the "spreading cone" concept explained in Sect. 2.5, as the trajectories of the heat run along a truncated pyramid from the small chip toward the significantly larger package base.

The end of the package can be identified as the divergence point at 0.6 K/W thermal resistance and 0.178 J/K thermal capacitance. Until this point, the heat propagates within the packaged device; the different TIM qualities still did not affect the spreading.

Beyond the identification of the structural elements within the package and the junction to case thermal resistance, also the thermal conductivity of the ceramics can be calculated from the chart. The inserted sheet with its 2.5 mm thickness added 0.84 K/W to the total junction to ambient thermal resistance. The effective cross-sectional area of the heat spreading was limited to the copper surface of the tab, which was 13 mm × 9 mm. According to (2.6), it follows from these geometrical data that the thermal conductivity of the ceramics is $\lambda = 25$ W/mK, a plausible value for sintered alumina material. The actual alumina sheet has a heat transfer coefficient of $h = 1/R_{th} \cdot A$, that is, $h = 7180$ W/m^{2}K.

The TDIM methodology, besides helping in structure identification, is the most broadly used methodology in finding the junction to case thermal resistance value for package characterization. As exposed in Sect. 2.4.2 and especially in Example 2.5, a change in the interface quality (such as thermal interface material composition, its thickness, etc.) changes the shape of all descriptive functions, including the Z_{th} curve and the structure functions.

The divergence of the Z_{th} curves, such as visible in Fig. 7.1, already gives a first estimate on the R_{thJC} value. However, the gradual growth of the difference makes the exact determination of the actual numerical value of the separation point uneasy.

In the terminology of the theory of LTI systems, the thermal system can be interpreted as an RC low pass filter in its Foster or Cauer form. The input to the filter is the sharp power step; the output signal is the Z_{th} curve. The Cauer form also enables an easy interpretation of transfer Z_{th} curves at a location farther from the excitation (driving point). The low pass character of the RC ladder causes the bumpiness and the smooth alteration in the Z_{th} curves which impedes the identification of the divergence point.

There are several signal processing techniques under the common term "edge enhancement" which improve the detection of differences in signals. Two such techniques were already referred in this book, the derivation and the deconvolution. A possible augmentation of changes in a signal is examining the derivatives of the transient, as defined in (2.15), or applying the numeric deconvolution of (2.21).

The JEDEC JESD51-14 standard [30] uses both methodologies and defines the parameters which govern the TDIM calculation algorithms. In the simpler approach (called Method 2 in the standard), the *difference of two cumulative structure functions* is calculated, and the curves are considered divergent when their difference exceeds a predefined *ε threshold*. In the alternative approach (Method 1), a formula with a *proportionality factor* and a *constant shift* defines the divergence point *in the derivatives* of the Z_{th} thermal impedances. The next example presents both methodologies.

Example 7.2: Identification of the Junction to Case Thermal Resistance of a Power Module

Previously in Chap. 6 Sect. 6.1.5, the measurement of a large power module (SEMIKRON SKM150GAL12T4) was presented, with many details of the powering and initial transient correction. It was stated that it is not straightforward to determine a junction to case thermal resistance value in the figures depicting the Z_{th} thermal impedances (Fig. 6.23) or structure functions (Fig. 6.24). First of all, the curves in those plots shrink toward lower R_{th} values at higher currents because of the "accordion" effect caused by the additional power dissipated in the internal wiring in the module, as explained in Sect. 6.1. Even selecting a pair of curves belonging to approximately the same power, as the Z_{th} plot in Fig. 7.4 which was measured at 80 A heating current and 1 A sensing current, the funnel between the two curves corresponding to the two boundary conditions grows slowly; a divergence point cannot be clearly identified. The point marked at 250 ms time and 0.37 K/W thermal impedance value is actually projected back from later results in this subsection.

The available software tools, such as the one in the configuration of [54] typically facilitate simultaneous calculation of the two algorithms defined in the standard. Fig. 7.5 shows the parameter entry window of the software, with a *proportionality factor* of 0.08 and a *constant shift* of 0.01 J/K for calculating the divergence point threshold of the derivatives with Method 1 of the standard and a threshold of 0.4 J/K as the *ε difference threshold* for the structure function, if we wish to use Method 2 of the standard.

(continued)

Example 7.2 (continued)

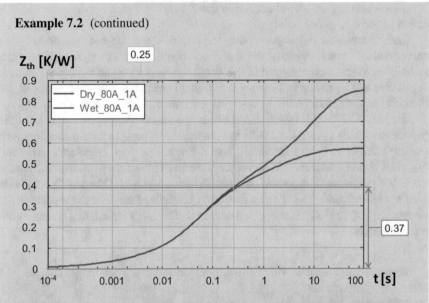

Fig. 7.4 Power module on cold plate, dry and wet boundary. Zth curves and an assumed divergence point shown

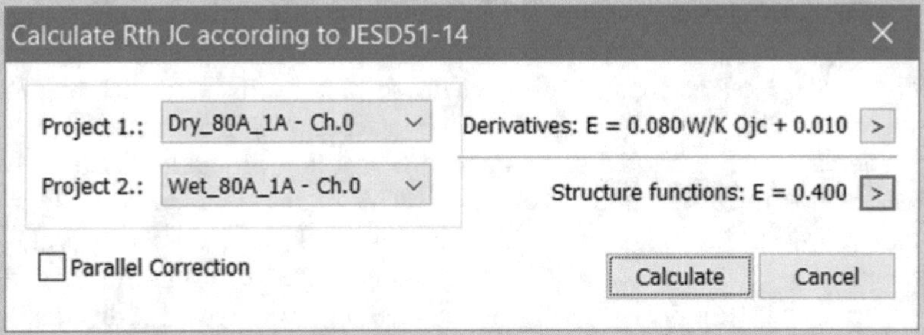

Fig. 7.5 Parameters for the calculation of the divergence point in a thermal transient evaluation software. The divergence threshold for the derivatives will be calculated with a proportionality factor of 0.08 and a shift of 0.01. The divergence threshold (E) for the structure functions is set to 0.4 Ws/K

It has to be noted that the parameters set in Fig. 7.5 differ from the default values specified in the JEDEC JESD51-14 standard [40], which have been defined and are valid for small discrete packages. Due to the large size of the module, the parameters have to be increased proportionally.

Let us examine first the case of using Method 2 of the standard, which means finding the separation point on the structure functions. The structure

(continued)

Example 7.2 (continued)

functions at the given powering are presented in Fig. 7.6. A long straight section can be observed between 0.25 and 0.45 K/W, which corresponds to the radial spreading in the base plate around the hot chips. The heat trajectories step out of the base plate and enter the TIM at various locations; this makes that the divergence of the curves is not very expressed.

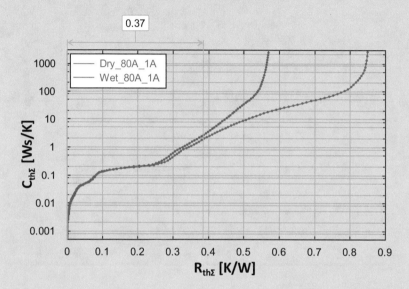

Fig. 7.6 Structure functions of the power module on cold plate, at dry and wet boundary

A clearer definition can be achieved by constructing the difference of the structure functions; see Fig. 7.7.

The fast growth in the difference of the structure functions starts after 0.32 K/W. The difference of $\varepsilon = 0.4$ J/K, which was defined in the parameter entry window before, is reached at $R_{thJC} = 0.37$ K/W.

The concept of Method 1 is similar, but it starts with finding the difference of the time derivatives of the Z_{th} thermal impedance curves (Fig. 7.8).

In the next step, the chart of the differences (green curve in Fig. 7.8) is recalculated, transforming the x axis scaled in time, to an x axis, scaled in the thermal impedance belonging to the "wet" boundary, in this case the Wet_80A_1A thermal impedance curve in Fig. 7.4.

The ε threshold is to be constructed in this transformed chart as a straight line determined by the proportional factor and the constant shift prescribed in Fig. 7.5. In Fig. 7.9, the threshold line intersects the transformed curve at $R_{thJC} = 0.4$ K/W.

(continued)

Example 7.2 (continued)

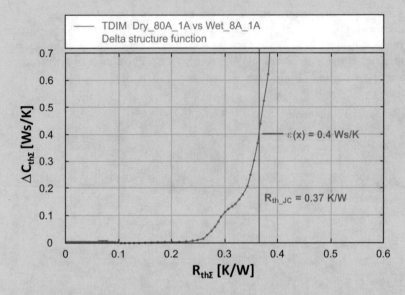

Fig. 7.7 Difference of the structure functions in Fig. 7.6. The predefined difference threshold of 0.4 J/K is reached at $R_{thJC} = 0.37$ K/W

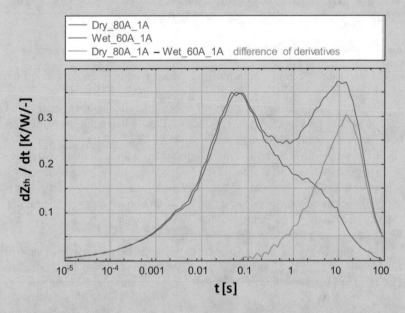

Fig. 7.8 Time derivatives of the Z_{th} curves in Fig. 7.4 and their difference. After 250 ms, the difference becomes more expressed

(continued)

Example 7.2 (continued)

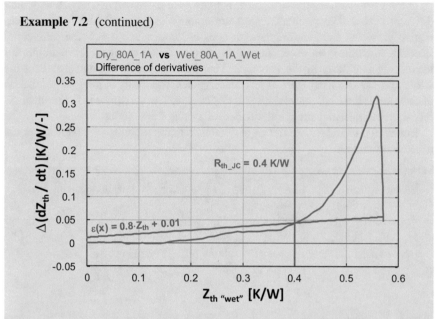

Fig. 7.9 The difference of time derivatives plotted against the thermal impedance values at the "wet" boundary in Fig. 7.4

As the above example confirms, after establishing a reasonable ε difference value for a device category, the method yields stable results. Although the ε threshold was chosen arbitrarily, it can be observed that in Fig. 7.7 with an ε value selected between 0.14 Ws/K and 0.7 Ws/K, which is of a ratio of 1:5; the deduced R_{thJC} value lies in the 0.33 K/W – 0.39 K/W range. This scatter is much lower than that of the typical spatial temperature difference measured separately at the junction and an ill-defined "case" location in the case of two point measurements.

A deep analysis was carried out at Infineon and presented in [96]. A round-robin test with the comparison of results from TDIM and two point measurements was carried out by different laboratories. Their conclusion is that a well-selected constant ε threshold for a device category ensures good repeatability among different laboratories. They also concluded that an appropriate threshold for Method 2, that is, for the difference of the structure functions, is to be determined for different device categories (SMD package, large module, etc.) separately.

7.2   Quality Testing by Finding Sample Differences with the Help of the Local Thermal Resistance Function

As it was mentioned in Chap. 2, the structure functions offer an excellent possibility to reveal differences between samples of assumed similar device structures. This capability of distinction makes the thermal transient methodology an excellent testing tool.

Different versions of the structure functions were presented in Chap. 2, with their different virtues. The cumulative structure function is best suited for solving quantitative problems and answers the questions of "how much it is." The differential structure function is an excellent tool for finding small changes, for example, for identifying interfaces in the structure. The version, which distinguishes best the stable and varying local thermal resistance values, especially in a larger population of samples, is the *local thermal resistance function*. The variations in these functions reveal often undesired structural differences among the samples.

Below two examples are shown to demonstrate the usability of the local thermal resistance function.

Example 7.3: Local Thermal Resistance Functions of a MOS Transistor at Different Boundary Conditions
The two structure function curves in Fig. 7.2 representing the structure of the MOSFET device with different attachment on cold plate are redrawn in the form of local thermal resistance functions in Fig. 7.10.

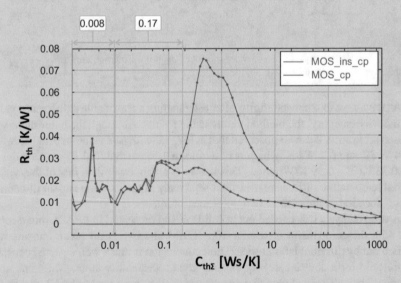

Fig. 7.10 Local thermal resistance functions of the MOSFET on cold plate with and without inserted ceramics sheet, redrawn from Fig. 7.2

The first section until 8 mJ/K thermal capacitance, identified as chip and die attach in Example 7.1, is almost identical in the two functions, which is expected as the same transistor was tested twice. Different samples could expose a conspicuous dissimilarity in this section because of the potential scatter in die attach quality. The steep step at 0.1 K/W of Fig. 7.2 corresponds to a peak in the first section of this chart.

(continued)

Example 7.3 (continued)

The similarity and dissimilarity in structural regions is even more expressed in the *difference of the local thermal resistance functions* as shown in Fig. 7.11. With the inserted alumina sheet, the trajectories of the heat spreading follow a very different path in the ceramics and also in the adjacent metal section of the cold plate. This is expressed in the large R_{th} growth of thermal resistance from 0.178 to 10 J/K. The thermal capacitance of 0.178 J/K can be converted to 50 mm^3 copper volume of the package, and 10 J/K corresponds to 4000 mm^3 of aluminum. Deeper in the metal, the difference vanishes, as the package and ceramics compound behaves as a single heater for the farther portions of the cold plate.

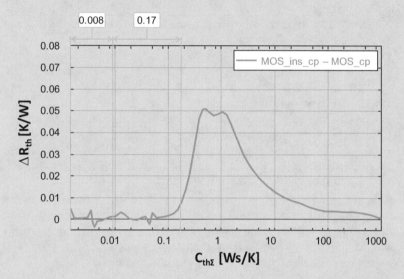

Fig. 7.11 Difference of the local thermal resistance functions of the MOSFET on cold plate with and without inserted ceramics sheet

It can be noted that the local thermal resistance function has less virtue when used for a single sample at various boundary conditions. It is best suited for volume testing or in-line testing on a number of samples where the technology variation can be followed and outlier samples can be selected and sorted out.

The next example presents how the methodology was used for finding the most variable parameters in an LED structure.

Example 7.4: Variance of the Material Properties in Interface Layers of Power LED Devices

In the European Delphi4LED project, an in-depth test was done on a broad type selection of power LED devices [152], mainly with the aim of providing valid models of packaged LEDs for luminaire construction, simulation, and electronic data sheets. The tests involved the investigation of electrical, thermal, optical, and mechanical properties and also the variance of the corresponding parameters [154].

One of the selected types was the Cree XP-E2 white medium power LED. In these devices, an internal III–V semiconductor chip emits monochromatic blue light which then excites a luminescent phosphor coating. The blue peak of the internal chip and the yellow light of broader visible spectral distribution from the phosphor merge into a white blend.

Some tests were carried out on packaged LED devices manufactured with a blue chip without phosphor coating.

Figure 7.12a shows a blue XP-E2 device in its standard packaged form; the light-emitting chip is attached to a ceramic heat spreader block and is covered with a silicone-filled dome.

Figure 7.12b presents the next assembly level; the packaged LED (now a white type with phosphor coating) is soldered on an aluminum starboard.

The devices were measured in an integrating sphere which facilitated the measurement of the emitted optical power. The starboard was mounted with no applied thermal paste to the temperature-controlled thermoelectric cooler (TEC) attached to the integrating sphere.

Thermal transient tests were carried out at various current levels and different plate temperatures. In Fig. 7.13, four structure functions of white LED packages, soldered to aluminum starboard and mounted on the thermally stabilized plate, are shown. The structure functions are calculated from cooling transients after heating at 700 mA, at 85 °C plate temperature.

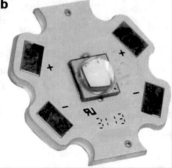

Fig. 7.12 Power LED assembly: (**a**) Blue XP-E2 chip attached to a ceramic heat spreader block, with a silicone-filled dome, (**b**) the ceramic block soldered to an aluminum starboard

(continued)

Example 7.4 (continued)

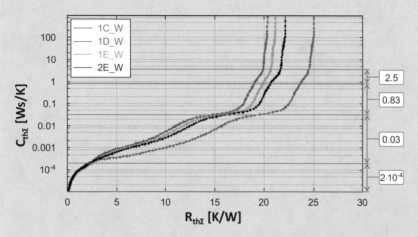

Fig. 7.13 Structure functions of four white LED packages, soldered to aluminum starboard and mounted on the temperature-controlled cold plate of an integrating sphere

Three samples, 1D_W, 1E_W, and 2E_W show only minor variations. Sample 1C_W is obviously faulty; it has a severe die delamination. Above 30 mJ/K, all structure functions are nearly identical; this can be proved by shifting them to the right and fitting them at their highest thermal capacitance.

This delamination of 1C_W is useful now in the sense that it unambiguously identifies the end of the LED chip section at 0.2 mJ/K, which corresponds to an intermetallic compound chip of 0.12 mm^3 volume.

Further structural elements can be recognized in the flatter and steeper structure function sections in Fig. 7.13 based on their material and dimensions.

The local thermal resistance function view of these structure functions (upper plot in Fig. 7.14) offers an easier way to identify components in the assembly. In this function, the higher or lower thermal conductivity of the material can be directly perceived as lower or higher R_{th} value, and the low or high scatter of curves informs whether the section is of "stable" or "varying" nature. This approach offers higher resolution than scrutinizing the variation of the steepness in the structure function. In the actual case, the following regions of heat spreading were identified, based on the value and variance of thermal resistance:

1. Compound semiconductor chip, low variance, 0.2 mJ/K (0.12 mm^3)
2. Conical spreading from the low cross-sectional area limited by chip size into the alumina block through die attach, high variance 0.03 J/K
3. 1D spreading in the alumina block, over full cross-sectional area, low variance, 0.01 J/K (10 mm^3)
4. Spreading from the cross-sectional area limited by alumina block size through soldered attachment, low variance, 0.06 J/K
5. Radial spreading in the aluminum starboard, low variance, 0.76 J/K (350 mm^3)

(continued)

Example 7.4 (continued)

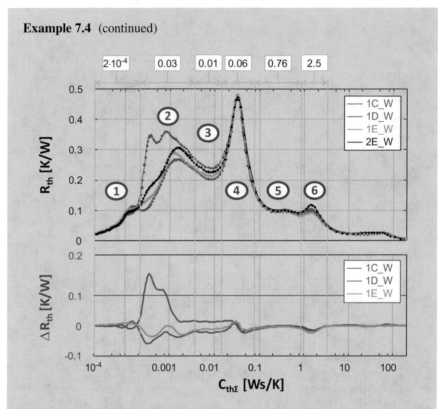

Fig. 7.14 Local thermal resistance functions of four white LED packages, soldered to aluminum starboard and mounted on the temperature-controlled cold plate, and their difference from the reference device 2E_W. Characteristic sections in the assembly are identified

6. Spreading through the dry boundary between the starboard and the thermostat plate

The stability of the different layers can be visualized plotting the difference of the local thermal resistance functions of each sample and of a reference device. The lower plot in Fig. 7.14 demonstrates that only the die attach shows significant variation.

Systematic analysis of the variance of power LED devices was presented in the studies of [153] and [154].

7.3 Structural Integrity Testing, TIM Testing

Structural integrity testing is probably the most important use case of thermal transient testing. It is based on the fact that a structure function is a signature of a given structure, and if the structure integrity changes for any reason, e.g., die attach delamination or any other change in the structure as a result or wear or ageing, the structure function taken from the same input point with the same boundary conditions will be different. From the location of the change in the structure function, we can even tell the place of the change in the physical structure that we test. Similarly to testing in electronics, we usually start from a "known good" sample or "golden" sample to which all other samples are compared. The known good sample can be calculated from a simulated model, but most of the time it is coming from verification measurements.

It is usually the TIM layer that is different in the manufactured samples that are supposed to be of the same structure. The manufacturing process itself can vary, the thickness of the TIM material can be different for various undesired reasons, and it often happens that the TIM material that is frequently a paste is not covering the entire surface that it is supposed to be covering. The effect of any change in the TIM layer structure can significantly change the value of the heat transfer properties of the package structure.

This is very well demonstrated in the following simulation experiment, which repeats the pioneering measurement experiment, presented in [157].

Example 7.5: Detecting Die Attach Defects with the Help of the Structure Functions

A MOSFET in a TO220 package is simulated with 1.3 mm die size, considering different strength of die attach delamination. This series of simulations is mimicking a die attach delamination process, starting from the upper left corner of the die. The first case shows the perfect die attach, and the following cases represent the gradually growing die attach delamination. The numbers under the die sketches in Fig. 7.15 give, respectively, the side lengths of the square-shaped delamination and the percentage that the defect represents of the total die area.

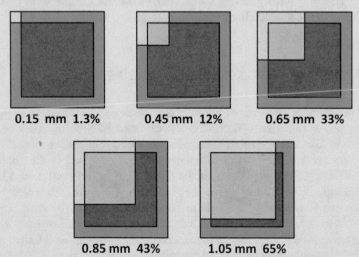

Fig. 7.15 Increasingly growing die attach delamination starting from the upper left corner of MOSFET in TO220 package. Blue square shows the chip; red denotes the active area. The orange squares highlight the considered defect

(continued)

Example 7.5 (continued)

As we can expect, the growing delamination results in growing die attach thermal resistance, which is well observable on the change of the structure functions referring to the different cases (Fig. 7.16).

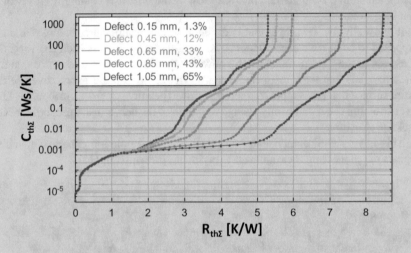

Fig. 7.16 The growing delamination appears in the structure functions as a growing die attach thermal resistance, shown by the parallel shift of the structure function curves from the point of the die

This well-observable change on the structure functions makes the method applicable for structure quality testing either after production, in the form of volume testing, or in the production line, enabling even in-line testing.

7.4 Reliability Testing, Ageing Monitoring

One of the primary purposes of the reliability testing of electronics components, modules, and systems is to determine the expected lifetime of a device, often described by the mean time to failure (MTTF) or mean time between failures (MTBF) parameters [158]. The main empirical method used is the accelerated lifetime testing [159], which exposes the tested device to excessive load beyond its normal application conditions in order to accelerate the main failure mechanisms, hence be able to conduct the tests in reasonable time frame. The most important output of these tests is the time to the failure of the tested devices; however, it always needs to be verified if the cause of the failure matches the expected failure mechanism. This is usually done by detailed characterization of the component before and

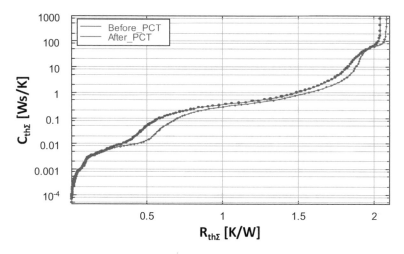

Fig. 7.17 Structure function of an IRF540N transistor in TO-220 package before (blue) and after (red) power cycling test (PCT)

after the tests with nondestructive test methodologies and by additional destructive testing after the device has been considered as failed (optical microscopy, x-ray microscopy, acoustic microscopy, cross-sectional analysis, etc.).

In case of thermally induced failure mechanisms affecting the structural integrity of the heat flow path and hence the thermal impedance of the structure, the thermal transient testing is a simple, quick, and nondestructive measurement option.

As it was discussed in Sect. 7.3, the structure function calculated from the measured thermal transient data can reveal structural differences between a golden sample with known structural and material parameters and an unknown sample. Similarly, the structural differences at various stages of the lifecycle of the same device can also be identified. A crack formed in a soldered layer due to the repeated load cycles or the degradation of a TIM material layer both lead to the increase of the thermal resistance of the affected layer. Consequently, they can be detected and located on the structure function plot by comparing the two curves corresponding to the new and the aged state of the same device. In Fig. 7.17, the structure function of an IRF540N transistor can be seen before and after power cycling test. A change in the structure can be clearly seen at about 0.3 K/W, which in this case corresponds to a die-attach delamination.

As the thermal transient testing requires only electrical access to the tested semiconductor device or module, intermediate tests can be added during the lifetime testing to evaluate the status of the degradation after a predefined amount of aging. Finally, by integrating the thermal transient testing and the accelerated lifetime testing systems, the thermal transient curves can be captured in regular intervals – enabling continuous monitoring of the degradation of the structural features in the device heat flow path.

This information can reveal the time-dependent progress of the degradation, e.g., ageing, and also a possible interaction of the different failure modes. The results can be fed back to the responsible manufacturing and development units, allowing the optimization of the relevant manufacturing process parameters or even modifying the device structure.

7.4.1 Active Power Cycling Combined with Thermal Transient Testing

For the accelerated lifetime testing of power transistors and transistor modules, active power cycling (APC) is a widely used method. During active power cycling, the power dissipated by the transistor itself is used to heat up the device, the chip, and the package, and after this the powering is turned off to let the device cool down.

The repeated heating and cooling cycles induce thermomechanical stresses in the package. These may either cause degradation in the heat conduction path, such as die attach delamination or crack of the base-plate solder, or the deterioration of the bond wires [160]. As the heating power is generated by the chip itself, the resulting temperature distribution inside the package can be very similar to the real application conditions.

In power converter and motor drive applications, the main switching transistors are driven at high frequencies (10..100 kHz or beyond) in order to generate the desired signal waveform, but due to the thermal mass of the chip, these short pulses cannot generate high temperature changes (see Sect. 2.8.2). Considerable temperature changes are induced by the changes in the load conditions, often described by mission profiles especially in the automotive applications [161].

As the temperature variation of the chip can be neglected, the power cycling is usually implemented with DC heating utilizing the conduction losses of the power transistors. These heating conditions are essentially identical to those used for the heating phase in thermal transient testing, which enables optimal combination of the two technologies. Theoretically, a thermal transient testing system can be used for active power cycling by repeating the same heating and measurement cycles until the device failure. However, while during thermal transient testing both the heating and the measurement need to be conducted until the system reaches the thermal equilibrium, in case of power cycling, usually much higher heating power (current) is applied to reach the desired temperature change in a much shorter time. Similarly, the cooling stage is shortened as well, in order to reduce the overall test duration.

In practice, the structural degradation is a slow process. For this reason, it is enough to insert a full-length thermal transient test measurement into the cycles after regular intervals, e.g., after every 1000 cycles, to enable monitoring the structural changes. The advantage of the integration of the two systems is that the intermediate tests can be run automatically without disassembling or even altering the tested structure. The measurement of regular monitoring parameters like on state voltage,

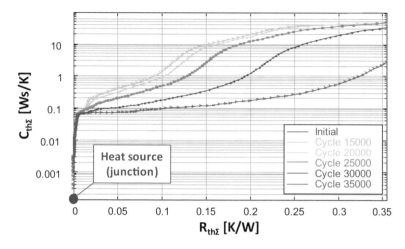

Fig. 7.18 Structure functions of an IGBT device recorded during a power-cycling test

minimum and maximum temperature, temperature swing, baseplate temperature, etc. can also help in separating the different failure modes. A detailed discussion about this is given in [162].

In Fig. 7.18, the structure functions generated from the regular thermal transient tests of an IGBT device exposed to power cycling are plotted after every 5000 heating cycles, starting from the 10,000th cycle.

The curves up to about 15,000 cycles fit nicely, indicating the structural integrity of the components. Starting from cycle 20,000, a gradually decreasing slope and increasing thermal resistance can be seen around 0.02 K/W corresponding to the die attach. As opposite to detailed testing after the device failure, this monitoring enables the investigation of the time dependence and propagation of the degradation as well [162]. Please, refer to [167] regarding further information on separation of failure modes in power-cycling tests of high-power transistor modules.

7.5   Determining Thermal Material Parameters

As many of the examples in this book have demonstrated, the heat dissipated in semiconductor components has to pass through a series of structural layers of different materials and geometry both inside and outside of the package. Although the structure functions calculated from the thermal transient response supply an accurate thermal RC equivalent model of the heat conduction path, the RC elements are not deterministic either for the geometry or the material properties of a certain layer. For example, a die attach with twice the thickness and twice the thermal conductivity would most likely appear similarly in the structure functions to its original counterpart. The Cauer model is usually sufficient for understanding the thermal behavior of a system, but for proper modeling and for generating material properties for the TIM (Thermal Interface Material) datasheets, additional analysis is necessary.

This section gives an overview of potential approaches to determine thermal material parameters using thermal transient testing.

7.5.1 Determining Thermal Data Directly from the Structure Functions

In quasi one-dimensional heat spreading cases, such as discrete power semiconductor components, LED-s, individually tested switch positions in a power module, etc., the RC elements resolved by the structure functions are in strong correlation to the real, physical RC values of the package layers. The less the heat spreads laterally in a material, the more this statement is valid. This makes it possible to derive material data from structure functions directly if the geometry is known. Equations (2.6) and (2.8) in Chap. 2 describe the basic formulae on how to convert thermal resistance or thermal capacitance information to material parameters, such as thermal conductivity coefficient or specific heat.

In our example below, we show how these latter two parameters can be determined from the measured structure functions.

Example 7.6: Determining Actual Material Parameters from the Structure Functions of an IGBT

In Fig. 7.19, both the (cumulative) structure function and its derivative, the differential structure function, are shown corresponding to an IGBT component on a DBC (direct bond on copper) substrate. The chip size was $7.3 \times 7.3 \times 0.3$ mm, indicated in area ①, and its thermal capacitance value was measured at 0.026 Ws/K. Based on these inputs to Eq. (2.8), the specific heat of the measured material is $1.62 \cdot 10^6$ Ws/m^{3}K, closely resembling the volumetric heat capacity of silicon. Areas ② and ③ correspond to the assumed die attach and top copper regions, respectively. The boundaries of these regions were identified by the slope changes in the (cumulative) structure functions and the corresponding inflexion points in the differential structure functions as discussed in Sect. 2.4.2. If the geometry is known, the effective thermal conductivity of these layers could be calculated based on Eq. (2.6).

(continued)

Example 7.6 (continued)

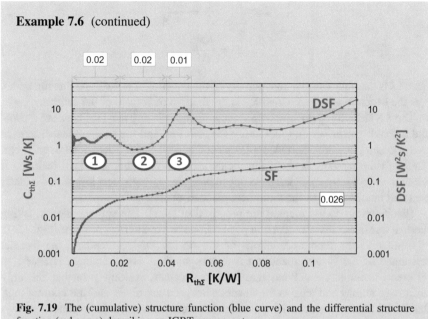

Fig. 7.19 The (cumulative) structure function (blue curve) and the differential structure function (red curve) describing an IGBT component

There are, however, some uncertainties in this method, as there is no clear boundary between the layers from the thermal perspective as multiple isotherms cross a single physical layer (see Fig. 3.16); also in most cases, it is hard to distinguish the thermal resistance of a structural element from the interfacial thermal resistances at its boundaries. For this reason, the above-explained method can only be applied to obtain effective thermal conductivity data.

7.5.2 Measuring Bulk Thermal Conductivity of TIM Materials

One of the biggest challenges of measuring the bulk thermal conductivity of TIM materials is eliminating the aforementioned interfacial thermal resistance values between the tested material and its environment. The ASTM D5470 standard [53] was designed to solve this problem by proposing a setup where the thermal resistance of the TIM is measured at multiple *bond line thickness* (BLT) values to cancel out the effect of the interfacial thermal resistance. It is assuming that the interfacial thermal resistance will not change with the BLT:

$$\lambda = \frac{\Delta L}{\Delta R_{\text{th}}} \cdot \frac{1}{A} = \frac{1}{m \cdot A} \tag{7.1}$$

$$m = \frac{\Delta R_{\text{th}}}{\Delta L} \tag{7.2}$$

where λ is the bulk thermal conductivity of the material, L is the bond line thickness, A is the heat spreading area, and R_{th} is the thermal resistance of the sample.

The ASTM D5470 document [53] recommends placing the TIM between a "hot" and a "cold" meter bar and measuring the heat flux and the temperature difference over the sample using thermocouples placed at well-defined locations in each bar. Although the method theoretically yields accurate measurement data, realizing it requires very careful considerations due to potential parasitic heat loss over the meter bars and inaccuracies of the thermocouple measurements.

Modifications to the method are proposed in [163]. The authors suggest taking thermal transient measurement of a power diode in a cylindrical package (Fig. 7.20).

Similar to the recommendations of the ASTM D5470 standard, the thermal resistance of the system measured from the junction to the ambient is tested: at different, accurately set bond line thickness values, making sure that the only difference among the tests is the distance between the diode and the cooling unit, that is, the BLT of the material. Although the measurement could be done in steady-

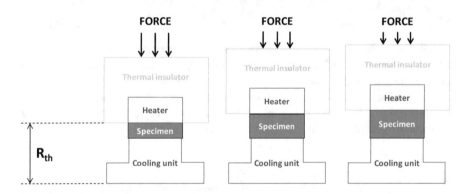

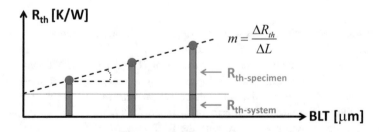

Fig. 7.20 Operation principle for bulk thermal conductivity measurement using thermal transient testing. The effect of the interface thermal resistances is eliminated by using the slope of the curve, inversely proportional to the thermal conductivity of the material

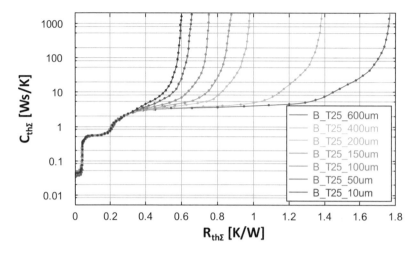

Fig. 7.21 Structure functions corresponding to the measurement of a thermal grease sample from 600 to 10 μm

state, it is recommended to do it in the transient space. In case of multiple transient measurements the first parts over the diode region of the structure functions are identical, also with prior reference measurements, proving the accuracy and consistency of the measured data [118].

In Fig. 7.21 all structure functions overlap up to approximately 0.4 K/W, indicating the R_{thJC} of the heater/sensor element. The flat plateaus after this point correspond to the thermal resistance of the TIM. The longer the plateau is, the higher the bond line thickness and the thermal resistance of the cooling unit.

This method has proven to work accurately for a wide range of TIM-s; however, originally it is aimed at characterizing highly conductive, low thermal resistance materials in order to avoid parasitic heat loss. For the measurement of more resistive materials, the parasitic heat loss can be accurately estimated using thermal simulation and can be compensated for during the test process.

Some examples of different test data measured this way are shown in Fig. 7.22.

7.6 3D Thermal Model Calibration for Simulations

Thermal modeling and simulation are of the most important design steps in product development. Creating a so-called digital twin of a real component or system allows engineers to make more solid design decisions, optimize design parameters, reduce cost, and improve time to market.

For thermal simulation, FEM and FVM (finite element method, finite volume method) solvers are both used. CFD (computational fluid dynamics) is perhaps the most common numerical method to analyze the behavior of fluids, most solvers

R_th [K/W]

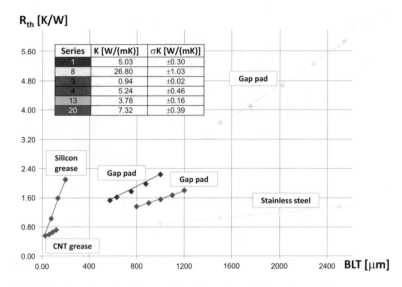

Series	K [W/(mK)]	σK [W/(mK)]
1	5.03	±0.30
8	26.80	±1.03
3	0.94	±0.02
4	5.24	±0.46
13	3.78	±0.16
20	7.32	±0.39

Fig. 7.22 Bulk thermal conductivity test data of a variety of thermal materials

being capable of handling all three major heat transfer methods as well. Simulations are however only as accurate as their input parameters, making it essential to create models which reflect the thermal behavior of the real components both in steady-state and in transient applications.

To create a detailed numerical model of a component, the exact geometry and the material properties of its structural elements have to be defined. Geometry is typically known for component makers; however, the bond-line thickness of a die attach material, the solder at the base plate, or any other TIM may vary with manufacturing. The same applies to material parameters, as the properties of metals and semiconductors are typically easy to measure in bulk with high accuracy, but most thermal interface materials behave differently in the real application. Even if the bulk thermal conductivity coefficient is known, e.g., applying the method introduced in Sect. 7.5.2, the interfacial thermal resistances between the layers can be significant, and they are not known in the design phase. These values can be determined only by calibrating the simulation model to real measured results. The calibrated models can be used later on in further design steps as the accurate, thermally verified digital twin of the structure.

To understand the need for calibrating the thermal model, let us have a look at Fig. 7.23. This figure shows a typical mismatch between an uncalibrated simulation model's thermal behavior and real test data.

Structure functions are ideal tools to serve as a test-based reference for model calibration. In one dimensional cases, they directly indicate the location and extent of the difference of the model parameters from reality. But even in truly three dimensional heat spreading cases, they can act as a reference by enhancing minor changes in the original time domain transient response [151, 123]. Some high-end CFD

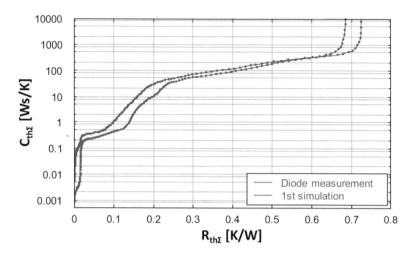

Fig. 7.23 Structure function based on an uncalibrated model (red) vs. real test data (blue)

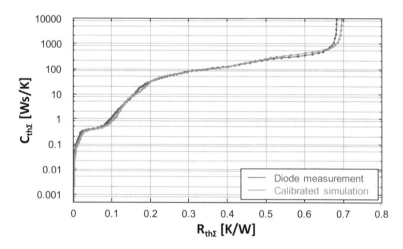

Fig. 7.24 Structure function based on a calibrated model (green) vs. real test data (blue)

solutions use structure functions to find calibrated model parameters as simulation input [145]. By selecting uncertain geometry or material parameters, setting up their probable variable range, the tools are capable of selecting the best matching parameter set after a number of iterative transient simulations.

Figure 7.24 shows the results of a calibration study where not only the component, but its near environment was also calibrated to a reference test data. In practical cases, changing the thermal conductivity values of TIM layers, introducing interfacial thermal resistances between key structural elements, and the proper selection of the semiconductor active area are the most sensitive variables.

Calibrating a thermal model is particularly important for transient simulation cases, where the individual RC values of each model elements play an important role unlike in steady-state situations. Transient simulations are also important in the cases where direct temperature measurements are either impossible or very hard to realize. These could be measuring the junction temperature of an IGBT in an electric vehicle in operation or simply understanding the temperature field in a multi-core SoC in complex, changing power scenarios.

As transient CFD simulations of a complex system are usually time consuming, some CFD tools allow the export of ROM-s (reduced-order network model) to be solved much quicker (over 40k times faster than CFD) [164, 165]. Of course ROM-s have to be generated based on calibrated CFD models. ROM-s not only allow rapid simulations, but they can be solved faster than real time, allowing them to be used in system level 1D simulations, control system design, or HIL (hardware-in-the-loop) systems [166].

7.7 Deriving Compact Models from Measured Data for Various Purposes

Generating compact models is a very frequent use of thermal transient measurements. Various methodologies exist for this. We have to distinguish them on the resulting models that they deliver: if these models are truly boundary condition independent models enabling fast 3D thermal simulation in the form of reduced-order models or delivering simple models enabling fast and easy "on the back of an envelope" calculations without needing a thermal simulator.

7.7.1 Simple Models Enabling Fast 1D Heat Flow Calculations

As introduced in Chap. 2, structure functions are essentially multi-element Cauer models of the heat conduction path between a semiconductor junction and the ambient. They contain typically over 100 thermal RC elements, describing both the internal features of the tested package and the thermal performance of the external cooling solutions. Using the TDIM method (introduced in Chap. 2, demonstrated in Chaps. 3 and 7), one can determine the junction-to-case thermal resistance of a component and potentially cut the chain of RC elements at that point (Figs. 7.25 and 7.26).

Figure 7.27 illustrates the composition of a one-dimensional RC compact model based on the TDIM methodology. A discrete IGBT component was tested at two different boundary conditions; the junction to case thermal resistance was determined by defining a difference of $\varepsilon = 0.04$ in the corresponding structure functions. An automated process in the software toolkit of [54] generated five RC ladder pairs until the junction to case thermal resistance point at 0.52 K/W.

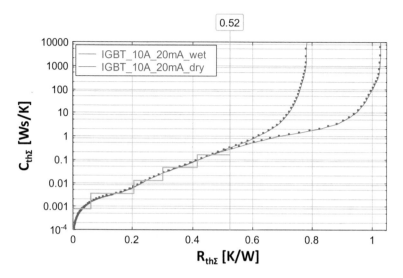

Fig. 7.25 Bifurcation point highlighting the R_{thJC} value of a discrete IGBT component. Five-stage compact thermal model is fitted on the same structure function between the junction and the selected "case" point

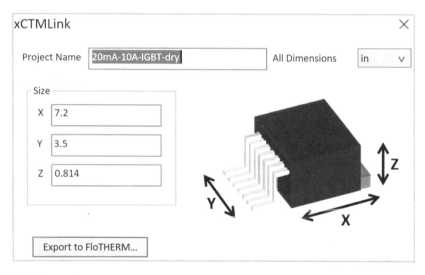

Fig. 7.26 Example export window from a thermal test environment to the Siemens SIMCENTER Flotherm tool

The resulting RC ladder may be a suitable thermal description of the package if the heat spreading problem is essentially 1D, and the temperature distribution over the package surface can be considered close to uniform. This may be the case for common package types such as TO-220, TO-263, and some LED packages, especially with larger chip sizes compared to the case area. The description will definitely

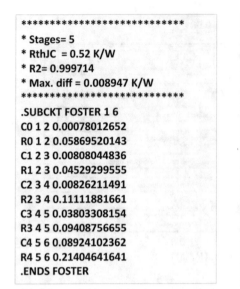
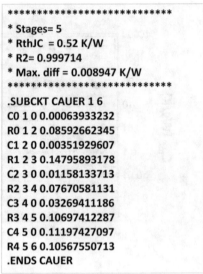

```
****************************
* Stages= 5
* RthJC  = 0.52 K/W
* R2= 0.999714
* Max. diff = 0.008947 K/W
****************************
.SUBCKT FOSTER 1 6
C0 1 2 0.00078012652
R0 1 2 0.05869520143
C1 2 3 0.00808044836
R1 2 3 0.04529299555
C2 3 4 0.00826211491
R2 3 4 0.11111881661
C3 4 5 0.03803308154
R3 4 5 0.09408756655
C4 5 6 0.08924102362
R4 5 6 0.21404641641
.ENDS FOSTER
```

```
****************************
* Stages= 5
* RthJC  = 0.52 K/W
* R2= 0.999714
* Max. diff = 0.008947 K/W
****************************
.SUBCKT CAUER 1 6
C0 1 0 0.00063933232
R0 1 2 0.08592662345
C1 2 0 0.00351929607
R1 2 3 0.14795893178
C2 3 0 0.01158133713
R2 3 4 0.07670581131
C3 4 0 0.03269411186
R3 4 5 0.10697412287
C4 5 0 0.11197427097
R4 5 6 0.10567550713
.ENDS CAUER
```

Fig. 7.27 Foster and Cauer network elements of the component examined in Fig. 7.25

not work for power modules with large surface area, multi heat-source problems, and most IC packages.

For the valid cases, the so-obtained compact thermal model is typically reduced to 3–5 RC stages, making it simpler and applicable for publication in product datasheets.

Beside their use in datasheets, compact models can also be used as simulation models in thermal simulator tools, exported directly from the test environment. To use them this way, the geometry of the package and the parallel thermal resistance between the junction and the top of the package have to be defined for the simulator tool to be able to match the model with the 3D environment and to be able to estimate the temperatures on the top surface of the package.

Due to their simplicity, control system design engineers also often use test-based compact thermal models for thermally aware system design of electric drives and motor controllers as an example. To support this application, the thermal network can be described in a SPICE netlist, either in a Foster or a Cauer network format, as shown in Fig. 7.27.

In more complex cases, such as LED-s, the compact thermal model can be combined with the electrical and optical characteristics of the component to create a multi-domain component description.

Although compact thermal models are often used and widely supported by thermal and network analysis tools, their limited applicability has encouraged researchers to find more generic modeling solutions.

7.7.2   *Boundary Condition Independent Model Generation with the Help of Transient Measurements*

Detailed 3D numerical models can be generally applied for most thermal problems; however, FVM or FEM solvers can be relatively slow especially at larger cell sizes. These models also reveal detailed structural and material information about the component, so they often cannot be shared among companies due to confidentiality issues.

Boundary-condition-independent reduced-order network models (BCI ROM-s) discussed in [164] offer a new possibility to transform detailed thermal models to an equivalent network which can be simulated in SPICE-like simulators faster than real time, while hiding the construction of the component. The method supports multi heat-source problems and can model components accurately in a wide range of predefined boundary conditions. BCI ROM-s should be created based on calibrated thermal models to make sure that their input data set is accurate. This workflow is now also implemented in a commercial toolset,[1] and it works as shown in Fig. 7.28.

In Fig. 7.29, an actual physical module is shown with its detailed model. In Step 1 of the workflow, the measured thermal transients on the physical sample are compared to simulated transients on the model with detailed geometry and assumed material parameters.

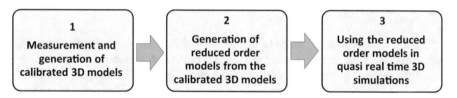

Fig. 7.28 Workflow to support the generation of reduced-order thermal models from simulation and measurement data

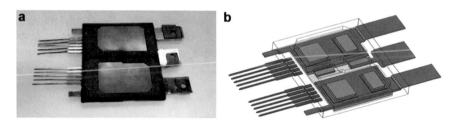

Fig. 7.29 (a) The physical module and (b) its model with detailed geometry and assumed thermal parameters of the materials

---

[1]The workflow is embraced by the SIEMENS SIMCENTER thermal test and simulation portfolio.

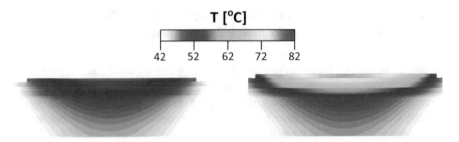

Fig. 7.30 Temperature distribution in the initial and in the calibrated detailed model

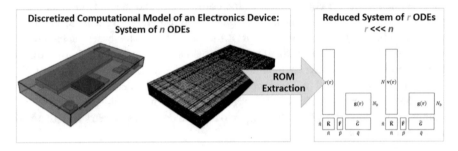

Fig. 7.31 Reduce-order model generated from the calibrated detailed 3D model

In the initial state of Step 1, before the calibration, the structure functions generated from the transient measurement do not match correctly, as indicated by the different steady-state temperature distributions in Fig. 7.23. Figure 7.30 illustrates the temperature distribution in the initial and in the calibrated detailed model at 100 W powering [168].

Carrying out the calibration procedure defined in Sect. 7.6, the match corresponding to Fig. 7.24 can be achieved. The calibrated model can now be used for generating the reduced-order model, as shown in Fig. 7.31.

The resulting models can be packaged in the so-called FMI (functional mockup interface) to be used with system level simulation or control system design tools while carrying the accuracy of a 3D simulation model with faster than real-time solver speeds.

7.7.3 Derivation of Boundary Condition Independent Models and Thermal Metrics by Measurements at Different Thermal Boundaries

In the previous section, we presented a way how thermal transient measurements can be used for compact model generation. As it was mentioned in Chap. 3, thermal transient measurements are often used also to determine thermal metrics. Here now,

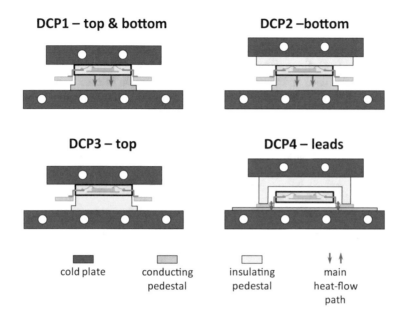

Fig. 7.32 The four DELPHI measurement setups

we present how the thermal transient methodology can be used for the characterization in the case of one of the latest and most advanced packaging technology of power devices, that is, devices with double-side cooling. To understand it, we have to go back to the DELPHI methodology [86, 138].

The DELPHI methodology has been elaborated to create simple, easy to be used, boundary condition independent compact models for packages. Several resistor arrangements are defined, representing the different heat flow paths of different package types, and 67 different boundary conditions are specified, with which the detailed model of a given package has to be first simulated. An optimization process has been established to select out of these results the best performing model of the package. The models are finally calibrated and verified with experimental methodologies in the predefined measurement setups of Fig. 7.32.

The DELPHI methodology can be well used both for device characterization and gaining parameters for numerical models. The four "hard" boundary conditions for testing R_{th} values at extreme points are shown in Fig. 7.32.

The four measurement setups are defined as follows:

- DCP1: No spacers between the package and the cold plates, mimics jet impingement tests
- DCP2: Insulating spacer on top, mimics use of a heat sink
- DCP3: Insulating spacer on bottom, representing a top heat sink
- DCP4: Insulating spacer on top, thin insulator on bottom, between leads. Represents component mounting without a heat sink.

These measurement setups are recommended also for the TSP calibration environment, getting rid of the uncertainties of single cold plate calibration [92].

The following example considers DCP1 arrangement for the more and more often used double-side cooling of power packages. In case of such packages, it is not easy to determine the place and value of the actual degradation in case of power cycling ageing testing. The conclusions of the calculation may help in understanding the results of ageing testing.

Example 7.7: Calculations on the R_{th} Change in Case of Double-Side Cooling

Let us assume that the thermal resistance toward "one side" is R_1; toward the "other side," it is R_2.

For reducing the number of variables to one, suppose there is a ratio of S between the two, $R_2 = S \cdot R_1$.

The resulting total thermal resistance is

$$R_{\text{sum}} = \frac{R_1 \cdot R_2}{R_1 + R_2} = \frac{S \cdot R_1^2}{(1+S) \cdot R_1} = \frac{S}{1+S} \cdot R_1 \qquad (7.3)$$

In the trivial case, when the two sides are symmetric, $S = 1$ and $R_{\text{sum}} = R_1/2$.

At $S < 1$, R_2 is the side with better cooling, at $S > 1$, R_2 is the worse side.

Now let us assume that the cooling on the side of R_2 degraded with a degradation rate of E, the thermal resistance has grown to $R_{2+} = E \cdot R_2$. Accordingly, $R_{2+} = S \cdot E \cdot R_1$.

R_{sum} also changed to $R_{\text{sum}+}$; the degradation has its effect in an attenuated way:

$$R_{\text{sum}+} = \frac{R_1 \cdot R_{2+}}{R_1 + R_{2+}} = \frac{E \cdot S \cdot R_1^2}{(1 + E \cdot S) \cdot R_1} = \frac{E \cdot S}{(1 + E \cdot S)} \cdot R_1 \qquad (7.4)$$

Instead of E, only an attenuated Z degradation factor is experienced, Z is the ratio of the total thermal resistance after/before degradation:

$$Z = \frac{R_{\text{sum}+}}{R_{\text{sum}}} = \frac{ES}{1 + ES} \cdot \frac{1+S}{S} = \frac{E(1+S)}{1+ES} \qquad (7.5)$$

(continued)

Example 7.7 (continued)

Figure 7.33 shows the resulting parallel thermal resistance of double-side cooling depending on the ratio of the thermal resistances toward the two sides. $S = 1$ corresponds to equal R_{th} into the two directions, $S > 1$ means that the change of E occurred on the worse side, and $S < 1$ means that the change of E occurred on the better side. Curve of 0% corresponds to no degradation; it just expresses the resulting parallel thermal resistance of the double-side cooling. The $S = 1$ and "$E-1$" $= 0\%$ point corresponds to symmetric thermal resistance with no degradation; accordingly, a half of the single-side thermal resistance is experienced due to parallel arrangement.

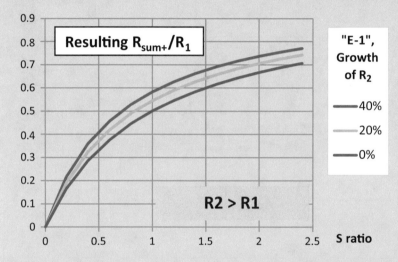

Fig. 7.33 Resulting parallel thermal resistance of double-side cooling depending on the ratio of the thermal resistances toward the two sides. Degradation of 20% and 40% in R_2 shown

A more interesting chart is shown in Fig. 7.34. It represents the experienced Z change in the parallel cooling at different S thermal resistance ratios of the sides and E degradation rate in R_2. At single-side cooling ($S = 0$), one experiences the full change, $Z = E$. Otherwise, the intact R_1 side causes an attenuation in the observed change, $Z < E$.

It can be seen that even at symmetric cooling a 40% degradation seems to be just around 17% Z change, because most of the heating power is directed toward the healthy side, demonstrating the efficiency of double-side cooling.

(continued)

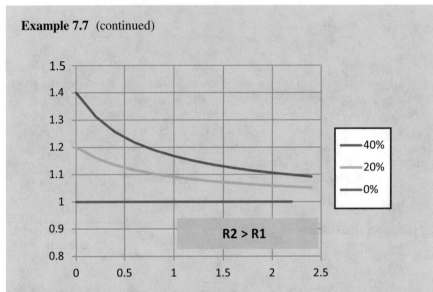

Example 7.7 (continued)

Fig. 7.34 Experienced Z change of the total thermal resistance at double-side cooling, depending on the ratio of the thermal resistances toward the two sides

Chapter 8
On the Accuracy and Repeatability of Thermal Transient Measurements

András Poppe and Márta Rencz

Uncertainty, accuracy, repeatability, etc. are among the most important, generic problems of metrology. In modern metrology, estimation of measurement uncertainties is a basic requirement. Therefore, in this chapter, we aim to introduce the concept of measurement uncertainties first in a wider context from metrology aspect. After this, we aim to provide an inventory of possible sources of uncertainties in thermal transient testing.

8.1 Introduction to the Concept of Measurement Uncertainty

International metrology guides suggest that for the measurement of a certain quantity, a model of the measurement has to be set up, ideally based on analytic model equations. In Chap. 2, the theoretical background of measuring thermal resistance/impedance was given, and Chap. 3 provided a short summary of the standardized versions, the thermal resistance/impedance measurements that form the daily thermal testing practice in the electronics industry. For example, Eq. (2.19) in Chap. 2 for the *thermal impedance in general* or Eq. (3.4) or (3.8) of Chap. 3 for the *junction-to-ambient thermal resistance* or Eq. (3.9) for the *junction-to-ambient thermal impedance in particular* can be considered as such model equations. For the *real thermal resistance* and real *thermal impedance measurement of LEDs* e.g. Eqs. (6.33) and (6.34) are the measurement model equations.

A. Poppe · M. Rencz (✉)
Siemens Digital Industry Software STS, Budapest, Hungary

Budapest University of Technology and Economics, Budapest, Hungary
e-mail: rencz.marta@vik.bme.hu

© The Author(s), under exclusive license to Springer Nature Switzerland AG 2022   353
M. Rencz et al. (eds.), *Theory and Practice of Thermal Transient Testing of Electronic Components*, https://doi.org/10.1007/978-3-030-86174-2_8

As it can be seen from these model equations, the thermal resistance is derived from the SI *base units* of *temperature* and *current* and from the SI *derived unit* of *voltage*, and thermal impedance measurements also rely on the measurement of *time*. In case of LEDs, the derived unit of the *emitted optical power (radiant flux)* is also involved.

The international definitions of these units are maintained by BIPM,[1] involving four international metrology areas, as follows:

- For measuring the electric *current, voltage,* and *power: international metrology in the field of electricity and magnetism*
- For measuring the *temperature: international metrology in the field of thermometry*
- For measuring the *time: international metrology in the field of time and frequency*
- For measuring the emitted *optical power: international metrology in the field of photometry and radiometry*

The *National Metrology Institutes*[2] (NMI-s) deal with the actual realizations of these units and maintain the corresponding etalons for the calibration of their own equipment and for the calibrations of the test equipment of *commercial calibration laboratories*.

These commercial laboratories provide calibration services for the industry, e.g., for *test equipment manufacturers*.

Ordinary *end-users of the test equipment* rely on the calibration certificates of their test equipment that are issued by the manufacturer. This way the values they measure can be traced back to the primary etalons of the NMIs.

In the above-unbroken chain of calibrations that is called *tractability or traceability chain*, there are multiple players involved. NMIs spend every effort in the implementation of their own measurement capabilities with the *highest possible accuracy* and *lowest possible uncertainty* available through the state of the art of the technology. From time to time, NMIs organize key comparisons of measurements of different units; therefore, one can speak about the NMI-level world average of *measurement uncertainty* (MU) of a given unit of measure.

The next level in the traceability chain is represented by the different commercial testing and calibration laboratories. Their etalons are based on regular calibrations performed by NMIs; thus, the uncertainty of their measurements adds to the uncertainty of the NMIs. Obviously, as further we are in the traceability chain from the primary standards of an NMI, the higher is the uncertainty of the measurement of a given unit.

In the assessment/calculation of the measurement uncertainty, one relies on the uncertainty of the measurements stated on the calibration reports of the metrological entity that precedes the given entity in the traceability chain.

---

[1] International Bureau of Weights and Measures (Bureau international des poids et mesures, BIPM), https://www.bipm.org

[2] Such as NIST in the USA, PTB in Germany, AIST in Japan, KRISS in South, etc.

For example, in the temperature measurements for obtaining the thermal resistance/thermal impedance of semiconductor device packages, the weakest element in the traceability chain is the "TSP calibration" performed by the user of the thermal test equipment in a field laboratory.

Regarding this, the uncertainty factors related to this device calibration process detailed in different sections of Chaps. 5 and 6 are added to the uncertainty stated on the temperature calibration certificate of the thermostat being used. The advantage of the differential formulation of the *junction-to-environment* X thermal resistance (see Sect. 3.1.5) is that no further item needs to be considered for the uncertainty calculation of the junction temperature change used for the calculation of the thermal resistance/impedance – see also Sect. 6.10.5.

If the absolute junction temperature is to be identified, for example, during the combined thermal and radiometric/photometric measurements of power LEDs using Eq. (6.37) of Sect. 6.10.6, then it is important to use the same temperature-controlled cold plate for the TSP calibration and for the actual measurement; otherwise, if two different thermostats are used, their individual uncertainties need to be combined to obtain the final, combined uncertainty of the LEDs' junction temperature setting/ measurement.

For the sake of a common understanding on the concepts and methods behind the calculation of measurement uncertainties, the *Joint Committee for Guides in Metrology (JCGM)* and their partner organizations who are stakeholders in different fields of international standardization (such as ISO, IEC) have published the *Guide to the Expression of Uncertainty in Measurement* often referred to as GUM [49].

Some NMIs like NIST and bigger organizations such as NASA derived and published their own handbooks on the assessment of measurement uncertainty [50, 51]. Some standardization bodies in specific fields of metrology also issued their own guides on the assessment/calculation of measurement uncertainties, such as CIE's guide on *determination of measurement uncertainties in photometry* [52]. Some specific measurement guidelines such as CIE's recent document on the measurement of power LEDs [45] provides examples on the assessment of the measurement uncertainties related to the quantities they deal with.

Unfortunately, in the field of measuring thermal resistance/thermal impedance of semiconductor device packages, the relevant stakeholders (such as the JEDEC JC15 committee) did not provide their specific guidelines about the assessment and quantification of the possible factors of the uncertainty of the measurement of thermal resistance and thermal impedance. The topic exceeds the limits of this book chapter, and the authors of this book do not dare to undertake the years' long task of a standardization committee in this regards. To highlight the complexity of the issue, in Fig. 8.1, we only quote a diagram from a temperature test chamber manufacturer that illustrates the factors of uncertainty of measuring the temperature of such chambers.

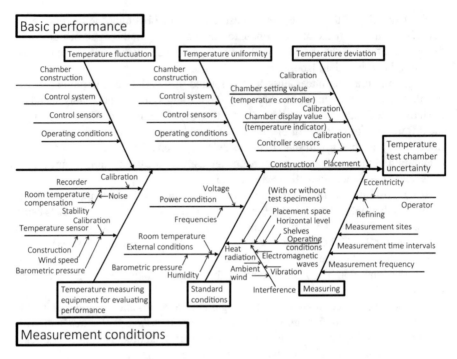

Fig. 8.1 An overview chart about different factors influencing the temperature measurement in a temperature chamber. (Source: [169])

8.2 Common Factors of Measurement Uncertainty

Uncertainty estimation has always been a major issue in modern metrology, and since the first publication of the GUM, it is better and better understood, though it is not a topic that is easy to master as some knowledge in statistical analysis is also needed. As a general introduction to the topic, it is worth quoting a few sentences from Ref. [51]:

> Measurement uncertainty is an estimation of the potential error in a measurement result that is caused by variability in the equipment, the processes, the environment, and other sources.
>
> Every element within a measurement process contributes errors to the measurement result, including characteristics of the item being tested.
>
> Evaluation of the measurement uncertainty characterizes what is reasonable to believe about a measurement result based on knowledge of the measurement process. It is through this process that credible data can be provided to those responsible for making decisions based on the measurements.

Besides the detailed handbooks already quoted [51], there are a few, more easy to read works also available [170, 171]. Based on these handbooks, hereby we provide a short overview of the factors that need to be considered related to the measurement of thermal resistance/impedance of semiconductor device packages.

8.2.1 Measurement Process Selection

In terms of the measurement process, there is little choice left. All thermal resistance/ impedance measurements of semiconductor device packages are based on JEDEC's JESD51-x series of standards [29], defining the so-called *electrical test method* [30], in case of certain device types like LEDs with a few further clarifications and additions [43, 44].

The choice left is between the JEDEC JESD51-1 *dynamic* or *static* test methods. In prior chapters of this book, several reasons have been given that suggest that the *static test method* with the extension of continuous transient measurement is the best choice; see Sect. 5.4 as well as Ref. [115]. Many aspects of these test methods have been discussed there except the question of measurement uncertainty. As in case of the JEDEC JESD51-1 dynamic test method, the complete heating curve of a device is composed from a series of measurements to single heating pulse; the uncertainty of a measurement based on this test method is obviously larger than the uncertainty of the measurement of the same device conducted according to the transient extension of static test method where the response to only a single switching in the heating power is involved.

The complete measurement process is best elaborated for LED devices. For this reason, we refer to the detailed example given in Sect. 6.10.3, which uses the transient extension of the static test method for power LEDs.

8.2.2 Measurement Error/Accuracy

All measurements are accompanied by error; one's lack of knowledge about the sign and magnitude of measurement error is the *measurement uncertainty*.

Prior to the publication of GUM [49], measurement errors were categorized as either random or systematic. The random components of the measurement error change when measurements are repeated; the systematic components remain constant in the repeated tests for the same quantity. According to the concepts of GUM, there are errors only, irrespective of their nature; therefore, each error is considered as a random variable contributing to the overall uncertainty of a measurement. Therefore, the different measurement process errors are the basic elements of uncertainty analysis.

- In the case of thermal transient testing, these different sources that disturb the measurement can originate from the transient behavior of the DUT, from the tester equipment and the related instrumentation, and/or from the ambient, not to mention the possible operator bias and the possible numerical and computational errors. As an example, the $Z_{th}(t)$ curves are measured and post-processed on a logarithmic time scale. The discretized logarithm of the elapsed time definitely differs from the real logarithm of the time. This adds to the overall, inherent noise present in the measured transients.

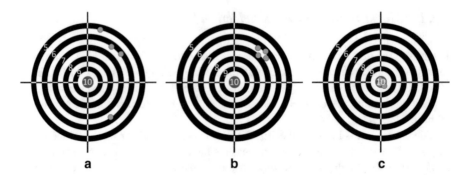

Fig. 8.2 Illustration of the concepts of accuracy and repeatability of measurements repeated within a short time period. (**a**) neither repeatable, nor accurate, (**b**) repeatable but not accurate, (**c**) repeatable and accurate

- The NID method that is used to obtain structure functions has its own theoretical and practical limits [58] that lead, for example, to limiting the structural details that can be resolved by the structure functions. These are factors that also contribute to the uncertainty budget for certain structure function-based measurement methods.
- As the GUM [49] recommends, all the above factors need to be quantified either based on a statistical approach (these are called Type A uncertainties) or have to be elaborated by well-established engineering estimates (Type B uncertainties).

Before discussing these with more details, let us specify what we mean by accuracy and repeatability of measurements from an engineering standpoint with the help of Fig. 8.2, where we use the following notions:

- *Accuracy* is the degree of closeness of measurements of a quantity to that quantity's true value. Note that accuracy can be declared only within a certain bound. As seen in Fig. 8.2c, even in case of the highest repeatability and accuracy, the subsequent "shots" differ a little from one another, though they are all located within the smallest, innermost circle that represents the targeted smallest region. The most important aspect of measurement uncertainty is how small the area of acceptance is around the targeted "true value."
- *Precision* is the degree to which repeated measurements under unchanged conditions show the same results; precision can relate to:

  - *Repeatability* – the variation of measurements with the same instrument and operator and repeated in a short time period
  - *Reproducibility* – the variation among different instruments and operators (even among, e.g., different laboratories) and over longer time periods

- *Resolution* is the smallest change which can be detected in the quantity that it is measured.

Reproducibility could be illustrated in the same way, but over longer time periods of time and eventually at different laboratories and instrumentation.

Resolution can be defined in case of measurements where the results are transformed into digital values. In this case, it may correspond to the thickness of the black and white rings of the target circles in Fig. 8.2; i.e., if a shot is anywhere in a thick ring, the digital value associated with that shot is the number of the ring.

8.3 Factors in the Uncertainty of Thermal Transient Measurements

Several factors contribute to the uncertainty of thermal transient measurements. These may origin from the measuring equipment, from the methodology, from the test environment, or from the uncertainties in the operation of the devices themselves. In this section, we shortly summarize the most important and most frequent ones.

8.3.1 The Transient Behavior of the Tested Electronic Devices

As it was discussed in Sect. 5.4.1 in details, in case of switching the point of operation of a semiconductor device, electrical and thermal transients occur simultaneously. The electrical transients are very fast but hide the first part of the measured thermal transient curve that has to be extrapolated. This extrapolation brings in certain amount of uncertainty, but this effects only the very first part of the measurements, and it may be corrected by the measuring instrument. The caused error depends on the construction of the device itself, but it is affecting less than the first microsecond of the measurement.

8.3.2 Effects of the Test Equipment

Let us shortly consider the most common equipment-related sources of measurement errors. In case of the thermal resistance/impedance measurement of semiconductor device packages, the *offset* and *scale/calibration errors* can be handled by the proper choice of the test method/procedure. As discussed in prior chapters, if the JEDEC JESD51-1 electrical test method is applied in a differential approach, using solely the temporal difference of the junction temperature (see Sect. 3.1.5) and if the same current sources and voltage meters are used both for the TSP calibration and the actual measurements, the offset and scale errors cancel out, or, at least, their effect can be reduced to the possible minimum (as discussed in details for LEDs in Sect. 6.10.5).

Linearity and *quantization errors* cannot be mitigated on the level of test procedures. These parameters should be provided on the specifications sheet of the test equipment to allow application level-combined uncertainty calculations by the end-user.

Properties of the Data Acquisition Channels

As discussed above, the data acquisition channels of the measurement instrument may have some *gain and offset errors*, and in case of transient measurements, the bandwidth of the channels limits the details of the thermal impedance that can be resolved with the measurement.

The finite resolution and the accuracy of the junction temperature measurement also depends on the device under test and on the properties of the thermostat used for the TSP calibration.

As a short example, if we assume that the junction temperature-induced change of a diode's forward voltage fits into a 50 mV differential voltage measurement range and the measured voltage change values are digitalized with 12 bits, the least significant bit corresponds to 12.2 μV. Assuming that the temperature sensitivity of the diode's forward voltage is cca. -2 mV/K, this would result in a temperature resolution below 0.01 °C.

The possible linearity error of the data acquisition channels can be checked with the help of an electronic only golden reference device discussed later.

Noise

In thermal transient measurements, we measure in most of the cases electrical signals, current, and voltage values. These value can be measured today with an extremely high accuracy/low uncertainty. In case of measuring with electrical signals, the fundamental source of uncertainty is the always present *noise* on the electrical signals. When the electrical signals are converted into digital ones, the conversion adds the quantization noise to the budget, coming from the finite bit length of the digital representation. The effect of noise on the electrical signals usually can be mended with high sampling rate and with repeating and averaging the measurements. The devices under test are also sources of noise. The different mechanisms in device operation result in different noise characteristics for diodes and transistors. These affect the choice of the measurement current. For further details, refer to Sect. 5.7.

Stability and Linearity of the Equipment and the Way of Testing It

Among the test equipment-related uncertainty issues, a very important factor is the stability of the test equipment. Stability of the test equipment means how the metrological properties of a test equipment remain constant in time, involving not only the actual readings of the values provided by the test equipment but including also the measurement process itself. Thus, equipment stability is also a factor to be considered in the overall uncertainty budget of a measurement.

The usual way of testing the stability of a testing apparatus is to maintain a physical object, an artifact, serving as an etalon that possesses a constant property that one aims to measure and use that physical object to check regularly how that object's property is measured by the test equipment. A widely known example for such an object was the international mass etalon, the 1 kg platinum-iridium cylinder known as the International Prototype Kilogram (IPK) maintained at BPIM in France that was used until recently as the base reference for measuring mass.[3] Similar to the IPK, one may think of establishing a physical prototype to realize a thermal resistance/thermal impedance etalon that could be used to check the stability of a thermal transient test equipment from time to time.

As described in Chaps. 3, 5, and 6, the thermal resistance/impedance measurement of semiconductor device packages is based on the electrical test method, meaning that the test equipment measures the changes of a temperature-sensitive parameter that indicates the change in the junction temperature. The measured change of the electrical signal is transferred to the temperature scale through the TSP calibration process that relies on the temperature control and the measurement capability of the calibration environment. Therefore, if this transfer to the temperature scale is separated from the measuring equipment, a real, physically realized thermal RC system as a thermal impedance etalon can be completely replaced by an electrical RC system that is built of components with precisely known resistance and capacitance values and with a long shelf lifetime.

This way, electronic golden reference devices can be built. Such a device is a linear, passive electrical RC circuit that comprises high-precision discrete resistors and capacitors. The schematic of such an electrical circuit was provided in Fig. 2.10 of Example 2.3 provided in Chap. 2.

A practical example of such a device attached to a test equipment is shown in Fig. 8.3. Besides getting rid of the need of the temperature calibration, the advantage of using such an electrical-only device for the stability checking of the thermal test equipment is that all uncertainties due to the thermal test environment are completely eliminated from these regular equipment stability control tests.

Another advantage of such electrical-only reference devices is that the sources of the I_H heating current and the I_M measurement current can be tested separately.

Besides checking the current sources and the voltage meters of the test equipment, its time base can also be tested by measuring the time-domain impedance function of the reference device and by identifying the time constant spectrum from this.

Figure 8.4a presents the unit step response of such an electrical-only golden reference device (translated to a quasi-thermal impedance by a dummy, constant temperature sensitivity). Applying the realization of the NID method with predefined numerical parameters yields the time constant spectrum of the circuit; see Fig. 8.4b.

---

[3] Effective from 20 May 2019, new definitions of the SI base units are used, no longer requiring physical artifacts to be used as prototypes for an SI unit.

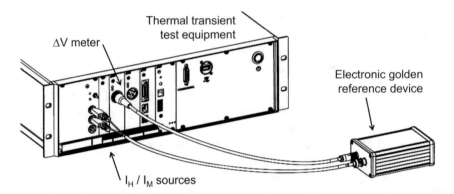

Fig. 8.3 A passive electric RC circuit built of high precision discrete resistors and capacitors with precisely known unit-step response (e.g., a four-stage Foster network) used as a golden reference device to test the stability of the impedance measurement capability of a thermal transient tester equipment

The electronic-only golden reference device shown here has four well-defined time constants. If the locations and the heights of the maxima are within a tolerance band specified by the equipment manufacturer, the current, voltage, and time measurement capability of the equipment can be considered unchanged. This condition can be identified through a manual checklist or by a fully computerized process.

The same golden reference device can also be used for checking the linearity of the measurement channels: the same impedance measurement should be performed at different current levels. If the resulting impedance curves and time-constant spectra are the same, then the linearity of both the measurement channel and the golden reference device should be assumed. For further insight, refer to Example 2.13 in Sect. 2.12.3.

8.3.3 Reproducibility Issues of the Test Environment

The R_{thJX} *junction to reference environment* thermal resistances are very sensitive to the applied test conditions. The two most common environments, still-air and infinite heat-sink (cold plate), represent practical limits of test environments (see Sect. 3.1). JEDEC's JESD51-x family of thermal testing standards provide specifications for the test environments with the aim of reducing the measurement uncertainties to the possible minimum and supporting the highest possible reproducibility of the measurements while allowing high level freedom for the physical realization of the environments.

For example, measuring IC packages in a natural convection test environment, i.e., in a JEDEC JESD51-2A [31], compliant still-air chamber also assumes using one of the JEDEC standard thermal test boards that would fit the package type.

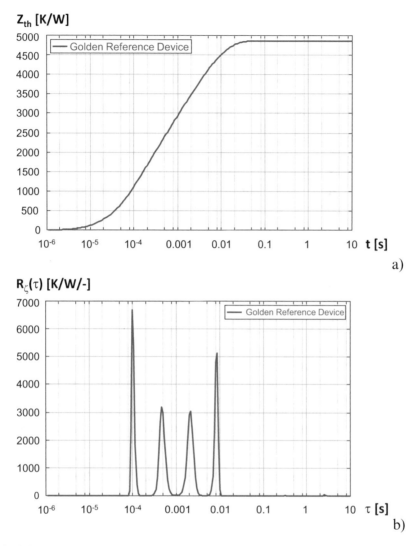

Fig. 8.4 The equipment stability test results obtained with the help of the golden reference device: (a) the unit step response, (b) the time constant spectrum, both obtained with prescribed current source setting and prescribed parameters of the numerical deconvolution process

General guidelines for the design of such test boards for the main IC package types are also provided in the different documents of the JESD51-x family [33, 35–38].

While a JEDEC JESD51-2A [31] compliant still-air chamber assures that the wider environment, i.e., the thermal testing laboratory, does not affect the R_{thJA} measurement results, the small variability among the standard test boards from arbitrary manufacturers will have an influence, especially on the *reproducibility* of a certain measurement type. The main reason is that in case of the junction to ambient thermal resistance, R_{thJA}, a large portion (e.g., 50..70%) of the total thermal resistance represents the test environment, mostly by the package attachment to the test board and the test board itself. This could be revealed by the structure functions, as illustrated by Example 3.1 in Sect. 3.1.2.

Using a cold plate as thermal test environment for power device packages considerably shortens the junction to ambient heat flow path, and most of the measured overall thermal resistance belongs to the package under test, though effect of the test environment also appears in the results.

The use of structure functions helps to separate the heat conduction path sections belonging to the package and belonging to the test environment (e.g., the thermal interface material between the package and the cold plate, the cold plate itself). The JEDEC JESD51-14 standard [40] defines the *transient dual thermal interface method* for this. Besides the definition of the test method, this document also provides hints on the proper construction of the cold plate to be used.

It has to be noted that despite such specifications, the realization of the cold plates, the different properties of the cold plates have a large scatter, since different laboratories apply different materials and geometries for the cold plate that is used in the measurement and use other formation of the liquid flow, and various surface roughness and planarities are used in the structures. Even if using the same equipment, the type and thickness of the applied thermal paste usually varies. Hints on the proper construction of cold plates are given in [40].

The above factors, however, usually do not affect the R_{thJC} junction to case thermal resistance values of the power semiconductor device packages identified this way. Most of the uncertainty factors of structure function-based methods are associated with the data processing procedures that are applied to the measured thermal $Z_{th}(t)$ and thermal impedance curves; see Sect. 8.4.

If in the thermal resistance measurement process external temperature sensors are used besides the DUT's temperature-sensitive parameter as a thermometer, further issues arise. For example, the type and position of external temperature sensors may be different, with different properties. Sources of inaccuracy related to the probe position at two point measurements are discussed in details in [97], with the help of a simulation experiment.

In a real measurement, further error sources can be identified, such as:

- The thermal contact resistance between case surface and probe tip can be quite large, especially since the contact area in case of a spherical probe is just a point.
- The heat flow from the tip through the thermally conductive material of a thermocouple diminishes the probe tip temperature.
- There is a temperature drop inside the alloy joint of the thermocouple, since the thermocouple does not measure the temperature at its tip but at the point where the two wires of different alloys separate, etc.

8.3.4 Uncertainty from the TSP Calibration of the Samples

Although the calibration of the temperature-sensitive parameter is not part of the measuring equipment, its accuracy is ultimately determining the accuracy of the measured thermal resistance/impedance results.

The uncertainty of the TSP calibration is primarily determined by the uncertainty stated on the calibration certificate of the temperature-controlled environment used for the TSP calibration. Other factors are the biasing of the device under test that is being calibrated for temperature sensitivity and the voltage meter(s) used.

The uncertainty is also determined by the authenticated accuracy of the thermal sensor that is used for the calibration of the device to be measured, but a number of inadvertent errors may further influence the accuracy of the calibration. The artifacts that may be created by the inappropriate calibration is presented in details in Sect. 2.12.2.

For details of the calibration process, refer to Sect. 5.6. Some specific recommendations for the TSP calibration of LEDs are provided in Sect. 6.10.5.

8.4 Uncertainty Issues Related to the Data Processing

In the prior chapters, many data processing aspects affecting the accuracy of the final results of thermal transient testing have been discussed already. In this section, we only recollect them.

8.4.1 Possible Uncertainty Related to the Initial Transient Correction

Depending on the stray electrical capacitances of the test setup, the DUT and test equipment, and the actual speed of switching on/off the heating power applied to the DUT, the initial part of the captured voltage transient is rather related to electrical changes than to the change of the DUT's junction temperature. This transient has to be discarded and replaced by an assumed $\Delta T_J(t)$ junction temperature transient that is proportional to the square root of the elapsed time t. Since usually neither the geometry nor the thermal parameters of the dissipating chip area are known, the $\Delta T_J(t)$ function corresponding to *Eq. (2.26)* *should be fitted to the measured data points* following the time instance before which data points have been discarded.

The more data points are involved in the regression calculation in the fitting, the better accuracy of the fitted curve is foreseen up to a certain point. That is, if a too long real $\Delta T_J(t)$ transient beyond the t_v time instance given by Eq. (2.28) is used for the fitting, where temperature data already correspond to a time when the dynamic heat propagation in the device already violates the assumptions made for Eq. (2.26), the quality of the fitted curve section would degrade. This happens, for example, when the heat generated at the junction would cross a boundary of two different, adjacent material layers.

It has to be mentioned that the quality of fitting, which depends on the operator's experience, affects the *reproducibility*. Table 2.2 provides practical hints about the t_v time limits in typical material layers up to which the assumptions made for the $\Delta T_J(t) \sim \sqrt{t}$ approximation are valid.

A possible way to avoid this problem is to create a detailed thermal simulation model of the chip/junction region of the DUT; fit this to a longer section of the measured transient and use the initial part of the $\Delta T_{J-sim}(t)$ transient provided by the simulation to replace the discarded portion of the measured one.

8.4.2 Properties of the Deconvolution Algorithms

One has to be aware of the properties of the deconvolution algorithm used for the implementation of the NID method when thinking about the final resolution of the time constant spectra and items at further stages of the data processing procedure yielding the structure functions.

To start with, let us refer to Example 2.4 of Sect. 2.4.1 where we had a known, lumped element model network (Fig. 2.16) with three distinct, discrete time constants. The unit-step response (see Fig. 2.17) of this model system was obtained by LTSpice simulation and was processed by the NID method, resulting in a time constant spectrum (Fig. 2.18).

All the three time constants are properly resolved, but instead of three Dirac–δlike single spectrum lines, we had a blurred, continuous distribution of the time constants with steep peaks having maxima corresponding to the discrete time constants calculated from the element values of the model network. Though in this example all time constants were nicely resolved, it is plausible to say that if two time constants are getting closer and closer, after a while they will not be distinguishable in the continuous spectrum calculated by the deconvolution process.

As a next example, let us consider another RC system with three discrete time constants at 0.5 ms, 1 ms, and 10 ms (see Fig. 8.5a), but in this case the distance between the first two time constants is small; it is only an octave of time, while the third time constant is in a decade distance from the middle one. Now, let us consider the first derivative of the unit-step response of this system, shown in Fig. 8.5b. This function has only two peaks. The second peak is exactly at the location of the 10 ms time constant while the first peak is located between 0.5 and 1 ms. That is, the effects of the first two time constants separated by an octave only are indistinguishable, but if the separation is already one decade, the effects of the two different time constants is clearly visible.

It is worth recalling Eq. (2.20) now. We can interpret it as if we had a linear system characterized by the $w_z(z)$ *weight function* (Fig. 8.6) to which an input with three Dirac-δ pulses with the R magnitudes shown in Fig. 8.5a have been applied,

δ-response R mangnitudes

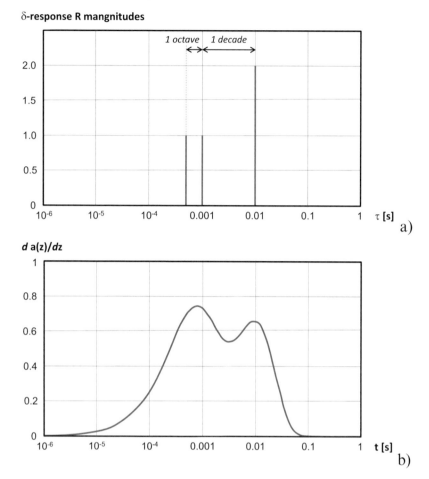

Fig. 8.5 RC system with three discrete spectrum lines: (**a**) the time constant distribution of the system (**b**) and the first derivative of the system unit-step response in logarithmic time. (Based on Székely [58])

and as a response, the function shown in Fig. 8.5b was produced. The function shown in Fig. 8.6 is a typical weight function of low-pass filters; therefore, the sharp input – the three discrete spectrum lines of Fig. 8.5a – gets blurred as seen in Fig. 8.5b.

This blurring effect means a fundamental limit in the theoretically achievable resolution of the time constant spectra obtained by deconvolution. This "blurring effect" is determined by the half-value width of the weight function. For the above function introduced as $w_z(z) = \exp[z - \exp(z)]$ in Sect. 2.4.1, this value is about 2.45. For a rigorous description of the these fundamental issues, refer to [58] or [5].

As a rule of thumb, one can say that spectrum lines separated by a time distance of a decade can be well resolved in the time constant spectra.

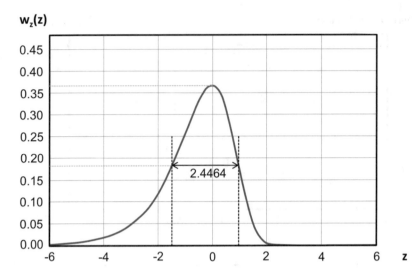

Fig. 8.6 The $w_z(z)$ function used in Eqs. (2.20) and (2.21) in Sect. 2.4.1

Further limitations in the resolution of the time constant spectra may come from the actual implementation of the deconvolution operation presented in Eq. (2.21). Both the numerical derivation and the deconvolution itself on the right hand side enhance the noise that is already inherently present in the unit-step response function.

In the implementation that is using the so-called *inverse Fourier filtering*, an interesting mitigation is applied in the frequency domain: adding some white noise to the frequency domain spectrum of the derivative of the unit-step response function and applying frequency domain filtering before the actual deconvolution step. For a detailed explanation, see the fundamental paper of [58] or the papers describing an actual numerical implementation of the method [172, 173]. The resulting resolution of the obtained time constant spectra also determines how detailed the structure functions can be, i.e. what are the smallest structural details resolved. There are a few, further fundamental papers that describe some improvement options of the thermal transient results evaluation process described in this book, both in terms of possible error corrections and definition of acceptance criteria of the final results [174–177].

This step inherently adds to the uncertainty regarding the resolution of the time constant spectra. The advantage of the inverse Fourier filtering-based deconvolution process is, however, that it works properly for transfer impedances as well, which also have negative values in their time constant spectra.

In contrast to the above process, iterative numerical deconvolution methods based, e.g., on Bayesian iteration [6] are deterministic in the sense that if the process is repeated with exactly the same parameters, the resulting time constant spectra are exactly the same. The uncertainty factor in this case is the actual parameter set with which the process is executed, e.g., the chosen number of iteration.

Note that in time constant spectra obtained by Bayesian iteration, all values are positive, thus, this deconvolution algorithm cannot be applied to processing thermal transfer impedances.

As a summary, one can state that in order to assure reproducibility, the type of the deconvolution algorithm and its major parameters need to be reported together with the thermal transient test results (e.g., Bayesian iteration, with iteration number X).

8.5 Final Remarks

At the end of this book, we wish to summarize what we need to know about the accuracy and repeatability of thermal transient measurements.

Thermal transient measurements are considered in general to be very accurate in the realm of thermal measurements as they operate with easily, accurately, well-measureable electrical signals. According to the experience of the authors of this book, the repeatability of the measurements is the best of all thermal measurements.

In this chapter, we have summarized all the factors that influence the actual accuracy of the measurements.

For the last 30 years of thermal transient measurements, the authors of this book were requested many times to calculate a number that gives the accuracy of thermal transient measurements, but we never agreed to do it. The discussion in this chapter presents all the factors that influence the accuracy of thermal transient measurements and demonstrates that the actual uncertainty of the measurement is influenced by several components.

We have to emphasize here again that the accuracy of these measurements strongly depends also on the care with which these measurements are carried out. The better we follow the prescriptions and recommendations of the measurement standards and the recommendation of this chapter, the more we can guarantee that our measurement is repeatable and accurate.

References

Books and Book Chapters, Data Compendiums

1. Fourier, J.J.: Théorie analytique de la chaleur. Paris (1822)
2. Ohm, G.S.: Die galvanische Kette, mathematisch bearbeitet. T. H. Riemann, Berlin (1827)
3. Incropera, F.P., DeWitt, D.P.: Introduction to Heat Transfer, 2nd edn. Wiley, New York (1990). ISBN: 0-471-51728-3
4. Doetsch, G.: Anleitung zum praktischen Gebrauch der Laplace-Transformation und der Z-Transformation, 3rd edn. R. Oldenbourg Verlag, München (1967)
5. Székely, V.: Distributed RC networks. In: Wai-Kai, C. (ed.) The Circuits and Filters Handbook, 2nd edn, pp. 1201–1222. CRC Press, Boca Raton/London/New York/Washington DC (2003). ISBN: 0-8493-0912-3. Available online: https://books.google.hu/books?hl=en&lr=&id=0SPNBQAAQBAJ. Accessed: 6 July 2021
6. Kennett, T.J., Prestwich, W.V., Robertson, A.: Bayesian deconvolution I: convergent properties. Nucl. Instrum. Methods. **151**(1–2), 285–292 (1978). https://doi.org/10.1016/0029-554X(78)90502-5
7. Lasance, C.J.M., Poppe, A.: Thermal Management for LED Applications. Springer (2014). https://doi.org/10.1007/978-1-4614-5091-7. ISBN (soft cover): 978-1-4939-4133-9
8. Farkas, G., Poppe, A.: Thermal testing of LEDs. In: Lasance, C.J.M., Poppe, A. (eds.) Thermal Management for LED Applications, pp. 73–165. Springer (2014). https://doi.org/10.1007/978-1-4614-5091-7_4
9. Lutz, J., Schlangenotto, H., Scheuermann, U., De Doncker, R.: Semiconductor Power Devices: Physics, Characteristics, Reliability, 2nd edn. Springer, Berlin (2018). https://doi.org/10.1007/978-3-319-70917-8. Section 11.4.2: One-Dimensional Equivalent Networks
10. Sze, S.M., Ng, K.K.: Physics of Semiconductor Devices, 3rd edn. John Wiley & Sons (2007). ISBN: 0-471-14323-5
11. Schubert, E.F.: Light-Emitting Diodes, 2nd edn. Cambridge University Press (2006). ISBN: 0-511-34476-7
12. Zeghbroeck, B.V.: Principles of Semiconductor Devices. Available online: http://ecee.colorado.edu/~bart/book. Accessed: 6 July 2021
13. Kittel, C.: Introduction to Solid State Physics, 8th edn. Wiley, New-York (2004). ISBN: 978-0-471-41526-8

M. Rencz et al. (eds.), *Theory and Practice of Thermal Transient Testing of Electronic Components*, https://doi.org/10.1007/978-3-030-86174-2

14. Föll, H.: "Semiconductors I", Course Handout. University of Kiel; Faculty of Engineering (2019). Available online: https://www.tf.uni-kiel.de/matwis/amat/semi_en/. Accessed: 6 July 2021

15. Farkas, G.: Appendix A: background in solid state physics of the heaters and sensors used in thermal transient testing. (2021). https://doi.org/10.13140/RG.2.2.26194.27843

16. Farkas, G., Hara, T., Rencz, M.: Thermal transient testing. In: Suganuma, K. (ed.) Wide Bandgap Power Semiconductor Packaging. Woodhead Publishing, Cambridge (2018). https://doi.org/10.1016/B978-0-08-102094-4.00008-6. Chapter 6

17. Online database of parameters of semiconductor materials. Available online: http://www.ioffe.ru/SVA/NSM/Semicond/. Accessed: 6 July 2021

Related Measurement Standards and Guidelines

18. Oettinger, F.F., Blackburn, D.L.: "Thermal Resistance Measurements", NIST Special Publication 400-86. NIST, Gaithersburg (1990). https://doi.org/10.6028/NIST.SP.400-86

19. IEC/EN 60747-2. Standard: Semiconductor Devices – Part 2: Discrete Devices – Rectifier Diodes. Available online: https://webstore.iec.ch/publication/24519. Accessed: 6 July 2021

20. IEC/EN 60747-15. Standard: Semiconductor Devices-Discrete Devices Part 15: Isolated Power Semiconductor Devices. Available online: https://webstore.iec.ch/publication/3255/. Accessed: 6 July 2021

21. IEC/EN 60749-34:2010: Standard: Semiconductor Devices – Mechanical and Climatic Test Methods – Part 34: Power Cycling. Available online: https://webstore.iec.ch/publication/3381/. Accessed: 6 July 2021

22. MIL-STD-750D: Test Methods for Semiconductor Devices. Available online: https://www.navsea.navy.mil/Portals/103/Documents/NSWC_Crane/SD-18/Test%20Methods/MILSTD750.pdf. Accessed: 13 Jan 2020

23. MIL-STD-750D Method 3101.3: Thermal Impedance (Response) Testing of Diodes. Available online: https://www.thermengr.net/PDF/MilStd750M3101_Diode.pdf. Accessed: 6 July 2021

24. MIL-STD-750F. Department of Defense (USA) Test Method Standard: Test Methods for Semiconductor Devices. Available online: http://everyspec.com/MIL-STD/MIL-STD-0700-0799/MIL-STD-750F_39654/. Accessed: 8 July 2019

25. ECPE Guideline AQG 324: Automotive Qualification Guideline. Available online: https://www.ecpe.org/research/working-groups/automotive-aqg-324/. Accessed: 13 Jan 2020

26. The JEDEC JC15 Standardization Committee on Thermal Characterization Techniques for Semiconductor Packages; Committee public webpage: https://www.jedec.org/committees/jc-15. Accessed: 29 Jan 2021

27. JEDEC Standard JESD15-3: Two-Resistor Compact Thermal Model Guideline (2008, July). Available online: https://www.jedec.org/standards-documents/docs/jesd-15-3. Accessed: 13 Jan 2020

28. JEDEC Standard JESD15-4: Delphi Compact Thermal Model Guidelines (2008, October). Available online: https://www.jedec.org/standards-documents/docs/jesd-15-4. Accessed: 13 Jan 2020

29. JEDEC Standard JESD51: Methodology for the Thermal Measurement of Component Packages (Single Semiconductor Devices) (1995, December). Available online: https://www.jedec.org/standards-documents/docs/jesd-51. Accessed: 13 Jan 2020

30. JEDEC Standard JESD51-1: Integrated Circuits Thermal Measurement Method – Electrical Test Method (Single Semiconductor Device) (1995, December). Available online: https://www.jedec.org/sites/default/files/docs/jesd51-1.pdf. Accessed: 6 July 2021

31. JEDEC Standard JESD51-2A: Integrated Circuits Thermal Test Method Environmental Conditions – Natural Convection (Still Air) (2008, January). https://www.jedec.org/system/files/docs/JESD51-2A.pdf. Accessed: 6 July 2021

32. JEDEC Standard JESD51-4A: Thermal Test Chip Guideline (Wire Bond and Flip Chip) (2019, June). Available online: https://www.jedec.org/system/files/docs/JESD51-4A.pdf. Accessed: 6 July 2021

33. JEDEC Standard JESD51-5: Extension of Thermal Test Board Standards for Packages with Direct Thermal Attachment Mechanisms (1999, February). Available online: https://www.jedec.org/sites/default/files/docs/jesd51-5.pdf. Accessed: 19 Feb 2021

34. JEDEC Standard JESD51-6: Integrated Circuit Thermal Test Method Environmental Conditions – Forced Convection (Moving Air) (1999, March). Available online: https://www.jedec.org/system/files/docs/jesd51-6.pdf. Accessed: 19 Feb 2021

35. JEDEC Standard JESD51-7: High Effective Thermal Conductivity Test Board for Leaded Surface Mount Packages (1999, February). Available online: https://www.jedec.org/sites/default/files/docs/jesd51-7.pdf. Accessed: 19 Feb 2021

36. JEDEC Standard JESD51-9: Test Boards for Area Array Surface Mount Package Thermal Measurements (2000, July). Available online: https://www.jedec.org/sites/default/files/docs/jesd51-9.pdf. Accessed: 19 Feb 2021

37. JEDEC Standard JESD51-10: Test Boards for Through-Hole Perimeter Leaded Package Thermal Measurements (2000, July). Available online: https://www.jedec.org/sites/default/files/docs/jesd51-10.pdf. Accessed: 19 Feb 2021

38. JEDEC Standard JESD51-11: Test Boards for Through-Hole Area Array Leaded Package Thermal Measurements (2001, June). Available online: https://www.jedec.org/sites/default/files/docs/jesd51-11.pdf. Accessed: 19 Feb 2021

39. JEDEC Standard JESD51-13: Glossary of terms and definitions (2009, June). Available online: https://www.jedec.org/system/files/docs/JESD51-13.pdf. Accessed: 19 Feb 2021

40. JEDEC Standard JESD51-14: Transient Dual Interface Test Method for the Measurement of Thermal Resistance Junction-to-Case of Semiconductor Devices with Heat Flow through a Single Path (2010, November). Available online: http://www.jedec.org/sites/default/files/docs/JESD51-14_1.pdf. Accessed: 13 Jan 2021

41. JEDEC Standard JESD51-31: Thermal Test Environment Modifications for Multi-Chip Packages (2008, July). Available online: https://www.jedec.org/system/files/docs/JESD51-31.pdf. Accessed: 13 Jan 2021

42. JEDEC Standard JESD51-32: Extension to JESD51 Thermal Test Board Standards to Accommodate Multi-Chip Packages (2010, December). Available online: https://www.jedec.org/system/files/docs/JESD51-32.pdf. Accessed: 13 Jan 2021

43. JEDEC Standard JESD51-51: Implementation of the Electrical Test Method for the Measurement of the Real Thermal Resistance and Impedance of Light-Emitting Diodes with Exposed Cooling Surface (2012, April). Available online: https://www.jedec.org/sites/default/files/docs/JESD51-51.pdf. Accessed: 13 Jan 2021

44. JEDEC Standard JESD51-52: Guidelines for Combining CIE 127-2007 Total Flux Measurements with Thermal Measurements of LEDs with Exposed Cooling Surface (2012, April). Available online: https://www.jedec.org/sites/default/files/docs/JESD51-52.pdf. Accessed: 13 January 2021

45. Zong, Y., Chou, P.T., Dekker, P., Distl, R., Godo, K., Hanselaer, P., Heidel, G., Hulett, J., Oshima, K., Poppe, A., Sauter, G., Schneider, M., Shen, H., Sisto, M.M., Sperling, A., Young, R., Zhao, W.: CIE 225:2017 Technical Report: "Optical Measurement of High-Power LEDs". CIE, Vienna. https://doi.org/10.25039/TR.225.2017. ISBN 978 3 902842 12 1

46. Muray, K., Goodman, T., Heidel, G., Ohno, Y., Sauter, G., Schanda, J., Steudtner, W., Young, R.: CIE 127:2007 Technical Report: "Measurement of LEDs". CIE, Vienna. ISBN: 978 3 901906 58 9. http://cie.co.at/publications/measurement-leds

47. Chou, P.-T., Bergen, T., Heidel, G., Jiao, J., Poppe, A., Ohno, Y., Zong, Y.: CIE 238:2020 Technical Report: "Characterization of AC-Driven LEDs for SSL Applications". CIE, Vienna. https://doi.org/10.25039/TR.238.2020. ISBN: 978-3-902842-64-0

48. Heidel, G., Cariou, N., Hou, P.T., Konjhodzic, D., Ng, K.F., Ohno, Y., Poppe, A., Sauter, G., Schneider, M., Sperling, Y.R., Zong, Y.: CIE 226:2017 Technical Report: "High-Speed Testing Methods for LEDs". CIE, Vienna. https://doi.org/10.25039/TR.226.2017. ISBN 978-3-902842-69-5

49. JCGM 100:2008 Document: Evaluation of Measurement Data – Guide to the Expression of Uncertainty in Measurement. Available online: http://www.bipm.org/en/publications/guides/gum. Accessed: 13 Jan 2021

50. Possolo, A.: Simple Guide for Evaluating and Expressing the Uncertainty of NIST Measurement Results. NIST Technical Note 1900, October 2015. https://doi.org/10.6028/NIST.TN.1900

51. NASA Handbook NASA-HDBK-8739.19-3: Measurement Uncertainty Analysis Principles and Methods. Available online: https://standards.nasa.gov/standard/osma/nasa-hdbk-873919-3. Accessed: 13 Jan 2021

52. CIE 198:2011 Technical Report: Determination of Measurement Uncertainties in Photometry. CIE, Vienna. ISBN: 978-3-902842-00-8. http://cie.co.at/publications/determination-measurement-uncertainties-photometry

53. ASTM D5470 Standard: "Standard Test Method for Thermal Transmission Properties of Thin Thermally Conductive Solid Electrical Insulation Materials", Designation D 5470-17. Available online: https://www.astm.org/Standards/D5470.htm. Accessed: 7 July 2021

Test Equipment Descriptions and Software Handbooks

54. SIEMENS Simcenter T3Ster®. Available online: https://www.plm.automation.siemens.com/global/en/products/simcenter/t3ster.html. Accessed: 6 July 2021

55. SIEMENS Simcenter Power Tester 1500A. Available online: https://www.plm.automation.siemens.com/global/en/products/simcenter/powertester.html. Accessed: 6 July 2021

56. SIMENS Simcenter Flotherm. Available online: https://www.plm.automation.siemens.com/global/en/products/simcenter/flotherm.html. Accessed: 6 July 2021

57. Linear Technology, "LTspice IV Getting Started Guide LTspice IV Getting Started Guide". Available online: http://cds.linear.com/docs/en/software-and-simulation/LTspiceGettingStartedGuide.pdf. Accessed: 7 July 2021

Journal and Conference Papers

58. Székely, V.: Identification of RC networks by deconvolution: chances and limits. IEEE Trans. Circuits Syst. I, Fundam. Theory Appl. 45(3), 244–258 (1998). https://doi.org/10.1109/81.662698

59. Székely, V.: On the representation of infinite-length distributed RC one-ports. IEEE Trans. Circuits Syst. 38(7), 711–719 (1991). https://doi.org/10.1109/31.135743

60. Székely, V., Rencz, M.: Thermal dynamics and the time constant domain. IEEE Trans. Compon. Packag. Technol. 23(3), 587–594 (2000). https://doi.org/10.1109/6144.868862

61. Székely, V., Bien, T.V.: Fine structure of heat flow path in semiconductor devices: a measurement and identification method. Solid State Electron. 31(9), 1363–1368 (1988). https://doi.org/10.1016/0038-1101(88)90099-8

62. Glavanovics, M., Zitta, H.: Thermal destruction testing: an indirect approach to a simple dynamic thermal model of smart power switches. In: Proceedings of the 27th European Solid-State Circuits Conference (ESSCIRC'01), pp. 221–224, Villach, Austria. Available online: https://ieeexplore.ieee.org/document/1471373. Accessed: 24 Jan 2021

63. Farkas, G., Zettner, J., Sárkány, Z., Rencz, M.: In-situ transient testing of thermal interface sheets and metal core boards in power switch assemblies. In: 2017 23rd International Workshop on Thermal Investigations of ICs and Systems (THERMINIC) (2017). https://doi.org/10.1109/THERMINIC.2017.8233799

64. Rencz, M., Szekely, V.: Non-linearity issues in the dynamic compact model generation. In: 2003 19th Annual IEEE Semiconductor Thermal Measurement and Management Symposium (SEMI-THERM), 11–13 March 2003, San Jose, CA, USA, pp. 263–270. https://doi.org/10.1109/STHERM.2003.1194372

65. Farkas, G., Rencz, M., Keranen, K., Heikkinen, V.: Joint characterisation of power semiconductors and their thermal boundary based on measurements range. In: 2020 26th International Workshop on THERMal INvestigation of ICs and Systems (THERMINIC), 24 September 2020, Berlin, Germany

66. Masana, F.N.: A closed form solution of junction to substrate thermal resistance in semiconductor chips. IEEE Trans. Compon. Packaging Manuf. Technol. Part A. **19**(4), 539–545 (1996). https://doi.org/10.1109/95.554935

67. Evdokimova, N.L., et al.: Examples of using the analytical structure functions for the thermal analysis of semiconductor devices. https://doi.org/10.13140/RG.2.2.19793.15209

68. Pape, H., Schweitzer, D., Janssen, J.H.J., Morelli, A., Villa, C.M.: Thermal transient modeling and experimental validation in the European project PROFIT. IEEE Trans. Compon. Packag. Technol. **27**(3), 530–538 (2004). https://doi.org/10.1109/TCAPT.2004.831791

69. Schweitzer, D., Ender, F., Hantos, G., Szabó, P.G.: Thermal transient characterization of semiconductor devices with multiple heat sources—fundamentals for a new thermal standard. Microelectron. J. **46**(2), 174–182 (2015). https://doi.org/10.1016/j.mejo.2014.11.001

70. Treurniet, T., Lammens, V.: Thermal management in color variable multi-chip LED modules. In: 2006 22nd Annual IEEE Semiconductor Thermal Measurement and Management Symposium (SEMI-THERM), 14–16 March 2006, Dallas, TX, USA, pp. 186–190. https://doi.org/10.1109/STHERM.2006.1625224

71. Poppe, A.: Simulation of LED based luminaires by using multi-domain compact models of LEDs and compact thermal models of their thermal environment. Microelectron. Reliab. **72**(5), 65–74 (2017). https://doi.org/10.1016/j.microrel.2017.03.039

72. Timár, A., Rencz, M.: Temperature dependent timing in standard cell designs. Microelectron. J. **45**(5), 521–529 (2014). https://doi.org/10.1016/j.mejo.2013.08.016

73. Poppe, A., Siegal, B., Farkas, G.: Issues of thermal testing of AC LEDs. In: 27th 2011 Semiconductor Thermal Measurement and Management Symposium (SEMI-THERM), 20–24 March 2011, San Jose, CA, USA, pp. 297–304. https://doi.org/10.1109/STHERM.2011.5767214

74. Rencz, M., Poppe, A., Farkas, G.: Determination of the complex thermal characteristics of discrete power devices and power modules. In: 2019 IEEE 21st Electronics Packaging Technology Conference (EPTC), 4–6 December 2019, Singapore. https://doi.org/10.1109/EPTC47984.2019.9026589

75. Rencz, M., Poppe, A., Kollár, E., Ress, S., Szekely, V.: Increasing the accuracy of structure function based thermal material parameter measurements. IEEE Trans. Compon. Packag. Technol. **28**(1), 51–57 (2005). https://doi.org/10.1109/TCAPT.2004.843204

76. Rencz, M., Poppe, A., Kollár, E., Ress, S., Székely, V., Courtois, B.: A procedure to correct the error in the structure function based thermal measuring methods. In: 2004 20th Annual IEEE Semiconductor Thermal Measurement and Management Symposium (SEMI-THERM), 9–11 March 2004, San Jose, CA, USA, pp. 92–97. https://doi.org/10.1109/STHERM.2004.1291307

77. Alexeev, A.: Characterization of light emitting diodes with transient measurements and simulations, Ph.D. Thesis; TU Eindhoven, Eindhoven, The Netherlands, 2020; print ISBN: 978-90-386-5035-7. Available online: https://research.tue.nl/en/publications/characterization-of-light-emitting-diodes-with-transient-measurem. Accessed: 21 Jan 2021

78. Alexeev, A., Onushkin, G., Linnartz, J.-P., Martin, G.: Multiple heat source thermal modeling and transient analysis of LEDs. Energies. **12**(10), 2019 (1860). https://doi.org/10.3390/en12101860

79. Luiten, W.: Length dimensions in thermal design. In: 2018 24th International Workshop on Thermal Investigations of ICs and Systems (THERMINIC), 26–28 September 2018, Stockholm, Sweden. https://doi.org/10.1109/THERMINIC.2018.8592877

80. De Mey, G., Torzewicz, T., Kawka, P., Czerwoniec, A., Janiczki, M., Napieralski, A.: Analysis of nonlinear heat exchange phenomena in natural convection cooled electronic systems. Microelectron. Reliab. **67**(12), 15–20 (2016). https://doi.org/10.1016/j.microrel.2016.11.003

81. Hantos, G., Hegedüs, J.: K-factor calibration issues of high power LEDs. In: 2017 23rd International Workshop on THERMal INvestigation of ICs and Systems (THERMINIC), 27–29 September, Amsterdam, The Netherlands. https://doi.org/10.1109/THERMINIC.2017.8233798

82. Hantos, G., Hegedüs, J., Poppe, A.: Different questions of today's LED thermal testing procedures. In: 2018 34th Annual IEEE Semiconductor Thermal Measurement and Management Symposium (SEMI-THERM), 19–23 March 2018, San Jose, CA, USA, pp. 63–70. https://doi.org/10.1109/SEMI-THERM.2018.8357354

83. Farkas, G., Sárkány, Z., Rencz, M.: Issues in testing advanced power semiconductor devices. In: 2016 32nd Annual IEEE Semiconductor Thermal Measurement and Management Symposium (SEMI-THERM), 14–17 March 2016, San Jose, CA, USA. https://doi.org/10.1109/SEMI-THERM.2016.7458458

84. Lasance, C.J.M.: The practical usefulness of various approaches to estimate heat spreading effects. In: The Second International Conference on Thermal Issues in Emerging Technologies (ThETA'08), 17–20 December 2008, Cairo, Egypt, pp. 149–158. https://doi.org/10.1109/THETA.2008.5190977

85. Lasance, C.J.M.: Heat spreading: not a trivial problem. Electronics Cooling Magazine **14**(2) (2008). Available online: https://www.electronics-cooling.com/2008/05/heat-spreading-not-a-trivial-problem/. Accessed: 7 July 2021

86. Rosten, H., Lasance, C.J.M.: DELPHI: the development of libraries of physical models of electronic components for an integrated design environment. In: Berge, J.-M., Levia, O., Rouillard, J. (eds.) Model Generation in Electronic Design, pp. 63–89. Kluwer Academic Press (1995). https://doi.org/10.1007/978-1-4615-2335-2_5

87. Németh, M., Poppe, A.: Parametric models of thermal transfer impedances within a successive node reduction based thermal simulation environment. Period. Polytech. Electr. Eng. Comput. Sci. **62**(1), 1–15 (2018). https://doi.org/10.3311/PPee.11058

88. Steffens, O., Szabó, P., Lenz, M., Farkas, G.: Thermal transient characterization methodology for single-chip and stacked structures. In: 2005 21st Annual IEEE Semiconductor Thermal Measurement and Management Symposium (SEMI-THERM), 15–17 March 2005, San Jose, CA, USA. https://doi.org/10.1109/STHERM.2005.1412198

89. Szabó, P., Steffens, O., Lenz, M., Farkas, G.: Transient junction-to-case thermal resistance measurement methodology of high accuracy and high repeatability. IEEE Trans. Compon. Packag. Technol. **28**(4), 630–636 (2005). https://doi.org/10.1109/TCAPT.2005.859768

90. Schweitzer, D., Pape, H., Chen, L.: Transient measurement of the junction-to-case thermal resistance using structure functions: chances and limits. In: 2008 24th Annual IEEE Semiconductor Thermal Measurement and Management Symposium (SEMI-THERM), 16–20 March 2008, San Jose, CA, USA. https://doi.org/10.1109/STHERM.2008.4509389

91. Schweitzer, D.: Transient dual interface measurement of the Rth-JC of power packages. In: 2008 14th Thermal Investigation of ICs and Systems (THERMINIC), 24–26 September 2008, Rome, Italy. https://doi.org/10.1109/THERMINIC.2008.4669871
92. Temmerman, W., Nelemans, W., Goosens, T., Lauwers, E., Lacaze, C.: "Experimental Validation Methods for Thermal Models", EUROTHERM Seminar No. 45, 20–22 September 1995, Leuven, Belgium, pp. 125–136. https://doi.org/10.1007/978-94-011-5506-9_12
93. Poppe, A., Zhang, Y., Wilson, J., Farkas, G., Szabó, P., Parry, J., Rencz, M., Székely, V.: Thermal measurement and modeling of multi-die packages. IEEE Trans. Compon. Packag. Technol. 32(2), 484–492 (2009). https://doi.org/10.1109/TCAPT.2008.2004578
94. Bein, M.C., Hegedüs, J., Hantos, G., Gaál, L., Farkas, G., Rencz, M., Poppe, A.: Comparison of two alternative junction temperature setting methods aimed for thermal and optical testing of high power LEDs. In: 2017 23rd International Workshop on Thermal Investigation of ICs and Systems (THERMINIC), 27–29 September 2017, Amsterdam, The Netherlands. https://doi.org/10.1109/THERMINIC.2017.8233838
95. Farkas, G.: Thermal transient characterization of semiconductor devices with programmed powering. In: 2013 29th Annual IEEE Semiconductor Thermal Measurement and Management Symposium (SEMI-THERM), 17–20 March 2013, San Jose, CA, USA. https://doi.org/10.1109/SEMI-THERM.2013.6526837
96. Schweitzer, D.: The junction-to-case thermal resistance: a boundary condition dependent thermal metric. In: 2010 26th Annual IEEE Semiconductor Thermal Measurement and Management Symposium (SEMI-THERM), 21–25 February 2010, Santa Clara, CA, USA. https://doi.org/10.1109/STHERM.2010.5444298
97. Farkas, G., Schweitzer, D., Sárkány, Z., Rencz, M.: On the reproducibility of thermal measurements and of related thermal metrics in static and transient tests of power devices. Energies. 13(3), 557 (2020). https://doi.org/10.3390/en13030557
98. Solar cell operational principles. In: TU Delft OpenCourseware "Introduction to Photovoltaic Solar Energy". Available online: https://ocw.tudelft.nl/wp-content/uploads/Solar-Cells-R4-CH4_Solar_cell_operational_principles.pdf. Accessed: 11 Apr 2021
99. A semiconductor device primer. In: Spieler, H. (ed.) "Introduction to Radiation Detectors and Electronics", Lecture Notes of Physics 198 Spring Semester 1999, UC Berkeley. Available online: https://www-physics.lbl.gov/~spieler/physics_198_notes/PDF/VIII-2-a-diodes.pdf. Accessed: 11 Apr 2021
100. Cristea, M. J.: The Shockley-type boundary conditions for semiconductor p-n junctions at medium and high injection levels. Available online: https://arxiv.org/ftp/cond-mat/papers/060 9/0609141.pdf. Accessed: 11 Apr 2021
101. Wachutka, G.: Rigorous thermodynamic treatment of heat generation and conduction in semiconductor device modeling. IEEE Trans. Comput. Aided Des. Integr. Circuits Syst. 9(11), 1141–1149 (1990). https://doi.org/10.1109/43.62751
102. Darwish, M., Al-Abassi, S.A.W., Neumann, P., Mizsei, J., Pohl, L.: Application of vanadium dioxide for thermal sensing. In: 2021 27th International Workshop on Thermal Investigations of ICs and Systems (THERMINIC), pp. 1–4 (2021). https://doi.org/10.1109/THERMINIC52472.2021.9626518
103. Kaminski, N., Hilt, O.: SiC and GaN devices – wide bandgap is not all the same. IET Circuits Devices Syst. 8(3), 227–236 (2014). https://doi.org/10.1049/iet-cds.2013.0223
104. Mishra, U.K., Parikh, P., Wu, Y.-F.: AlGaN/GaN HEMTs-an overview of device operation and applications. Proc. IEEE. 90(6), 1022–1031 (2002). https://doi.org/10.1109/JPROC.2002.1021567
105. Hamady, S.: "New Concepts for Normally-Off Power Gallium Nitride (GaN) High Electron Mobility Transistor (HEMT)", Ph.D. Thesis, Micro and Nanotechnologies/Microelectronics. Université Toulouse III Paul Sabatier, 2014, HAL Id: tel-01132563. Available online: https://tel.archives-ouvertes.fr/tel-01132563/document. Accessed: 7 July 2021
106. Szabó, P., Rencz, M., Farkas, G., Poppe, A.: Short time die attach characterization of LEDs for in-line testing application. In: 2006 8th Electronics Packaging Technology Conference

(EPTC), vol. 1, 6–8 December 2006, Singapore, pp. 360–366. https://doi.org/10.1109/EPTC. 2006.342743. ISBN: 1-4244-0664-1

107. Szabó, P.; Poppe, A.; Rencz, M.: Studies on the possibilities of in-line die attach characterization of semiconductor devices. In: 2007 9th Electronics Packaging Technology Conference (EPTC), 10–12 December 2007, Singapore, pp. 779–784. https://doi.org/10.1109/EPTC.2007. 4469707. ISBN: 978-1-4244-1324-9

108. Székely, V.: Evaluation of short pulse thermal transient measurements. In: Proceedings of the 2008 14th International Workshop on Thermal Investigations of ICs and Systems (THERMINIC), 24–26 September 2008, Rome, Italy, pp. 20–25. https://doi.org/10.1109/ THERMINIC.2008.4669872

109. Székely, V.: Evaluation of short pulse and short time thermal transient measurements. Microelectron. J. **41**(9), 560–565 (2010). https://doi.org/10.1016/j.mejo.2009.12.006

110. Singh, S., Proulx, J., Vass-Várnai, A.: Measuring the RthJC of power semiconductor components using short pulses. In: Proceedings of the 2021 27th International Workshop on Thermal Investigations of ICs and Systems (THERMINIC), 23–23 September 2021, Berlin, Germany. https://doi.org/10.1109/THERMINIC52472.2021.9626498

111. Poppe, A., Vass-Várnai, Sárkány, Z., Rencz, M., Hantos, G., Farkas, G.: Suggestions for extending the scope of the transient dual interface method. In: Proceedings of the 2021 27th International Workshop on Thermal Investigations of ICs and Systems (THERMINIC), 23–23 September 2021, Berlin, Germany. https://doi.org/10.1109/THERMINIC52472.2021. 9626508

112. Siegal, B.: Measuring thermal resistance is the key to a cool semiconductor. Electronics. **51**(14), 121–126 (1978)

113. Siegal, B.: An alternative approach to junction-to-case thermal resistance measurements. Electronics Cooling Magazine. **7**(2) (2001). Available online: https://www.electronics-cooling.com/2001/05/an-alternative-approach-to-junction-to-case-thermal-resistance-measurements/. Accessed: 7 July 2021

114. Sofia, J.W.: Analysis of thermal transient data with synthesized dynamic models for semiconductor devices. IEEE Trans. Compon. Packag. Manuf. Technol. Part A. **18**(1), 39–47 (1995). https://doi.org/10.1109/95.370733

115. Vass-Várnai, A., Parry, J., Tóth, G., Ress, S., Farkas, G., Poppe, A., Rencz, M.: Comparison of JEDEC dynamic and static test methods for thermal characterization of power LEDs. In: 2012 14th Electronics Packaging Technology Conference (EPTC), 5–7 December, Singapore, pp. 594–597. https://doi.org/10.1109/EPTC.2012.6507151

116. Farkas, G., van Voorst Vader, Q., Poppe, A., Bognár, G.: Thermal investigation of high power optical devices by transient testing. IEEE Trans. Compon. Packag. Technol. **28**(1), 45–50 (2005). https://doi.org/10.1109/TCAPT.2004.843197

117. Farkas, G., Sárkány, Z., Rencz, M.: Structural analysis of power devices and assemblies by thermal transient measurements. Energies. **12**(14), 2696 (2019). https://doi.org/10.3390/ en12142696

118. Vass-Várnai, A., Sárkány, Z., Barna, CS., Laky, S., Rencz, M.: A possible method to assess the accuracy of a TIM tester. In: 2013 International Conference on Electronics Packaging (ICEP), 10–12 April 2013, Osaka, Japan, Paper B11

119. Schoiswoh, J.: "Linear Mode Operation and Safe Operating Diagram of Power-MOSFETs", Infineon application note AP99007. Available online: https://www.infineon.com/dgdl/ Infineon-ApplicationNote_Linear_Mode_Operation_Safe_Operation_Diagram_MOSFETs-AN-v01_00-EN.pdf?fileId=db3a30433e30e4bf013e3646e9381200. Accessed: 7 July 2021

120. Bornoff, R., Vass-Várnai, A., Blackmore, B., Wang, G., Wong, V.H.: Full-circuit 3D electro-thermal modeling of an IGBT Power Inverter. In: 2017 33rd Annual IEEE Thermal Measurement, Modeling & Management Symposium (SEMI-THERM), 13–17 March 2017, San Jose, CA, USA, pp. 29–35. https://doi.org/10.1109/SEMI-THERM.2017.7896904

121. Schmidt, R., Scheuermann, U.: Using the chip as a temperature sensor – the influence of steep lateral temperature gradients on the VCE(T)-measurement. EPE J. **21**(2), 5–11 (2011). https://doi.org/10.1080/09398368.2011.11463790

122. Vass-Várnai, A., Cho, Y.J., Farkas, G., Rencz, M.: An alternative method to accurately determine the thermal resistance of SiC MOSFET structures with discrete diodes. In: 2018 International Power Electronics Conference (IPEC-Niigata 2018 -ECCE Asia), 20–24 May 2018, Niigata, Japan, pp. 137–141. https://doi.org/10.23919/IPEC.2018.8507995

123. Bornoff, R., Vass-Varnai, A.: A detailed IC package numerical model calibration methodology. In: 2013 29th Annual IEEE Semiconductor Thermal Measurement and Management Symposium (SEMI-THERM), 17–21 March 2013, San Jose, CA, USA. https://doi.org/10.1109/SEMI-THERM.2013.6526807

124. Imbruglia, A., et al.: WInSiC4AP: Wide Band Gap Innovative SiC for Advanced Power. In: 2019 AEIT International Conference of Electrical and Electronic Technologies for Automotive (AEIT AUTOMOTIVE), 2–4 July 2019, Turin, Italy. https://doi.org/10.23919/EETA.2019.8804586

125. Adamowicz, M., Giziewski, S., Pietryka, J., Krzeminski, Z.: Performance comparison of SiC Schottky diodes and silicon ultra fast recovery diodes. In: 2011 7th International Conference-Workshop Compatibility and Power Electronics (CPE), 1–3 June 2011, Tallinn, Estonia, pp. 144–149. https://doi.org/10.1109/CPE.2011.5942222

126. Elasser, A., Kheraluwala, M., Ghezzo, M., Steigerwald, R., Krishnamurthy, N., Kretchmer, J., Chow, T.P.: A comparative evaluation of new silicon carbide diodes and state-of-the-art silicon diodes for power electronic applications. In: Conference Record of the 1999 IEEE Industry Applications Conference. Thirty-Forth IAS Annual Meeting (Cat. No.99CH36370), vol. 1, pp. 341–345, Phoenix, AZ, USA, 3–7 October 1999. https://doi.org/10.1109/IAS.1999.799976

127. Gurfinkel, M., Xiong, H.D., Cheung, K.P., Suehle, J.S., Bernstein, J.B., Shapira, Y., Lelis, A.J., Habersat, D., Goldsman, N.: Characterization of transient gate oxide trapping in SiC MOSFETs using fast I–V techniques. IEEE Trans. Electron Devices. **55**(8), 2004–2012 (2008). https://doi.org/10.1109/TED.2008.926626

128. Funaki, T., Fukunaga, S.: Difficulties in characterizing transient thermal resistance of SiC MOSFETs. In: 2016 22nd International Workshop on Thermal Investigations of ICs and Systems (THERMINIC), 21–23 September 2016, Budapest, Hungary, pp. 141–146. https://doi.org/10.1109/THERMINIC.2016.7749042

129. Roccaforte, F., Greco, G., Fiorenza, P., Iucolano, F.: An overview of normally-off GaN-based high electron mobility transistors. Materials. **12**, 1599 (2019). https://doi.org/10.3390/ma12101599

130. Saito, W., Takada, Y., Kuraguchi, M., Tsuda, K., Omura, I.: Recessed-gate structure approach toward normally off high-voltage AlGaN/GaN HEMT for power electronics applications. IEEE Trans. Electron Devices. **53**(2), 356–362 (2006). https://doi.org/10.1109/TED.2005.862708

131. Hamady, S., Morancho, F., Beydoun, B., Austin, P., Gavelle, M.: A new concept of enhanced-mode GaN HEMT using fluorine implantation in the GaN layer. In: 2013 15th European Conference on Power Electronics and Applications (EPE), pp. 1–6 (2013). https://doi.org/10.1109/EPE.2013.6631904

132. Huang, S., et al.: Ultrathin-barrier AlGaN/GaN heterostructure: a recess-free technology for manufacturing high-performance GaN-on-Si power devices. IEEE Trans. Electron Devices. **65**(1), 207–214 (2018). https://doi.org/10.1109/TED.2017.2773201

133. He, J., Cheng, W., Wang, Q., Cheng, K., Yu, H., Chai, Y.: Recent advances in GaN-based power HEMT devices. Adv. Electron. Mater. **7**, 2001045 (2021). https://doi.org/10.1002/aelm.202001045

134. Alexeev, A., Martin, G., Onushkin, G.: Multiple heat path dynamic thermal compact modeling for silicone encapsulated LEDs. Microelectron. Reliab. **87**(8), 89–96 (2018). https://doi.org/10.1016/j.microrel.2018.05.014

135. Pohl, L., Hantos, G., Hegedüs, J., Németh, M., Kohári, Z.S., Poppe, A.: Mixed detailed and compact multi-domain modeling to describe CoB LEDs. Energies. **13**(16), 4051 (2020). https://doi.org/10.3390/en13164051

136. Tang, Y.: A modified single pulse method for transient thermal impedance (TTI) measurement of VDMOSFET relates gate bias to the TTI results. J. Semicond. Technol. Sci. **18**(3), 383–391 (2018). https://doi.org/10.5573/JSTS.2018.18.3.383

137. Protonotarios, E.N., Wing, O.: Theory of nonuniform RC lines, part I: analytic properties and realizability conditions in the frequency domain. IEEE Trans. Circuits Theory. **14**(1), 2–12 (1967). https://doi.org/10.1109/TCT.1967.1082650

138. Schweitzer, D., Pape, H.: Boundary condition independent dynamic thermal compact models of IC packages. In: 2003 9th International Workshop on Thermal Investigation of ICs and Systems (THERMINIC), 24–26 September 2003, Aix-en-Provence, France, pp. 225–230.-ISBN: 2-84813-020-2

139. Vermeersch, B., De Mey, G.: A fixed-angle heat spreading model for dynamic thermal characterization of rear-cooled substrates. In: 2007 23rd Annual IEEE Semiconductor Thermal Measurement and Management Symposium (SEMI-THERM), 18–22 March 2007, San Jose, CA, USA. https://doi.org/10.1109/STHERM.2007.352393

140. Onushkin, G., Bosschaart, K.J., Yu, J., van Aalderen, H.J., Joly J., Martin, G., Poppe, A.: Assessment of isothermal electro-optical-thermal measurement procedures for LEDs. In: 2017 23rd International Workshop on THERMal INvestigation of ICs and Systems (THERMINIC). https://doi.org/10.1109/THERMINIC.2017.8233796

141. Public deliverable of the Delphi4LED project, "Deliverable 5.6 D5.6 XML Schema for Compact Model Data Exchange, Created and Described". Available online: https://delphi4led.org/pydio/public/6d3437. Accessed: 7 July 2021

142. Poppe, A., Farkas, G., Gaál, L., Hantos, G., Hegedüs, J., Rencz, M.: Multi-domain modelling of LEDs for supporting virtual prototyping of luminaires. Energies. **12**(10), 2019 (1909). https://doi.org/10.3390/en12101909

143. van der Schans, M., Yu, J., Martin, G.: Digital luminaire design using led digital twins—accuracy and reduced computation time: a Delphi4LED methodology. Energies. **13**(16), 4979 (2020). https://doi.org/10.3390/en13164051

144. Martin, G., Marty, C., Bornoff, R., Poppe, A., Onushkin, G., Rencz, M., Yu, J.: Luminaire digital design flow with multi-domain digital twins of LEDs. Energies. **12**(12), 2389 (2019). https://doi.org/10.3390/en12122389

145. Bornoff, R.: Extraction of boundary condition independent dynamic compact thermal models of LEDs—a Delphi4LED methodology. Energies. **12**(9), 1628 (2019). https://doi.org/10.3390/en12091628

146. Poppe, A., Farkas, G., Szabó, F., Joly, J., Thomé, J., Yu, J., Bosschaartl, K., Juntunen, E., Vaumorin, E., di Bucchianico, A., et al.: "Inter Laboratory Comparison of LED Measurements Aimed as Input for Multi-Domain Compact Model Development within a European-Wide R&D Project" Conference on "Smarter Lighting for Better Life" at the CIE Midterm Meeting 2017, 23–25 October 2017, Jeju, Korea, pp. 569–579. https://doi.org/10.25039/x44.2017.PP16

147. Martin, G., Poppe, A., Schöps, S., Kraker, E., Marty, Ch., Soer, W., Yu, J.: AI-TWILIGHT: AI-digital TWIn for LIGHTing – a new European project. In: Proceedings of the 2021 27th International Workshop on Thermal Investigations of ICs and Systems (THERMINIC), 23–23 September 2021, Berlin, Germany. https://doi.org/10.1109/THERMINIC52472.2021.9626541

148. Website of the AI-TWILIGHT H2020 ECSEL Research Project of the European Union. Available online: https://ai-twilight.eu/. Accessed: 25 Feb 2022

149. Terms of Reference of the CIE TC2-91 Technical Committee. Available online: http://cie.co.at/technicalcommittees/optical-measurement-methods-led-packages-and-led-arrays. Accessed: 25 Feb 2022

150. Yake, F., Gang, W., Chen, X., Hon, W.V., Fu, X., Vass-Várnai, A.: Detailed analysis of IC packages using thermal transient testing and CFD modelling for communication device applications. In: 2016 22nd International Workshop on Thermal Investigations of ICs and

Systems (THERMINIC), 21–23 September 2016, pp. 164–168. https://doi.org/10.1109/THERMINIC.2016.7749046

151. Vass-Várnai, A., Bornoff, R., Sárkány, Z., Ress, S., Rencz, M.: Measurement based compact thermal model creation – accurate approach to neglect inaccurate TIM conductivity data. In: 2011 13th Electronics Packaging Technology Conference (EPTC), 7–9 December 2011, Singapore, pp. 67–72. ISBN: 978-1-4577-1981-3. https://doi.org/10.1109/EPTC.2011.6184388

152. Public Deliverable of the Delphi4LED Project, "Deliverable 2.1: Report on Round-Robin Measurement of the Thermal, Electrical and Light Output Properties of Selected LED Packages". Available online: https://delphi4led.org/pydio/public/2f72dd. Accessed on Oct 2019

153. Bornoff, R., Farkas, G., Gaál, L., Rencz, M., Poppe, A.: LED 3D thermal model calibration against measurement. In: 2018 19th International Conference on Thermal, Mechanical and Multi-Physics Simulation and Experiments in Microelectronics and Microsystems (EuroSimE), 15–18 April 2018, Toulouse, France. https://doi.org/10.1109/EuroSimE.2018.8369929

154. Bornoff, R., Mérelle, T., Sari, J., Di Bucchianico, A., Farkas, G.: Quantified Insights into LED variability. In: 2018 24th International Workshop on Thermal Investigations of ICs and Systems (THERMINIC), 26–28 September 2018, Stockholm, Sweden. https://doi.org/10.1109/THERMINIC.2018.8593315

155. Poppe, A., Farkas, G., Székely, V., Horváth, Gy., Rencz, M.: Multi-domain simulation and measurement of power LED-s and power LED assemblies. In: 2006 22nd Annual IEEE Semiconductor Thermal Measurement and Management Symposium (SEMI-THERM), 14–16 March 2006, Dallas, TX, USA, pp. 191–198. https://doi.org/10.1109/STHERM.2006.1625227

156. Alexeev, A., Martin, G., Hildenbrand, V.: Structure function analysis and thermal compact model development of a mid-power LED. In: 2017 33rd Annual IEEE Thermal Measurement, Modeling & Management Symposium (SEMI-THERM), 13–17 March 2017, San Jose, CA, USA, pp. 283–289. https://doi.org/10.1109/SEMI-THERM.2017.7896942

157. Rencz, M., Székely, V., Morelli, A., Villa, C.: Determining partial thermal resistances with transient measurements, and using the method to detect die attach discontinuities. In: Proceedings of the 2002 18th Annual IEEE Semiconductor Thermal Measurement and Management Symposium (SEMI-THERM), March 12–14 2002, San Jose, CA, USA, pp. 15–20. https://doi.org/10.1109/STHERM.2002.991340

158. Krasich, M.: How to estimate and use MTTF/MTBF would the real MTBF please stand up? In: 2009 Annual Reliability and Maintainability Symposium, 26–29 January 2009, Fort Worth, TX, USA, pp. 353–359. https://doi.org/10.1109/RAMS.2009.4914702

159. Lall, P.: Challenges in accelerated life testing. In: The Ninth Intersociety Conference on Thermal and Thermomechanical Phenomena in Electronic Systems (IEEE Cat. No.04CH37543, ITherm), vol. 2, p. 727, Las Vegas, NV, USA, 1–4 June 2004. https://doi.org/10.1109/ITHERM.2004.1318375

160. Stockmeier, T.: From packaging to "Un"-packaging – trends in power semiconductor modules. In: 20th International symposium on Semiconductor Devices & IC's, 18–22 May 2008, Orlando, FL, USA. https://doi.org/10.1109/ISPSD.2008.4538886

161. Ciappa, M.: Lifetime prediction on the base of mission profiles. Microelectron. Reliab. 45(9–11), 1293–1298 (2005). https://doi.org/10.1016/j.microrel.2005.07.060

162. Sárkány, Z., Vass-Várnai, A., Rencz, M.: Analysis of concurrent failure mechanisms in IGBT structures during active power cycling tests. In: 2014 16th Electronics Packaging Technology Conference (EPTC), 3–5 December 2014, Singapore, pp. 650–654. https://doi.org/10.1109/EPTC.2014.7028349

163. Vass-Várnai, A., Sárkány, Z., Rencz, M.: Characterization method for thermal interface materials imitating an in-situ environment. Microelectron. J. 43(9), 661–668 (2012). https://doi.org/10.1016/j.mejo.2011.06.013

164. Codecasa, L., D'Alessandro, V., Magnani, A., Rinaldi, N., Zampardi, P.: Fast novel thermal analysis simulation tool for integrated circuits (FANTASTIC). In: 2014 20th International

Workshop on Thermal Investigations of ICs and Systems (THERMINIC), 24–26 September 2014, Greenwich, UK. https://doi.org/10.1109/THERMINIC.2014.6972507

165. Codecasa, L., Magnani, A., D'Alessandro, V., Niccolò, R., Metzger, A., Bornoff, R., Parry, J.: Novel MOR approach for extracting dynamic compact thermal models with massive numbers of heat sources. In: 2016 32nd Annual IEEE Thermal Measurement, Modeling & Management Symposium (SEMI-THERM), 14–17 March 2016, San Jose, CA, USA, pp. 218–223. https://doi.org/10.1109/SEMI-THERM.2016.7458469

166. Parry, J.: "Simcenter Flotherm 2020.2: What's New", Blog Post, available online: https://blogs.sw.siemens.com/simcenter/simcenter-flotherm-2020-2-whats-new/.   Accessed: 7 July 2021

167. Sarkany, Z., Rencz, M.: Methods for the separation of failure modes in power-cycling tests of high-power transistor modules using accurate voltage monitoring. Energies. **13**(11), 2718 (2020). https://doi.org/10.3390/en13112718

168. Hara, T., Aoki, Y., Funaki, T.: Thermal fluid simulation modeling based on thermal transient test and fatigue analysis of asymmetric structural double-sided cooling power module. J. Jpn. Inst. Electron. Packag. **24**(1), 130–142 (2021)

169. Nakahama, H.: Estimation Method for Temperature Uncertainty of Temperature Chambers (JTM K 08), ESPEC Technology Report No. 26. Available online: https://www.test-navi.com/eng/report/pdf/EstimationMethodForTemperatureUncertaintyOfTemperatureChamber_JTM_K_08.pdf. Accessed: 7 July 2021

170. Muelaner, J.: Calculating Your Uncertainty Budget in Manufacturing: Quantifying Sources of Uncertainty in Measurement. Available online: https://www.engineering.com/story/calculating-your-uncertainty-budget-in-manufacturing. Accessed: 14 Feb 2021

171. Hogan, R.: "Sources of Uncertainty in Measurement For Every Uncertainty Budget", ISOBudgets LLC, Yorktown, VA, USA (2015). Available online: https://nfogm.no/wp-content/uploads/2016/03/8-Sources-of-Uncertainty-in-Measurement-for-Every-Uncertainty-Budget-by-Rick-Hogan.pdf. Accessed: 14 Feb 2021

172. Székely, V.: THERMODEL: a tool for compact dynamic thermal model generation. Microelectron. J. **29**(4–5), 257–267 (1998). https://doi.org/10.1016/S0026-2692(97)00065-7

173. Székely, V., Rencz, M., Poppe, A., Courtois, B.: THERMODEL: a tool for thermal model generation, and application for MEMS. Analog Integr. Circ. Sig. Process. **29**(1–2), 49–59 (2001). https://doi.org/10.1023/A:1011226213197

174. Székely, V., Szalai, A.: Measurement of the time-constant spectrum: systematic errors, correction. Microelectron. J. **43**(11), 904–907 (2012). https://doi.org/10.1016/j.mejo.2012.05.011

175. Székely, V., Szalai, A.: Measurement of the time-constant spectrum: systematic errors, correction. In: 2011 17th International Workshop on THERMal INvestigation of ICs and Systems (THERMINIC), 27–29 September 2011, Paris, France, pp. 45–48. Available online: https://ieeexplore.ieee.org/document/6081013. Accessed: 7 July 2021

176. Szalai, A., Székely, V.: Possible acception criteria for structure functions. Microelectron. J. **43**(2), 164–168 (2012). https://doi.org/10.1016/j.mejo.2011.08.010

177. Szalai, A., Székely, V.: How do we know if a structure function is correct? In: 2010 16th International Workshop on THERMal INvestigation of ICs and Systems (THERMINIC), 6–8 October 2010, Barcelona, Spain, pp. 80–83. Available online: https://ieeexplore.ieee.org/document/5636321. Accessed: 7 July 2021

178. Dibra, D., Stecher, M., Decker, S., Lindemann, A., Lutz, J., Kadow, C.: On the origin of thermal runaway in a trench power MOSFET. IEEE Trans. Electron Devices. 58(10), 3477–3484 (2011). https://doi.org/10.1109/TED.2011.2160867

179. Reggiani, S., Valdinoci, M., Colalongo, L., Rudan, M., Baccarani, G.: An analytical, temperature-dependent model for majority- and minority-carrier mobility in silicon devices. VLSI Des. 4(10), 467–483 (2000). Available online: https://pdfs.semanticscholar.org/2aec/e6c9b6863bb314d5a875f375dc1a022eca4c.pdf. Accessed: 7 July 2021

Index